Advanced Technologies for Sensor Systems

Advanced Technologies for Sensor Systems

Editor: Bob Tucker

NY RESEARCH PRESS

New York

Published by NY Research Press
118-35 Queens Blvd., Suite 400,
Forest Hills, NY 11375, USA
www.nyresearchpress.com

Advanced Technologies for Sensor Systems
Edited by Bob Tucker

International Standard Book Number: 978-1-63238-567-3 (Hardback)

Cataloging-in-Publication Data

    Advanced technologies for sensor systems / edited by Bob Tucker.
       p. cm.
    Includes bibliographical references and index.
    ISBN 978-1-63238-567-3
    1. Detectors. 2. Engineering instruments. I. Tucker, Bob.
TK7871.674 .A38 2018
621.381 536--dc23

Contents

Preface

This book has been a concerted effort by a group of academicians, researchers and scientists, who have contributed their research works for the realization of the book. This book has materialized in the wake of emerging advancements and innovations in this field. Therefore, the need of the hour was to compile all the required researches and disseminate the knowledge to a broad spectrum of people comprising of students, researchers and specialists of the field.

Sensors are objects that are used to detect events or changes which occur in the surrounding environment. It helps in converting real world information into data that computers can understand. They can be objects that are used on a daily basis, such as touch-sensitive elevator buttons, flow measurement devices, etc. Some widely used types of sensors are biosensors, chemical sensors, nanosensors, etc. This book is an essential guide for both academicians and those who want to develop a better understanding of the subject matter. In this book, using case studies and examples, constant effort has been made to make the understanding of the concept of sensor systems as easy and informative as possible.

At the end of the preface, I would like to thank the authors for their brilliant chapters and the publisher for guiding us all-through the making of the book till its final stage. Also, I would like to thank my family for providing the support and encouragement throughout my academic career and research projects.

Editor

An electrical characterisation system for the real-time acquisition of multiple independent sensing parameters from organic thin film transistors

A. Dragoneas, L. Hague, and M. Grell

Physics and Astronomy, The University of Sheffield, Hicks Building, Hounsfield Road, S3 7RH, Sheffield, UK

Correspondence to: A. Dragoneas (a.dragoneas@mpic.de)

Abstract. The presence of multiple independent sensing parameters in a single device is the key conceptual advantage of sensor devices based on an organic thin film transistor (OTFT) over simple organic chemiresistors. Practically, however, these multiple parameters must first be extracted from the electrical characteristics of the OTFTs and, thus, they are not immediately apparent. To exploit the advantage of OTFT sensors, we require a measurement technology to extract these parameters in real time. Here, we introduce an efficient, cost-effective system that is a faster and more compact alternative to the expensive and cumbersome laboratory-based instruments currently available. The characterisation system presented here records the electric behaviour of OTFTs in the form of its "saturated transfer characteristics" multiple times per second for virtually unlimited periods of time, with the option to multiplex up to 20 devices in parallel. By applying a bespoke algorithm to the measured transfer characteristics, the system then extracts, in real time, several underlying transistor parameters (on- and off-current, threshold voltage, and charge carrier mobility). Tests were conducted on the example of a poly(thieno[3,2-b]thiophene) (PBTTT) OTFT exposed to ethanol vapour. The system extracts the underlying OTFT parameters with very low noise without introducing apparent correlations between independent parameters as an artefact.

1 Introduction

Organic semiconductors (OSCs) are generally more susceptible to environmental factors than traditional (inorganic) semiconductors, so their electronic properties, and consequently the device performance, often strongly decay under ambient atmosphere. This may be due to oxygen, humidity, or pollutants in the air. While such sensitivity is a disadvantage in the context of many applications, it may be an asset for sensor technology. A prominent example is the poly(triaryl amine) (PTAA) family of compounds, which proved unsuitable as realistic photoconductors due to their strong vulnerability to the common air pollutant NO_2, yet qualified them for use in NO_2 sensors (Das et al., 2007).

A further attraction of OSCs for sensor technology is that their response can be read electrically, which allows for straightforward transduction. A number of "chemiresistors" have been demonstrated, wherein the interaction between an analyte and an OSC is converted to a change in electrical resistance. This may be a drop from a high resistance when OSCs are exposed to dopants, e.g. poly(aniline) (PAni) to acidic vapours (Huang et al., 2003), or polythiophene exposed to humidity (Dragoneas et al., 2013). Alternatively, there may be an increase in resistance in previously doped organic semiconductors ("synthetic metals") exposed to analyte odour that acts as de-doping agents, e.g. doped PAni under amines (Sotzing et al., 2000).

To improve on the simple chemiresistor concept, sensitive OSCs have also been used in organic thin film transistors (OTFTs). In an OTFT, conductivity between two terminals of the device, the "source" (S) and "drain" (D), is enhanced by a sufficiently high voltage of the appropriate polarity applied to a third, insulated terminal, the "gate" (G) (Horowitz, 2000). Under a high gate voltage (V_G), the induced "saturated drain current" ($I_{D,sat}$), hereafter referred to as the "on-current" (I_{ON}), can be orders of magnitude higher than the current of a non-gated chemiresistor, which typically is similar to the OTFTs "off-current", when no V_G is applied. This brings advantages in terms of signal-to-noise ratio (SNR) and thus the limit of detection (LoD); a large number of OTFT sensors have already been reported in the literature (e.g. Crone et al., 2001; Das et al., 2007; Guo et al., 2006; Hague et al., 2011b; Street et al., 2007; Torsi et al., 2008, 2009).

There is also a conceptual advantage of OTFT sensors over chemiresistors: the resistance of a chemiresistor is a single parameter, given simply by the product of charge carrier density and mobility, which are themselves correlated (Hague et al., 2011b). The characteristics of an OTFT (e.g., the saturated transfer characteristics, $I_{D,sat}(V_G)$) may reveal multiple uncorrelated parameters, i.e. carrier mobility (μ), threshold voltage (V_{th}) and on/off ratio; each of these parameters may respond independently and differently to different analytes. OTFT sensors can therefore provide richer information than chemiresistors for the identification of different analytes based upon their effects on several parameters of the same sensor device. Bayn et al. (2013), Wang and Haick (2013a, b), Ermanok et al. (2013), and Wang et al. (2014) have recently shown impressive gas sensing performance of organic and organic/inorganic hybrid TFTs under multiparametric characterisation.

However, to take full advantage of the multiparametric nature of OTFT sensors, we first need to delineate these parameters from measured OTFT characteristics. While the extraction of mobility, threshold, and on/off ratio from OTFT saturated transfer characteristics is well established in principle (Horowitz, 2000), it is a substantial technological challenge to implement this in an automated and portable real-time OTFT measurement system. Bayn et al. (2013), Wang and Haick (2013a, b), Ermanok et al. (2013), and Wang et al. (2014) relied on conventional, laboratory-based semiconductor parameter analysers, which limit the scope for practical applications. Alternative real-time separation between mobility and threshold in sensor OTFTs has been attempted before (Das et al., 2009). However, when mobility and threshold data were fed separately into a genetic algorithm for analyte recognition (Wedge et al., 2009), they proved to be no more effective than using "on-current" I_{ON} alone. It is not known whether this was due to the noisy nature of extracted parameters, or because they became correlated as an artefact of the system's algorithms.

The advanced measurement solution we present here meets the requirements of thin film transistor sensors. The system is capable of acquiring OTFT saturated transfer ($I_{D,sat}(V_G)$) characteristics; by the application of a bespoke numerical data evaluation scheme, it can then extract charge carrier mobility and threshold voltage as uncorrelated parameters with very low noise in real time. To test this instrument, we characterised an OTFT with polymer OSC "PBTTT" (poly(2,5-bis(3-hexadecylthiophen-2-yl)thieno(3,2-b)thiophene) under exposure to ethanol (EtOH)

odour. This is not meant to lead to a practical sensor device – EtOH odour sensors are well established (fuel-cell based "breathalysers"; Leonard, 2012) – but to provide a realistic yet generic proof-of-concept. However, given the strong scientific interest that PBTTT has attracted since its introduction in 2006 (Mcculloch et al., 2006), studying its behaviour under EtOH has scientific merit in itself: it is known that exposure to EtOH may lead to morphological changes in other thiophene polymers (Jiang-Feng et al., 2011; Nam et al., 2010), but this has not yet been investigated for PBTTT.

2 Methods

2.1 OTFT preparation

OTFTs were built on heavily n-doped Si substrates with a thermally grown 300 nm thick oxide. The substrates were cleaned with an alkaline solution and isopropyl alcohol before a UV-ozone treatment was applied. For the sake of reproducibility and simplicity, as well as the exclusion of any additional interactions, no surface treatment was applied to the gate oxide. We used the OSC PBTTT-C16, supplied by Ossila Ltd (product number M141), dissolved in 1,2-dichlorobenzene ($7.5\,mg\,mL^{-1}$) at $100\,°C$. PBTTT is more stable under ambient conditions than the common organic semiconductor poly(3-hexyl thiophene) (P3HT), which has well-documented problems with oxygen and humidity (Hoshino et al., 2004). The solution was spincast onto the substrates in a spin coater with a nitrogen purge, at 1500 rpm for 60 s; the substrates were then annealed at $95\,°C$ for 45 min in dynamic vacuum. Gold was thermally evaporated through shadow masks in a high vacuum of 5×10^{-7} Torr at a rate of $0.03\,nm\,s^{-1}$. The shadow masks patterned 20 pairs of source-drain contacts on each substrate. The channel dimensions of the devices were $5\,\mu m$ (length) by $1000\,\mu m$ (width). The solution preparation and deposition, substrate annealing, gold deposition, as well as all device measurements, were conducted under either dimmed yellow light or in the dark.

2.2 Odour exposure

Gas flow and mixing ratios were regulated by two voltage-controlled mass-flow controllers (Tylan FC-260) with a full-scale throughput of 500 sccm (standard cubic centimetres per minute), which were addressed by a computer using a bespoke LabVIEW application and a NI-USB 6008 device. A solenoid valve was placed before each controller to block the small but undesirable gas flow at the minimum setting of their scales. One mass flow controller regulated the flow of N_2 through a "bubbler" bottle to generate saturated analyte odour; the other regulated the flow of pure N_2 to dilute the saturated odour when both lines joined at a mixing point. In this way, vapour atmospheres in the range (1 to 0.01) p/p_{sat} could be generated, where p_{sat} stands for the saturated vapour pressure. p_{sat} itself depends on the particular analyte (here, ethanol), and the analyte temperature, which can be

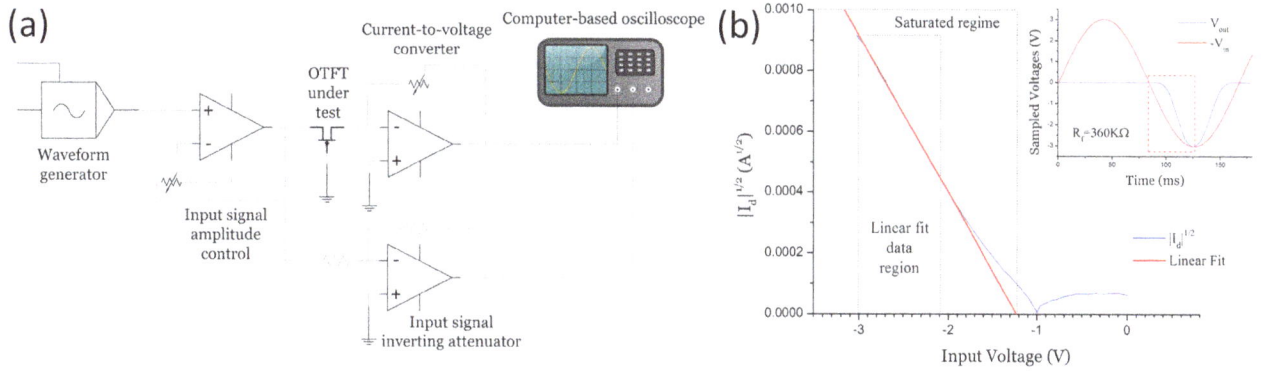

Figure 1. (**a**) A simplified schematic of the circuit for real-time measurements. (**b**) Inset: plot of $-V_{in}$ (red sine) and V_{out} (blue) versus time, showing "on" and "off" half-cycles. Main plot: saturated transfer characteristic (blue) in the form square root of modulus of drain current versus applied voltage ($-V_{in}$). Same data as in the inset, but time is eliminated. Red straight line: linear fit to high drive voltage data for evaluation of mobility and threshold.

controlled by immersion of the bubbler bottle into a thermostatically controlled water bath. All pipework and the exposure chamber were thoroughly flushed with nitrogen before each test. Transistor devices were placed inside the Teflon-lined exposure chamber and contacted by an array of spring-loaded Au contact pins, which are shown in Fig. 2b.

2.3 Real-time electrical characterisation

We designed and built a bespoke, fully automated, computer-controlled, real-time, multiparametric electrical characterisation system for OTFTs. Its operating principle is based upon our previous report on a OTFT characterisation system employing an operational amplifier (op-amp) (Hague et al., 2011a). The portable system was interfaced to a PC via a Pico Technology Picoscope 2204 device, which acts as both a two-channel digital oscilloscope and a waveform generator. The circuit is mainly built around three op-amps with different specifications. A high-voltage op-amp (Texas Instruments OPA445AP) is used to amplify a zero-offset, low-frequency ($f = 6$ Hz) sine drive provided by the Picoscope waveform generator, to extend the limited voltage amplitude of its output. A zero-offset sine waveform is chosen to minimise gate dielectric stress (Zschieschang et al., 2009) thanks to its symmetry, and to avoid higher harmonics. The amplified sinusoidal drive signal, V_{in} ($= V_{MAX} \sin(\omega(t))$), is fed into the source of the OTFT under test, while the gate is connected to the electrical ground and the drain is connected to virtual ground (see below). Therefore, this drive configuration is equivalent to connecting the source to ground and applying $-V_{in}$ to both the drain and gate. A hole (or electron) transporting OTFT will turn "on" for a sufficiently large positive (or negative) V_{in}, and deliver a saturated drain current, $I_{D,sat}$ ($V_G = V_D = -V_{in}$). V_{in} is "sufficiently" large when it has the same polarity as, and its modulus exceeds the modulus of, threshold, V_{th}. The resulting saturated drain current is fed into the inverting input of a high-input impedance op-

amp (Analog Devices AD549JH), which is configured as a current / voltage (I/V) converter by feeding back the output voltage (V_{out}) to the inverting input via a high-precision feedback resistor, R_f (here, $R_f = 1$ MΩ). The non-inverting input of the same op-amp is grounded, thus establishing virtual ground at the inverting input. V_{out} is related to the $I_{D,sat}$ of the OTFT via the equation $V_{out} = -R_f \, I_{D,sat}$. The third op-amp (Linear Technology LT1677) is configured as a low-noise attenuator and inverter of the input signal (V_{in}), which provides a reference voltage ($V_{ref} = -0.1 \, V_{in}$) which can be safely sampled by the oscilloscope and used to trigger the measurement. The output voltage (V_{out}), which is proportional to $I_{D,sat}$, and the attenuated and inverted input voltage (V_{ref}), are connected to the two oscilloscope channels of the Picoscope, which carry out the analogue-to-digital conversion. Together, $V_{out}(t)$ and $10 \, V_{ref}(t) = -V_{in}(t)$ represent the saturated transfer characteristic of the OTFT, parametric in time. Figure 1a schematically illustrates the system.

A National Instruments LabVIEW application has been developed to fully automate the characterisation process. This application collects the sampled data from the oscilloscope and plots the input and output voltages versus time as shown in the inset to Fig. 1b. Low-pass digital filters can also be applied. As highlighted by the dashed red rectangle in the inset to Fig. 1b, for all calculations presented here, the third quadrant of the sinusoidal signal period was used; i.e. the rising flank of the input voltage (V_{in}). Drain current at every sampling point in time (t) can be calculated according to

$$I_d(V_{in}(t)) = -\frac{V_{out}(t)}{R_f}. \tag{1}$$

Then, the root of drain current is plotted against V_{in} directly, eliminating time. This agrees with the common $I_{D,sat}^{1/2}$ versus V_G plot of saturated OTFT transfer characteristics, which is routinely used to evaluate charge carrier mobility (μ) and threshold voltage (V_{th}) (Locklin and Bao, 2006) from slope and intercept. Figure 1b shows an exam-

(a)

(b)

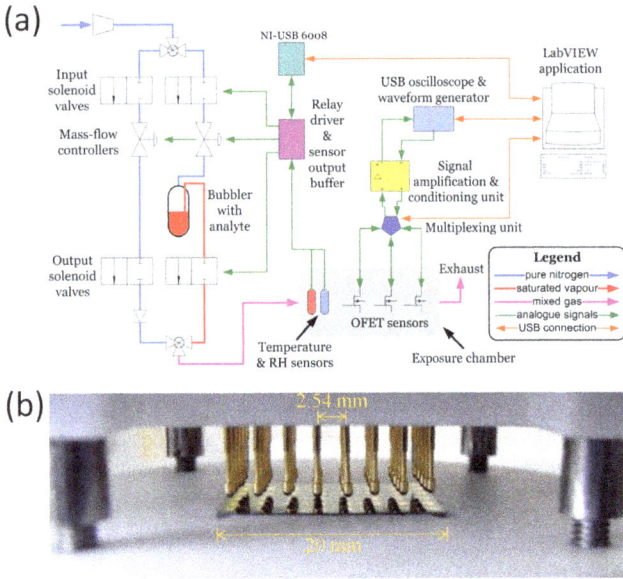

Figure 2. (a) Schematic representation of the computer-controlled pneumatic analyte odour delivery system, and multiplexed electrical OTFT drive and measurement system. **(b)** Array of 40 spring-loaded Au contact pins, matching the S/D contacts for the 20 OTFTs on the OTFT substrate below, shortly before making contact inside the exposure chamber.

ple: at high V_{in} we indeed find a good fit to a straight line. It is worth noting that here, drain voltage (V_{DS}) is swept simultaneously with the gate voltage (V_{GS}) rather than being kept constant, as in conventional transfer sweeps. This different convention results in the uncommon subthreshold behaviour shown in Fig. 1b. However, for high V_{in}, the curve coincides with the conventional saturated transfer characteristics. We therefore apply a simple linear regression fit in the high V_{in} regime of the $(I_{D,sat})^{1/2}$ versus V_{in} plot; in this study, the data range for the fit was in the range $V_{in} = -(0.707 \text{ to } 1) V_{MAX}$. Threshold voltage is extracted as the intercept of the extrapolated straight line with the V_{in} axis. We note that this procedure is robust against a transition from "normally off" to "normally on" behaviour, which we did in fact sometimes observe under exposure (see Sect. 3.2). Carrier mobility (μ) was evaluated from the slope (s) of the fit line, according to

$$\mu = \frac{2L}{WC_i}s^2. \tag{2}$$

wherein W and L stand for the width and length of the transistor's channel, respectively, and C_i is the gate insulator's specific capacitance, here $C_i = 10\,\text{nF cm}^{-2}$. Equation (2) derives from the common equation for saturated drain current in an OTFT (Horowitz, 2000),

$$I_{d,sat} = \frac{W}{2L}\mu C_i(V_G - V_{th})^2. \tag{3}$$

I_{ON} and I_{OFF} are simply defined as the currents recorded at $V_{in} = \pm V_{MAX}$, i.e. maximum/minimum I_d. At a scanning

speed $f = 6\,\text{Hz}$, it is possible to get six sets of data (I_{ON}, I_{OFF}, μ, V_{th}) per second, which are subsequently filtered by a median filter. The system is capable of supporting multiple sensors to allow for simultaneous comparative measurements; this attribute makes the system appropriate for applications such as "electronic noses". This is achieved by the use of one of two bespoke computer-controlled multiplexers; one of them is a relay-based circuit which is controlled by the NI-USB 6008 unit, whereas the other is a stand-alone microcontroller-based system, utilising solid-state analogue switches, which communicates with the computer over a USB connection. The maximum supported drive voltages are ± 45 and ± 8 V for the relay-based and solid-state units, respectively. A schematic representation of the described odour exposure and electrical characterisation scheme is shown in Fig. 2a.

OTFT threshold (V_{th}) and mobility (μ) are the subject of theoretical treatments, e.g. by Horowitz et al. for V_{th} (Horowitz, 2000; Horowitz et al., 1998), and Bässler for μ (Bässler, 1993). Extracting these from measured saturated transfer characteristics, therefore, in principle provides an opportunity to relate observed OTFT response under odour exposure to the physical interactions between analyte and OSC. From the band model for carrier injection into OTFTs, Horowitz et al. (1998) give Eq. (4) for V_{th}:

$$V_{th} = \pm\frac{qn_0 d_s}{C_i} + V_{FB}, \tag{4}$$

wherein $\pm$ applies to hole/electron transporting semiconductors, q is the elementary charge, n_0 the doping level in the semiconductor, d_s the thickness of the semiconducting film, and V_{FB} the flat-band voltage. V_{FB} contains a number of contributions, including the metal work function, Φ_m.

Equation (4) suggests how odour can influence V_{th}: either by doping (via n_0) or by adsorption of polar molecules on an interface, which changes Φ_m and, hence, V_{FB}. Bässler's model of organic mobility is more intricate, and involves semi-empirical "disorder parameters", which may also be impacted by organic odour. However, there may be further interactions not explicitly covered by these models (e.g. deep carrier traps).

3 Results and discussion

3.1 PBTTT OTFT long-term stability

Figure 3 shows both the on- and off-currents of a PBTTT OTFT recorded over several days using the characterisation system described above with $V_{MAX} = 20$ V, while the exposure chamber was continuously purged with dry nitrogen.

An initial off-current (possibly due to unintentional doping during device fabrication) was minimised over time, dropping from 200 to 65 nA over 3 days. The on-current also drops over time (possibly because charge carrier mobility also decreases with decreasing doping levels; Hague

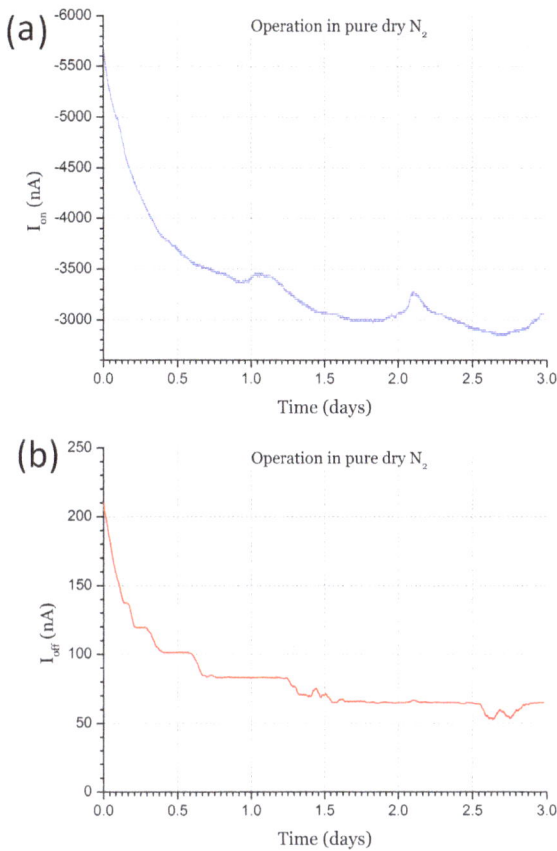

Figure 3. (a) Maximum drain current (I_{ON}); **(b)** minimum drain current (I_{OFF}) for a PBTTT OTFT under nitrogen purge over the course of 3 days. Data for $V_{DS} = V_{GS} = -20\,V$ (I_{ON}), $V_{DS} = V_{GS} = +20\,V$ (I_{OFF}).

Figure 4. (a) I_{ON} ($V_{DS} = V_{GS} = -20\,V$) under EtOH exposure/recovery cycles. All white regions depict purge under 500 sccm pure dry nitrogen. Shaded regions indicate exposure to different concentrations of EtOH vapour. From left to right: 5 % p_{sat} (equivalent to 2750 ppm), 15 % p_{sat} (8250 ppm) and 45 % p_{sat} (24 750 ppm). **(b)** Similar for I_{OFF} ($V_{DS} = V_{GS} = +20\,V$).

et al., 2011b), with some undulations on a $\sim 24\,h$ timescale, which could be attributed to the effects of daily room temperature fluctuations on charge carrier mobility. Later experiments were taken on a transistor pre-conditioned by a 3-day dry N_2 purge, after current drift with time had largely ceased. The initial study presented in Fig. 3 confirms that PBTTT OTFTs display sufficient stability against spontaneous drift on the timescale ($\sim 1\,h$) of the EtOH exposure/recovery tests reported here.

3.2 Testing the system with PBTTT OTFTs under exposure to ethanol odour

To test the multiparametric data acquisition system described above on a practical example, we ran exposure/recovery cycles of PBTTT OTFTs to ethanol (EtOH) odour. In Fig. 4, we present the observed on- and off-current, I_{ON} and I_{OFF}. We selected EtOH concentrations of 5, 15 and 45 % of p_{sat} at an ambient temperature of $\sim 20\,°C$, corresponding to 2750, 8250 and 24 750 ppm, beginning with the lowest concentration. Both I_{ON} and I_{OFF} display a distinct response to EtOH exposure. For the two lowest concentra-

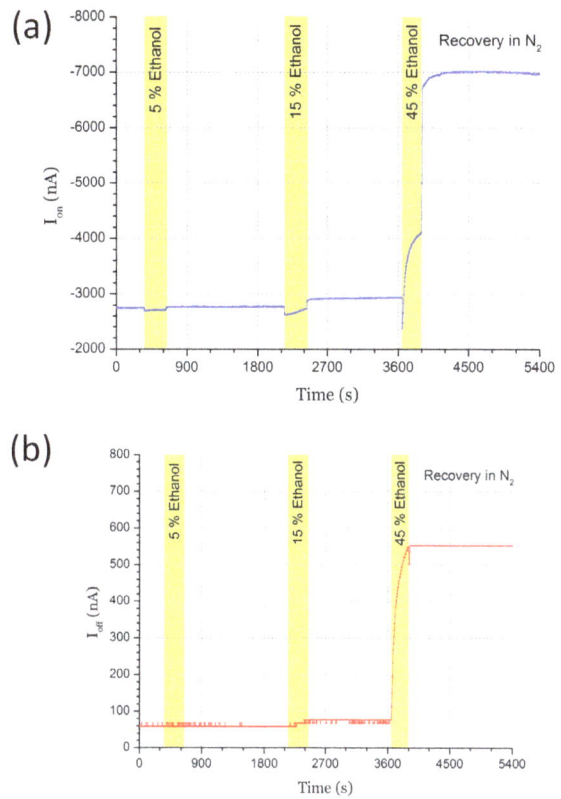

tion levels (5 and 15 % p_{sat}), the I_{OFF} showed a small rise, whereas I_{ON} instantly dropped but slowly increased again over time during 15 % p_{sat} exposure. Immediately after exposure, an instant recovery to a level slightly above the initial I_{ON} was recorded. The change in drain current under 5 % EtOH can be expressed as a sensitivity $S = 16\,pA\,ppm^{-1}$. We note that current versus time traces prior to exposure display very low noise: the scatter of measured data is limited to a single digit in the "quantisation" of analogue-to-digital (ADC) conversion, which at $R_f = 1\,M\Omega$ corresponds to approximately 16 nA. From current noise, we calculate the limit of detection (LoD) as LoD $\sim 16\,nA\,/\,(16\,pA\,ppm^{-1})$ $= 1000\,ppm = 0.1\,\%$ vol/vol, compared to a lower explosive limit (LEL) of 3.5 % vol/vol for EtOH (http://www.distill.com/materialsafety/msds-eu.html).

The most intriguing behaviour was observed under 45 % p_{sat}. I_{OFF} rose steadily over time without showing any recovery after the period of exposure; the behaviour of I_{ON} seemed erratic, with an instant drop at the beginning of exposure, followed by a steady increase with time during exposure and a

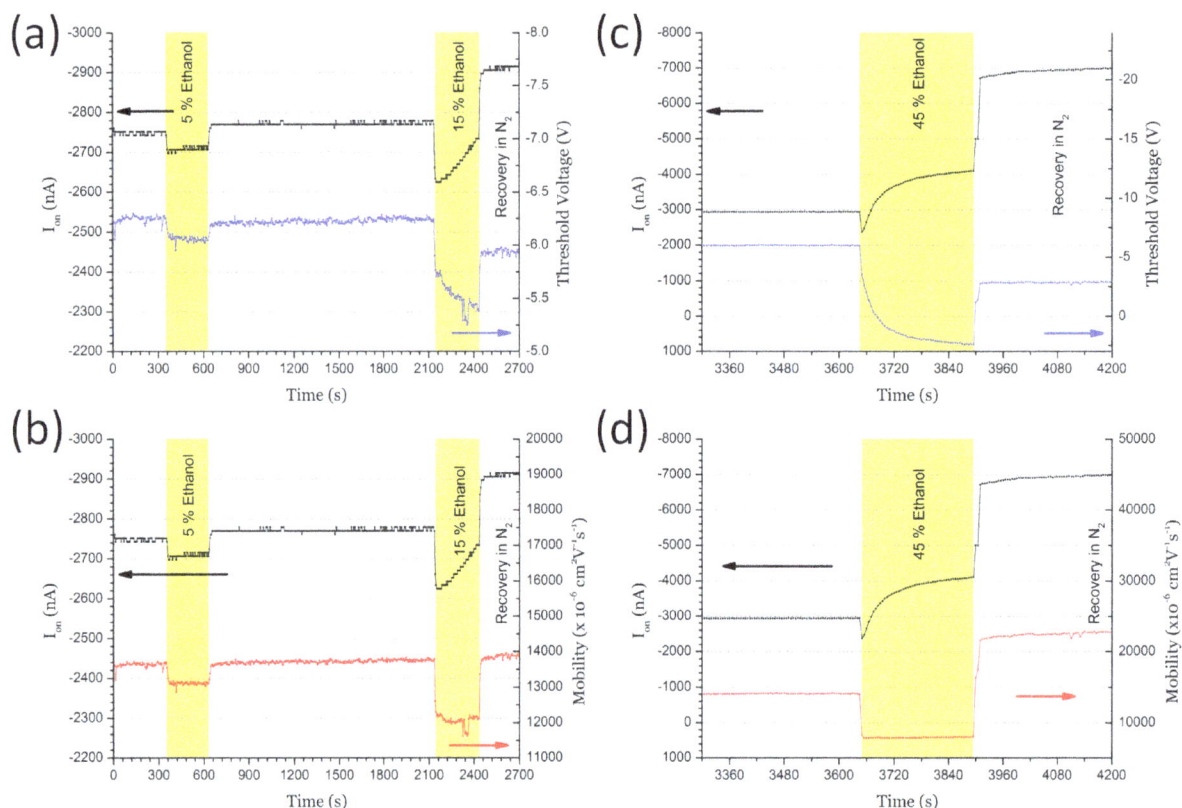

Figure 5. Exposure of PBTTT OTFTs to EtOH. (**a**) I_{ON} (top, black curve) and V_{th} (bottom, blue curve) versus time under 5/15 % p_{sat} EtOH. (**b**) I_{ON} (top, black curve) and μ (bottom, red curve) versus time under 5/15 % p_{sat} EtOH. White regions depict dry nitrogen purge under 500 sccm. Shaded regions indicate exposure to different concentrations of ethanol vapour. Left region: 5 % p_{sat} (2750 ppm) EtOH; right region: 15 % p_{sat} (8250 ppm) EtOH. (**c**) Similar to (**a**) but for exposure to 45 % p_{sat} (24 750 ppm) EtOH. (**d**) Similar to (**b**) but under 45 % p_{sat} (24 750 ppm) EtOH.

further step-like increase at the end of exposure to more than twice the prior level, with no subsequent recovery.

The I_{ON} / I_{OFF} data shown in Fig. 4 could have also been recorded with a less sophisticated data acquisition/evaluation system; however, the "zigzag" behaviour of I_{ON} under 45 % p_{sat} is particularly difficult to rationalise without further information. For the intended purpose of demonstrating the technical quality, and scientific merit, of this multiparametric characterisation system, a PBTTT OTFT under EtOH is thus a fortunate choice: only a sophisticated system, like the one developed here, can deliver separate mobility and threshold data with real-time resolution, which will be needed to gain a deeper understanding of such rapid and complex sensor responses.

Figure 5 shows the individual evolution of threshold voltage (V_{th}) and carrier mobility (μ) under exposure to 5, 15, and 45 % p_{sat} EtOH, as evaluated in real time by the algorithm described above. Note the good quality (low noise) of the V_{th}/μ data. For reference, I_{ON} is also shown.

It can be seen in Fig. 5 that both V_{th} and μ are affected by EtOH, showing stronger responses for higher EtOH concentrations. In all cases, mobility immediately drops to a lower

level and then remains constant throughout each exposure period; the modulus of V_{th} drops instantly and then gradually reduces further throughout the exposure period, in an approximately exponential curve. Under 45 % p_{sat}, V_{th} even turns slightly positive, indicating a transition from a "normally off" (enhancement-mode) to a "normally on" (depletion-mode) type of OTFT. Significantly, under EtOH exposure, one parameter (V_{th}) is shown to slowly evolve in time, while another (μ) remains constant after a rapid, initial change. This underscores the independence of the extracted parameters: this data analysis system does not introduce correlations between parameters as an artefact, when they are physically independent. Otherwise, real drift in one parameter would necessarily lead to apparent drift in the other.

According to Eq. (3), the trends in V_{th} and μ have opposite effects on I_{ON}: a drop in (the modulus of) V_{th} leads to an increase, whereas a drop in μ leads to a decrease. The initial drop/later gradual rise of I_{ON} under EtOH can thus be explained by the different dynamics of opposing trends.

At the end of analyte exposure, both threshold voltage and mobility responded quickly. For exposure to low EtOH odour concentration (5 % p_{sat}), recovery is almost complete, $i.e.$

sensor response is almost fully reversible. However, under exposure to higher EtOH concentrations, PBTTT OTFTs undergo irreversible changes. After exposure to 45 % p_{sat}, V_{th} returned to a "normally off" state, yet with a lower modulus than prior to exposure; meanwhile, mobility showed a reduction under exposure, which was followed by a recovery to a level higher than its pre-exposure value. Immediately after each exposure period, both reduced threshold voltage and increased mobility contributed to a higher overall I_{ON}, as compared to its value prior to exposure (I_{ON} more than doubles after 45 % p_{sat} EtOH exposure). I_{ON} does not return to its previous value, even when purged under dry nitrogen for more than 2 h after the last exposure. Exposure to EtOH thus leads to both reversible, and irreversible, changes in mobility and threshold, which we will discuss separately.

The results for PBTTT OTFTs under low EtOH concentrations are similar to those from previous work on a polythiophene derivative OTFT by Torsi et al. (2004), who reported a quick and reversible reduction of I_{ON} by ~ 0.2 % under 700 ppm = 1.3 % p_{sat} EtOH, albeit without delineating into mobility and threshold voltage changes. This is comparable to the observation, given here, of a largely reversible ~ 1.8 % reduction of I_{ON} under 5 % p_{sat} (2750 ppm) EtOH. Such reversible sensing may result from a manifold of possible interactions between analytes and the different components of an OTFT, e.g. doping or de-doping the OSC (Dragoneas et al., 2013; Sotzing et al., 2000), swelling (AlQahtani et al., 2012), assisted tunnelling at OSC grain boundaries (Hague et al., 2011a), changing work function by interfacial adsorption of polar molecules, and introduction of trap sites at the insulator surface. However, none of these lead to permanent morphological changes in the PBTTT film, and as EtOH escapes under nitrogen purge, electrical characteristics largely return to their values prior to exposure.

In contrast, exposure of thiophene OTFTs to more concentrated EtOH odour up to 45 % p_{sat} leads to irreversible changes that persist under subsequent purge. Such changes must therefore be assigned to morphological changes under EtOH, a process which is referred to as "vapour annealing". Exposure of PBTTT devices under highly concentrated EtOH odour has not previously been reported in the literature. However, it is known that the treatment of the chemically related OSC P3HT with liquid EtOH leads to irreversible morphological changes with improved carrier mobility, either by adding EtOH as a non-solvent to P3HT processing solution prior to casting (Jiang-Feng et al., 2011), or by application of a liquid droplet to a film after casting (Nam et al., 2010). We believe that exposure of PBTTT to concentrated EtOH vapour leads to similar morphological improvements, which partly explain the observed irreversible changes of OTFT parameters. Detailed X-ray diffraction studies of other conjugated polymers also have shown that solvent vapour annealing can significantly and irreversibly increase crystallinity, often more than thermal annealing (e.g. Grell et al., 1999). At this point, it is worth noting that these changes can compromise the sensing capabilities of these transistors and, thus, these devices should not be reused as sensors after exposure to such high analyte concentrations. However, irreversible morphological changes under concentrated odours are specific to a particular OSC–analyte combination, while the focus of this study is the characterisation system that can be applied to all odour-sensing OTFTs.

4 Conclusions

"Electronic noses" require the acquisition of manifold information-rich sensor data, and powerful data analysis algorithms to identify and quantify analytes. We here present a comprehensive and reliable system for the acquisition of low-noise, multiparametric, mutually independent OTFT sensor data in real time, which can optionally be used for sensor arrays via multiplexing. The bespoke solution presented here comes with a much lighter footprint and significantly lower cost than commercially available semiconductor parameter analysers; this system enables the user to fully exploit the conceptual advantages of OTFT-based sensors over organic chemiresistors.

Tests of this measurement system were conducted by exposing PBTTT OTFTs to ethanol (EtOH) vapour. I_{ON} displayed apparently erratic behaviour, particularly under high EtOH concentrations. However, we succeeded in extracting threshold voltage and mobility as independent, uncorrelated parameters with low noise. When inspected separately, mobility and threshold voltage display a more straightforward behaviour, which, taken together, accounted for the observed I_{ON} behaviour.

While we are confident that morphological changes under concentrated EtOH account in part for the observed behaviour, we cannot provide a full explanation on a molecular level. However, practical "electronic noses" do not rely on detailed interpretations of analyte/sensitiser interactions either, but pragmatically use measured exposure data as a "feedstock" to train advanced pattern recognition algorithms; we refer to the review by Scott et al. (2006). Such an approach is particularly appropriate in the case of OTFT sensors, where the theory of carrier mobility and threshold voltage is complex, and still involves empirical parameters, even in the absence of analytes [5, 21, 22].

The measurement technology solution for the acquisition of multiple independent sensing parameters from organic thin film transistors described here is well suited to deliver a rich data feedstock for such algorithms.

Acknowledgements. A. Dragoneas would like to acknowledge the European Commission for the provision of an Early Stage Researcher (ESR) Marie Curie fellowship within the FlexSmell ITN of the 7th Framework Programme (FP7). L. Hague would like to thank the Engineering and Physical Sciences Research Council (EPSRC) for the provision of his Doctoral Prize Fellowship.

References

AlQahtani, H., Alduraibi, M., Richardson, T., and Grell, M.: Manifold sensitivity improvement of swelling-based sensors, Phys. Chem. Chem. Phys., 14, 5558–5560, doi:10.1039/C2cp00003b, 2012.

Bässler, H.: Charge transport in disordered organic photoconductors – a monte-carlo simulation study, Phys. Status Solidi B, 175, 15–56, doi:10.1002/pssb.2221750102, 1993.

Bayn, A., Feng, X. L., Mullen, K., and Haick, H.: Field effect transistors based on polycyclic aromatic hydrocarbons for the detection and classification of volatile organic compounds, Acs. Appl. Mater. Inter., 5, 3431–3440, doi:10.1021/Am4005144, 2013.

Crone, B., Dodabalapur, A., Gelperin, A., Torsi, L., Katz, H. E., Lovinger, A. J., and Bao, Z.: Electronic sensing of vapors with organic transistors, Appl. Phys. Lett., 78, 2229–2231, doi:10.1063/1.1360785, 2001.

Das, A., Dost, R., Richardson, T., Grell, M., Morrison, J. J., and Turner, M. L.: A nitrogen dioxide sensor based on an organic transistor constructed from amorphous semiconducting polymers, Adv. Mater., 19, 4018–4023, doi:10.1002/adma.200701504, 2007.

Das, A., Dost, R., Richardson, T. H., Grell, M., Wedge, D. C., Kell, D. B., Morrison, J. J., and Turner, M. L.: Low cost, portable, fast multiparameter data acquisition system for organic transistor odour sensors, Sensor. Actuat. B-Chem., 137, 586–591, doi:10.1016/j.snb.2009.01.006, 2009.

Dragoneas, A., Grell, M., Hampton, M., and Macdonald, J. E.: Morphology-driven sensitivity enhancement in organic nanowire chemiresistors, Sensor Lett., 11, 552–555, doi:10.1166/Sl.2013.2917, 2013.

Ermanok, R., Assad, O., Zigelboim, K., Wang, B., and Haick, H.: Discriminative power of chemically sensitive silicon nanowire field effect transistors to volatile organic compounds, Acs. Appl. Mater. Inter., 5, 11172–11183, doi:10.1021/Am403421g, 2013.

Grell, M., Bradley, D. D. C., Ungar, G., Hill, J., and Whitehead, K. S.: Interplay of physical structure and photophysics for a liquid crystalline polyfluorene, Macromolecules, 32, 5810–5817, doi:10.1021/Ma990741o, 1999.

Guo, X., Myers, M., Xiao, S., Lefenfeld, M., Steiner, R., Tulevski, G. S., Tang, J., Baumert, J., Leibfarth, F., Yardley, J. T., Steiger-wald, M. L., Kim, P., and Nuckolls, C.: Chemoresponsive monolayer transistors, P. Natl. Acad. Sci. USA, 103, 11452–11456, doi:10.1073/pnas.0601675103, 2006.

Hague, L., Puzzovio, D., Dragoneas, A., and Grell, M.: Simplified real-time organic transistor characterisation schemes for sensing applications, Sci. Adv. Mat., 3, 907–911, doi:10.1166/sam.2011.1216, 2011a.

Hague, L., Puzzovio, D., Richardson, T. H., and Grell, M.: Discovery of a new odour sensing mechanism using an n-type organic transistor, Sensor Lett., 9, 1692–1696, doi:10.1166/Sl.2011.1734, 2011b.

Horowitz, G.: Physics of organic field-effect transistors, in: Semiconducting polymers, edited by: Hadziioannou, G. and Malliaras, G. G., 463–514, 2000.

Horowitz, G., Hajlaoui, R., Bouchriha, H., Bourguiga, R., and Hajlaoui, M.: The concept of "threshold voltage" in organic field-effect transistors, Adv Mater, 10, 923–927, doi:10.1002/(Sici)1521-4095(199808)10:12<923::Aid-Adma923>3.0.Co;2-W, 1998.

Hoshino, S., Yoshida, M., Uemura, S., Kodzasa, T., Takada, N., Kamata, T., and Yase, K.: Influence of moisture on device characteristics of polythiophene-based field-effect transistors, J. Appl. Phys., 95, 5088–5093, doi:10.1063/1.1691190, 2004.

Huang, J. X., Virji, S., Weiller, B. H., and Kaner, R. B.: Polyaniline nanofibers: Facile synthesis and chemical sensors, J. Am. Chem. Soc., 125, 314–315, doi:10.1021/Ja028371y, 2003.

Jiang-Feng, Y., Su-Ling, Z., Zheng, X., Jiang-Feng, Y., Fu-Jun, Z., and Xue-Yan, T.: Non-solvent addition induced self-organization for enhancement of performance of poly(3-hexylthiophene) organic field-effect transistors, Acta Phys. Sin., 60, 37201–037201, doi:10.7498/aps.60.037201, 2011.

Leonard, R. J.: Evaluation of the analytical performance of a fuel cell breath alcohol testing instrument: A seven-year comprehensive study, J. Forensic Sci., 57, 1614–1620, doi:10.1111/j.1556-4029.2012.02146.x, 2012.

Locklin, J. and Bao, Z. N.: Effect of morphology on organic thin film transistor sensors, Anal. Bioanal. Chem., 384, 336–342, doi:10.1007/s00216-005-0137-z, 2006.

Mcculloch, I., Heeney, M., Bailey, C., Genevicius, K., MacDonald, I., Shkunov, M., Sparrowe, D., Tierney, S., Wagner, R., Zhang, W. M., Chabinyc, M. L., Kline, R. J., Mcgehee, M. D., and Toney, M. F.: Liquid-crystalline semiconducting polymers with high charge-carrier mobility, Nat. Mater., 5, 328–333, doi:10.1038/Nmat1612, 2006.

Nam, S., Chung, D. S., Jang, J., Kim, S. H., Yang, C., Kwon, S. K., and Park, C. E.: Effects of poor solvent for solution-processing passivation of organic field effect transistors, J. Electrochem. Soc., 157, H90–H93, doi:10.1149/1.3251337, 2010.

Scott, S. M., James, D., and Ali, Z.: Data analysis for electronic nose systems, Microchim. Acta, 156, 183–207, doi:10.1007/s00604-006-0623-9, 2006.

Sotzing, G. A., Phend, J. N., Grubbs, R. H., and Lewis, N. S.: Highly sensitive detection and discrimination of biogenic amines utilizing arrays of polyaniline/carbon black composite vapor detectors, Chem. Mater., 12, 593–595, doi:10.1021/Cm990694e, 2000.

Street, R. A., Chabinyc, M. L., and Endicott, F.: Chemical impurity effects on transport in polymer transistors, Phys. Rev. B, 76, 045208, doi:10.1103/Physrevb.76.045208, 2007.

Torsi, L., Tanese, M. C., Cioffi, N., Gallazzi, M. C., Sabbatini, L., and Zambonin, P. G.: Alkoxy-substituted polyterthiophene thin-film-transistors as alcohol sensors, Sensor. Actuat. B-Chem., 98, 204–207, doi:10.1016/j.snb.2003.10.007, 2004.

Torsi, L., Farinola, G. M., Marinelli, F., Tanese, M. C., Omar, O. H., Valli, L., Babudri, F., Palmisano, F., Zambonin, P. G., and Naso,

F.: A sensitivity-enhanced field-effect chiral sensor, Nat. Mater., 7, 412–417, doi:10.1038/Nmat2167, 2008.

Torsi, L., Marinelli, F., Angione, M. D., Dell'Aquila, A., Cioffi, N., De Giglio, E., and Sabbatini, L.: Contact effects in organic thin-film transistor sensors, Org. Electron., 10, 233–239, doi:10.1016/j.orgel.2008.11.009, 2009.

Wang, B. and Haick, H.: Effect of functional groups on the sensing properties of silicon nanowires toward volatile compounds, Acs. Appl. Mater. Inter., 5, 2289–2299, doi:10.1021/Am4004649, 2013a.

Wang, B. and Haick, H.: Effect of chain length on the sensing of volatile organic compounds by means of silicon nanowires, Acs. Appl. Mater. Inter., 5, 5748–5756, doi:10.1021/Am401265z, 2013b.

Wang, B., Cancilla, J. C., Torrecilla, J. S., and Haick, H.: Artificial sensing intelligence with silicon nanowires for ultraselective detection in the gas phase, Nano Lett., 14, 933–938, doi:10.1021/Nl404335p, 2014.

Wedge, D. C., Das, A., Dost, R., Kettle, J., Madec, M. B., Morrison, J. J., Grell, M., Kell, D. B., Richardson, T. H., Yeates, S., and Turner, M. L.: Real-time vapour sensing using an ofet-based electronic nose and genetic programming, Sensor. Actuat. B-Chem., 143, 365–372, doi:10.1016/j.snb.2009.09.030, 2009.

Zschieschang, U., Weitz, R. T., Kern, K., and Klauk, H.: Bias stress effect in low-voltage organic thin-film transistors, Appl. Phys. A-Mater., 95, 139–145, doi:10.1007/s00339-008-5019-8, 2009.

Influence of operation temperature variations on NO measurements in low concentrations when applying the pulsed polarization technique to thimble-type lambda probes

S. Fischer[1,2]**, D. Schönauer-Kamin**[1]**, R. Pohle**[2]**, M. Fleischer**[2]**, and R. Moos**[1]

[1]Department of Functional Materials, University of Bayreuth, Bayreuth, Germany
[2]Corporate Technology, Siemens AG, Munich, Germany

Correspondence to: R. Moos (functional.materials@uni-bayreuth.de)

Abstract. By applying the pulsed polarization technique, a thimble-type lambda probe can be used as a NO_x sensor in the low ppm range. Due to the robustness of the sensor in harsh exhaust gas environments, this approach has many opportunities for application. The temperature operating range for best NO sensing properties is a crucial parameter. The sensor temperature changes with the ambient gas temperature, but can be stabilized actively by internal heating in a certain temperature range. This study evaluates in detail the temperature influence on NO sensitivity, so that an optimum operating point can be derived from these results using a dynamic measurement technique. Stepwise NO concentration changes between 0 and 12.5 ppm in synthetic exhausts demonstrate the potential of the concept.

1 Introduction

Environmental issues like acid rain and smog are caused by oxides of nitrogen (NO_x), so that strong emission regulations have been introduced for automotive emissions. For active emission control of automotive exhausts, NO_x detection in the low ppm range is important – without cross-sensitivities to other exhaust components. Despite many approaches in the past years, which are reviewed in the literature (Zhuiykov and Miura, 2007; Fergus, 2007; Guth and Zosel, 2004), only one type of NO_x sensor has made it to series production. This sensor is a derivative of the planar oxygen sensor manufactured in planar zirconia ceramic technology (Riegel et al., 2002; Moos, 2005). Due to the complex setup of this amperometric double pumping cell (Siegberg and Kilinc, 2014; Kato et al., 1996), cost issues must always be considered, especially with respect to the low sensitivies of the sensor raw signal of only a few nA ppm^{-1} NO (Kato et al., 1996), which requires costly circuitry and may also lead to a reduced accuracy (Kim and Van Nieuwstadt, 2006).

The thimble-type lambda probe is a well-known robust and reliable sensor that can be operated especially in the exhausts of internal combustion engines between a few hundred °C and almost 1000 °C (Baunach et al., 2006). As shown in previous works, it is possible to operate a commercial thimble-type lambda probe as a promising NO sensor just by applying a so-called pulsed polarization technique (Fischer et al., 2010). For this method, voltage pulses of alternating signs with intermediate depolarization phases (Fischer et al., 2010, 2012a, 2013) are applied. The self-discharge voltage (open circuit voltage) after each polarization pulse is strongly affected if NO is present in the exhaust. Since the properties of this sensor element of oxygen ion conducting yttria stabilized zirconia (YSZ) with porous platinum electrodes also change widely with temperature, "temperature" is a parameter with a large influence on the sensor response to NO.

Thimble-type lambda probes can be heated internally by the already available internal rod-type heater so that the sensor temperature can be controlled independently of the outer

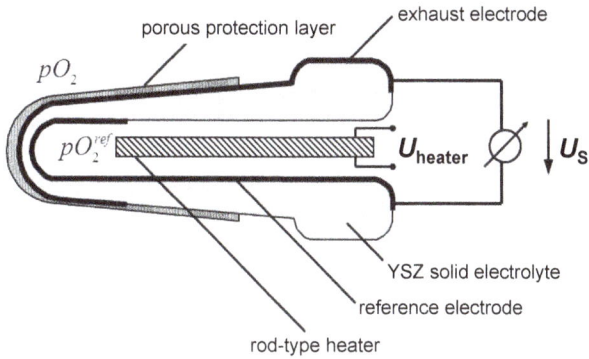

Figure 1. Thimble-type lambda probe (scheme, adapted from Moos, 2006).

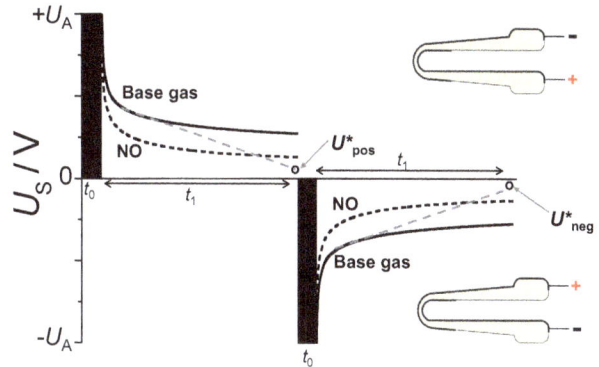

Figure 2. Scheme of the pulsed polarization technique with corresponding signs of electrode polarization of the thimble-type lambda probe and illustration of how U^* is evaluated.

exhaust gas temperature (in the case of a lower outer temperature compared to the desired sensor temperature).

This study raises and answers the question of whether the lambda probe can be applied as a NO sensor in a large temperature range only by controlling the heater voltage depending on ambient temperature. This information is essential for the development of an operating strategy and a possible sensor application. The target operation temperature is in the range of about 400 °C.

For that purpose, the discharge curves of the sensor after polarization pulses as well as polarization currents during voltage pulses are recorded at different external gas temperatures and internal heater voltages.

2   Experimental

2.1   Measurement setup

Thimble-type lambda probes are well known and have been described by many authors (Riegel et al., 2002; Moos, 2005; Baunach et al., 2006). The illustration of a commercial thimble-type sensor is shown in Fig. 1.

The porous Pt reference and exhaust electrodes on yttria stabilized zirconia (YSZ), which is an oxygen ion conductor at temperatures above 350 °C, separate the reference and the exhaust side. Due to the thimble-type design, the atmosphere of the inner electrode is on a well-known oxygen partial pressure reference level (pO_2^{ref}). Considering the changing oxygen content at the outer exhaust electrode (pO_2^{exhaust}), a sensor voltage according to Nernst forms. It depends on the ratio of the oxygen partial pressure at both electrodes (Riegel et al., 2002):

$$U_{\mathrm{Nernst}} = \frac{RT}{4F} \ln \frac{pO_2^{\mathrm{exhaust}}}{pO_2^{\mathrm{reference}}}, \tag{1}$$

with the known temperature T, the universal gas constant R and Faraday's constant F. The outer electrode is coated with a thick porous protection layer, mainly consisting of

$MgAl_2O_4$ spinel to ensure long-term stability in the harsh exhaust gas environment. The minimum sensor temperature of 350 °C for sufficient oxygen ion mobility is controlled by applying a heater voltage (U_{heater}) to an internal rod-type heater, so that a metering voltage (U_S) between the outer sensing and inner reference electrode occurs according to Eq. (1).

To evaluate the temperature range for NO sensing by applying the dynamic measurement technique, various temperatures are adjusted by different gas temperatures T_{gas} from 300 to 450 °C in 50 °C steps without internal heating. Additionally, the sensor response is investigated at different heating voltages. For that purpose, U_{heater} was varied from 0 to 10 V in 1 V steps, which are applied at a constant gas temperature $T_{\mathrm{gas}} = 350$ °C.

The gas composition used as a baseline for the measurements contains 10 % oxygen with a humidity level of 3 % absolute and a flow rate of 1 L min^{-1}. The NO response is evaluated by adding 12.5 ppm of this gas.

2.2   Pulsed polarization technique

The pulsed polarization technique is a dynamic measurement method. Voltage pulses of alternating signs but equal amplitudes and discharges pauses in between are applied (Fischer et al., 2010, 2012b). After polarization, the open circuit discharge curves are recorded during the discharge phase. Figure 2 illustrates a typical single-measurement process. The discharge curves are strongly dependent on the NO concentration in the exhaust. This is also schematically depicted in Fig. 2. The signs of the polarization pulses as related to the design of the thimble-type lambda probe are depicted as well. During the negative voltage pulse (as it is defined here), the negatively charged oxygen ions are pumped to the outer atmosphere driven by the higher potential of this electrode. During opposite polarization sign, the ions are transported to the inner reference electrode.

To obtain a single NO concentration value, for example at defined parameters T_{gas} and U_{heater} or at a defined gas composition, the evaluation of the whole discharge curve with many measured points would be too time-consuming. Therefore, only one parameter characterizing the discharge curve is used. The slope of the discharge curves dU/dt, which is fitted between 2 and 3 s after the polarization pulse, is used as an indicator for the analyte concentrations in the investigated exhausts. The extrapolation to 10 s defines the sensor output signal, the voltage U^*. The process of how U^* is evaluated is also illustrated in Fig. 2 for positive and negative polarization.

Many parameters of the dynamic measurement technique can be varied to optimize the setup for NO sensing. The optimum parameter for NO detection by the pulsed polarization method was found in a previous study by applying a polarization amplitude of $U_A = 2.5$ V with a pulse duration of $t_0 = 1$ s and a subsequent discharge time duration of $t_1 = 10$ s. Using these defined voltage pulses of alternating sign, high NO sensitivities in a low ppm range are measured as already shown previously (Fischer et al., 2010, 2012a, 2013).

In addition to discharge voltage evaluation between alternating polarizations, the polarization current (I_{pol}) during the polarization pulses is recorded. The current value is temperature-dependent, since the resistance of the oxygen ion conducting YSZ changes. Therefore, this parameter can be used as a good temperature indicator.

3   NO sensitivity at temperature variation

As described in Sect. 2.1, the sensor temperature is varied actively using different heating voltages and passively only by changing the gas temperatures. The NO sensitivities are evaluated for each of these methods.

3.1   Sensor operation temperature dependence due to gas temperature variations

When increasing the ambient exhaust temperature, the sensor temperature changes accordingly. The polarization current curves during positive and negative polarization pulses are shown in Fig. 3 for gas temperatures in the range from $T_{gas} = 300$ to $450\,°$C. For that investigation, the heater voltage U_{heater} was set to zero, so that no additional heating occurred. Due to heat loss through the sensor mount, the sensor temperature is roughly 10–20 °C less than the gas temperature. As already known from the literature (Fischer et al., 2013), the pulse voltage was $U_A = 2.5$ V for a pulse duration of $t_0 = 1$ s.

The final values after 1 s were further evaluated. Due to high resistance at low temperatures, a current of only 35 μA at 300 °C flows, which is too low for electrode polarization including charge separation. Only if the temperature exceeds about 350 °C is the oxygen ion mobility of YSZ sufficiently high enough so that a noticeable and time-dependent polar-

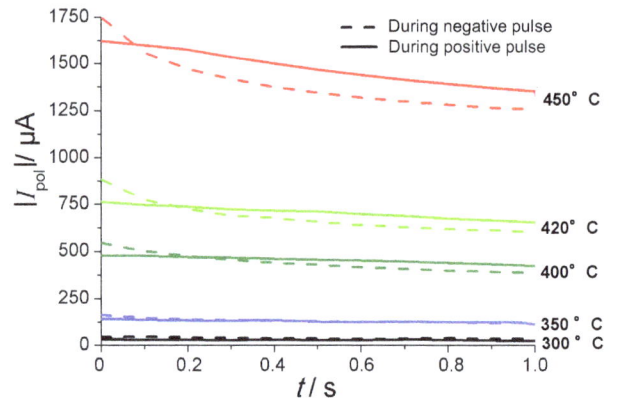

Figure 3. Polarization currents I_{pol} during positive and negative pulses at various gas temperatures (T_{gas}) as indicated.

ization current occurs. At 450 °C, the current is increased by a factor of 40 compared to 300 °C.

The current of both polarization processes with opposite signs can be separated into two phases. After a time-dependent phase at the beginning, the current becomes almost stationary. The time dependency can be seen above 350 °C. It is assumed that starting from that point, a sufficient electrode polarization with associated charge separation occurs. During the stationary current at the end of the polarization process, only oxygen ions were pumped through the ion-conducting YSZ without an effect on charge separation and polarization magnitude.

The temperature-dependent polarization currents for positive and negative voltage pulses are almost the same, but a closer look reveals differences in their time-dependent behavior. At the beginning of pulses, the absolute value of the current during negative polarization pulses is higher level and decreases more strongly with time compared to the opposite sign. The stationary current at the end of polarization is, on the other hand, clearly lower. This may be caused by two effects. The stationary current at the end of polarization, which can be interpreted as oxygen pumping through the ion conductor YSZ without an additional effect of polarization, is dependent on transport of oxygen to the electrode surface and removal of oxygen from the surface at the opposite electrode. These differences in oxygen transport during both polarization signs can be explained by different gas volumes, which are available for oxygen exchange. The reference atmosphere at the inside of the thimble-type sensor contains less gas volume so that the oxygen transport is limited compared to oxygen transport from the outer gas atmosphere to the inner reference electrode. The current differences at the beginning of the polarization, however, are caused by different depolarization voltages at the end of the depolarization pause. This is reported in detail in Fig. 5a and b below. After a positive polarization pulse, the voltage is at a higher voltage level (e.g., at 420 °C: $\sim +300$ mV) compared to the opposite polariza-

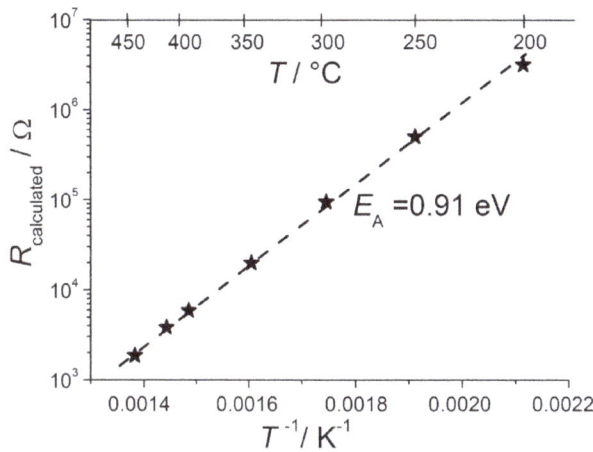

Figure 4. Arrhenius-like dependency of the calculated resistance with an activation energy of 0.91 eV.

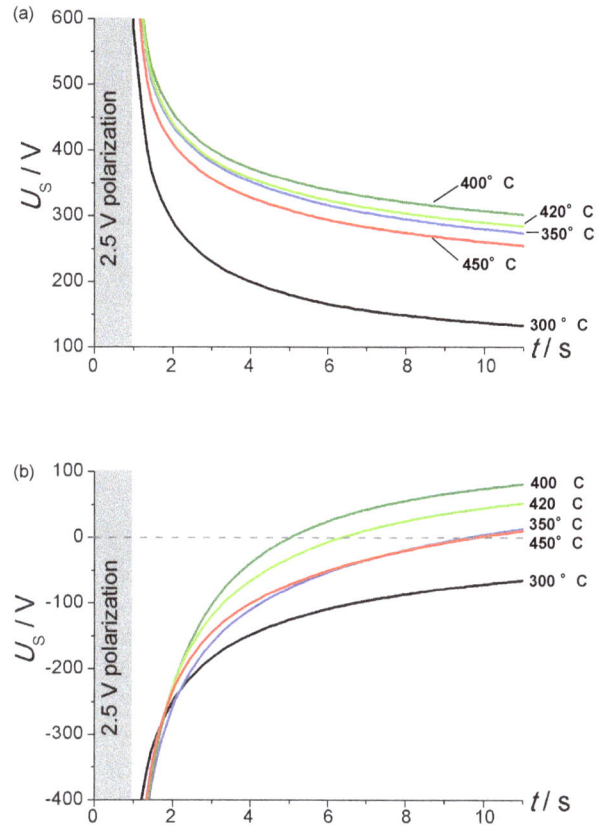

Figure 5. (a) Discharge voltage after positive voltage pulse at various temperatures. (b) Discharge voltage after negative voltage pulse at various temperatures.

tion sign (e.g., at 420 °C: $\sim +50$ mV). The effective voltage difference $|U_{\text{end of epolarization}} - U_{\text{subsequent voltage pulse}}|$ to the subsequent 2.5 V voltage pulse is higher at negative polarization [$+300$ mV $-(-2.5$ V$) = 2.8$ V] compared to positive polarization [50 mV -2.5 V $= -2.45$ V]. With this consideration the higher polarization current of 120 μA at 420 °C at polarization with a negative voltage pulse can be explained by the Ohmic law $\Delta I = \Delta U / R$ with respect to a resistance of 3 kΩ[$\Delta I = 350$ mV $/3$ k$\Omega \sim 120\,\mu$A].

Due to the Ohmic law $U = R \cdot I$, the temperature-dependent current correlates with a decreasing resistance. During polarization pulses with applied voltage amplitudes $U_A = 2.5$ V each, the following resistances calculated from the final values of the polarization current (I_{pol}) after 1 s were further evaluated, and their temperature dependency is shown in the Arrhenius-like diagram of Fig. 4 according to $R_{\text{calculated}} = U_A / I_{\text{pol}}$. For that evaluation the mean values after both polarization signs are used.

The activation energy of 0.91 eV agrees with the literature and is typical for 8 mol % stabilized zirconia (Badwal, 1992).

After a polarization pulse of 1 s, the self-discharge curves (open circuit voltages) are recorded for 10 s between the pulsed voltages with alternating sign as shown in Fig. 5a and b.

At temperatures below 300 °C, discharging after applied voltage pulses is very fast after both polarization signs, so that already 1 s after the applied voltage pulse voltages, only a few 10 mV occur. These measurements are not shown here, because at this low temperature regime almost no oxygen ion conductivity exists and polarization is negligible. At 300 °C, a polarization including charge separation occurs, so that a distinct discharge voltage is measured after both polarization signs, whereby the voltage after a positive polarization sign is 50 mV higher. This polarization correlates with a significant polarization current as already discussed.

After positive voltage pulses, the voltage is shifted to a more positive voltage up to 400 °C, so that the discharging is delayed further on by increasing temperature. However, it is apparent from observing alternating polarization signs that there is another underlying effect due to the completely different behavior after negative voltage pulses. Due to oxygen contents of 10 % at the outer electrode and 20 % at the reference electrode, an offset voltage occurs starting from this temperature range. After negative voltage pulse, the offset voltage is visible at 350 °C and especially at 400 °C, with a corresponding voltage of 80 mV after 10 s.

This offset voltage occurs due to the Nernst Eq. (1) $U_{\text{Nernst}} = \frac{RT}{4F} \ln \frac{p\text{O}_2^{\text{exhaust}}}{p\text{O}_2^{\text{ref}}}$, which indicates the oxygen gradient between the electrodes and increases with temperature. Thus, the same voltage in sign and amplitude overlies the whole discharge curve at temperatures at 350 °C and higher.

The slowest discharge occurs at 400 °C with a corresponding voltage of 300 mV after 10 s. At higher temperatures (420 and 450 °C) the sample discharges faster for both polarization signs. The resistance R decreases by further increasing temperatures (cf. Fig. 4), which leads to the observed faster discharge. At these high temperatures, the voltage decay is

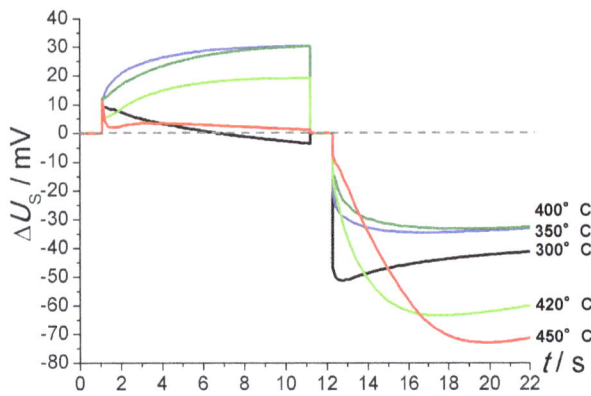

Figure 6. Voltage differences ΔU_S with 12.5 ppm compared to a measurement without NO in the base gas. 1–11 s: discharge after positive polarization. 12–22 s: voltage difference after negative polarization. Parameter: T_{gas}.

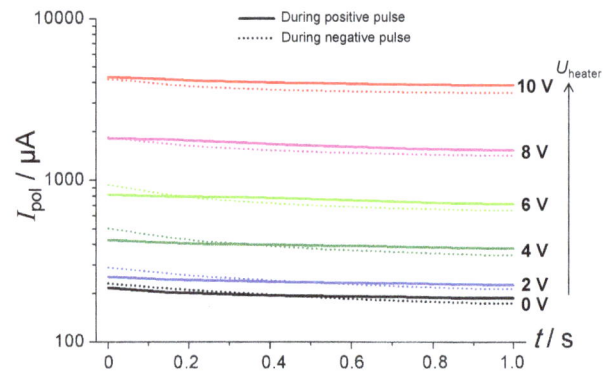

Figure 7. Polarization currents I_{pol} (logarithmic scale) at various heater voltages and $T_{gas} = 350\,°C$.

accelerated despite the strong polarization and overlying offset voltage. Thus, the decreasing resistance dominates the discharge behavior and the Nernst voltage is only less visible.

The main question is, however, how the temperature variation influences the NO sensitivity. For that purpose, 12.5 ppm NO are dosed additionally to the base gas. To evaluate the NO effect on the discharge voltage, the voltage differences $\Delta U_S(t) = U_S^{base\ gas}(t) - U_S^{12.5\ ppm\ NO}(t)$ between the curves with NO and without NO in the base gas are calculated for the whole discharge process as shown in Fig. 6.

The voltage differences regarding 12.5 ppm NO are different for both polarization signs and are dependent on the discharge time. At the lowest investigated temperature of 300 °C, almost no effect of NO after positive polarization occurs, whereas after opposite polarization sign, a voltage difference of around 40–50 mV occurs. The maximum signal is measured at the beginning of the discharge and decreases later slightly. At higher temperatures (350–400 °C) the time-dependent voltage differences ΔU_S are similar in curve shape and amplitude for both polarizations of opposite sign. This corresponds to accelerated discharging in NO-containing atmospheres, in contrast to the base gas.

After positive voltage, the response to NO decreases when the temperature is further increased, so that at the highest investigated temperature of 450 °C there is only an almost constant voltage difference ΔU_S of a few mV. In contrast to that, the NO effect after negative polarization voltage is markedly increased at temperatures above 400 °C, so that a strong time-dependent voltage response is measured. The maximum voltage difference regarding 12.5 ppm NO occurs at 450 °C and it is about 70 mV at the end of the discharge time.

3.2 Sensor operation temperature dependence due to heater voltage variations

Applying a defined heater voltage is an alternative method to vary the sensor operation temperature. The active heating of the sensor causes a temperature difference ΔT between the inner and outer electrodes, in contrast to passive heating by changing the gas temperature as described in Sect. 3.1. To reduce this temperature difference ΔT and an associated thermoelectric voltage overlaying the discharge curves, the ambient gas temperature was set to 350 °C and the heater voltage was varied in 1 V steps between 0 and 10 V. The polarization current was measured during polarization pulses of 1 s ($U_A = 2.5\,V$). The results are plotted in Fig. 7; for clarity reasons, only results for even heater voltages are displayed.

With increasing heater voltages, the current increases as well, from almost $200\,\mu A$ without additional heating at a gas temperature of 350 °C up to about 4 mA with an additional heating voltage of 10 V. In agreement with Fig. 3, the polarization current is time-dependent at temperatures of 350 °C, which corresponds to the measurement without active heating. By increasing the temperature, the time dependency becomes stronger, so that a current difference of $800\,\mu A$ is recorded between the beginning and the end of polarization pulse at a heating voltage of 10 V (due to the logarithmic scaling of the ordinate, this effect is barely visible in Fig. 7). The difference between the opposite polarization signs is less marked compared to the gas temperature influences, but still noticeable. Thus, the same explanation as before can be supposed: the various gas volumes, which are available at outer and inner electrodes, and the different amount of pumped oxygen ions causes the differences.

As before, the resistance R is calculated from the polarization current during the polarization pulse.

Due to the temperature increase, the resistance $R_{calculated}$ drops down from nearly 13 kΩ without active heating to 630 Ω at $U_{heater} = 10\,V$ (Fig. 8). If one compares the resistances with gas temperature variation with the one shown in Fig. 4, the highest investigated temperature of 450 °C corre-

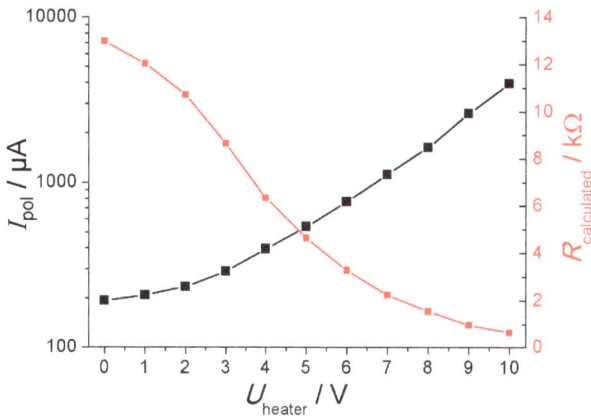

Figure 8. Influence of the heater voltage on the polarization current I_{pol} and the calculated resistance $R_{calculated}$ at $T_{gas} = 350\,°C$.

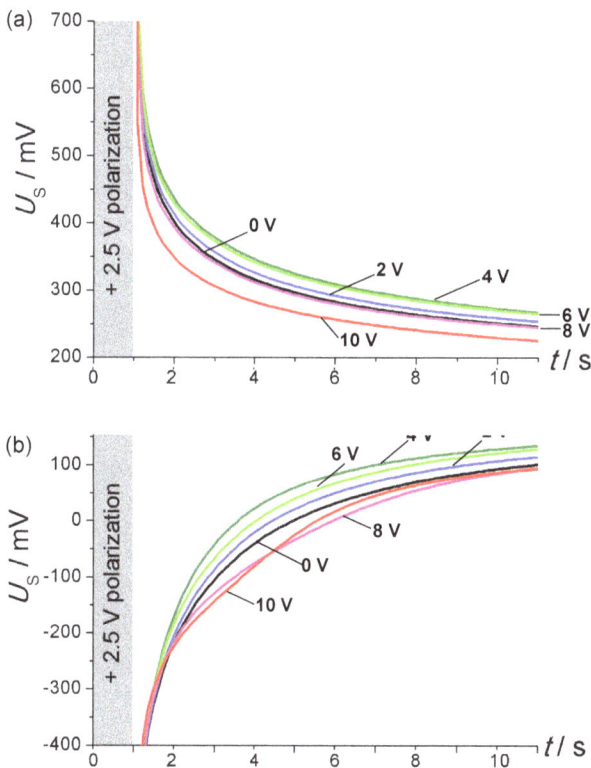

Figure 9. (**a**) Discharge voltage after positive voltage pulse at various heater voltages and $T_{gas} = 350\,°C$. (**b**) Discharge voltage after negative voltage pulse at various heater voltages and $T_{gas} = 350\,°C$.

sponds to an additional heating voltage of about 8 V, in conjunction with a gas temperature of 350 °C.

The influence of the active heating on the discharge curves at base gas is shown in Fig. 9a and b for both polarization signs.

As shown in the section before, the discharge curves after opposite signs are not symmetrical with the zero line due to the non-symmetric sensor design and the overlaying offset

voltage according to Nernst. Without additional active heating, the discharge curves are equal to the curve for 350 °C shown in Fig. 5a and b, including strong polarization due to sufficiently high charge transport. The discharge curve of 4 V is on the most positive voltage level and almost equal to the results obtained for 400 °C. Furthermore, the polarization currents also agree quite well.

At heater voltages up to 4 V both discharge curves are shifted to more positive voltages. With increasing sensor temperature, a Nernst voltage according to Eq. (1) forms due to the oxygen partial pressure difference between both electrodes. It adds to the open circuit voltage.

In contrast to temperature variation only by gas temperature, an additional offset voltage caused by active heating can be seen. We assume that a thermoelectric voltage occurs due to the temperature difference between inner and outer electrodes generated by the rod-type heater at the inside of the thimble-type lambda probe. Furthermore, heating voltages above 4 V cause a faster discharge after both polarization signs, which agrees with the aforementioned results. The decreasing resistance of the ion conductor with higher temperatures causes this accelerated discharging. Only at the highest voltage of 10 V is the curve shape after negative voltage pulse changed compared to 8 V, but at the end of discharging the voltages are both about 90 mV.

Again, the sensor response to NO when applying different heater voltages is evaluated by the voltage difference $\Delta U_S(t) = U_S^{base\ gas}(t) - U_S^{12.5\,pp\,NO}(t)$. The results are shown in Fig. 10 (analogously to Fig. 6).

In good agreement with temperature variation results that were obtained when solely the gas temperature was varied, discharging in a NO-containing atmosphere is also faster and the temperature dependencies are also almost the same. After positive voltage pulses, the voltage difference between 12.5 ppm NO in the base gas and the base gas continuously decreases with increasing heater voltages. At heater voltages of 8 and 10 V, almost no NO effect can be seen, so that the discharge curves are almost equal, except for a few mV.

After negative polarization pulses, the NO effect is present for all investigated heating amplitudes. Without active heating, a maximum voltage difference of 20 mV occurs, which increases up to almost 120 mV at 8 V. All curves in NO discharge faster compared to the base gas. The NO effect increases with temperature up to a heating voltage of 8 V. Additionally, the evaluated voltage difference increases with discharging time at higher temperatures. Applying higher heater voltages (10 V), the NO influence is lower and the maximum voltage difference is shifted to shorter discharging times.

All in all, the highest NO sensitivity is achieved by applying an additional heating voltage of 8 V at a gas temperature of 350 °C, so that a sensor response of about 120 mV occurs as a response of 12.5 ppm NO.

The NO sensitivity to different NO concentrations is evaluated for 0.5–12.5 ppm NO. The sensor characteristics were

Table 1. Dependency of the sensitivity to NO for several heater voltages, each after positive and negative voltage pulse. $T_{gas} = 350\,°C$.

U_{heater}/V	0	2	4	6	8	10
NO sensitivity after positive pulse in mV decade^{-1}	20.0	22.0	24.9	15.9	−3.1	−4.8
NO sensitivity after negative pulse in mV decade^{-1}	−12.0	−14.0	−21.5	−56.7	−88.0	−50.7

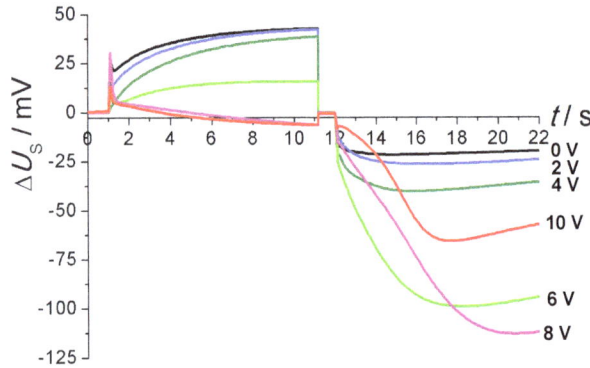

Figure 10. Voltage differences ΔU_S with 12.5 ppm NO in the base gas compared to a measurement without NO in the base gas. 1–11 s: discharge after positive polarization. 12–22 s: voltage difference after negative polarization. $T_{gas} = 350\,°C$. Parameter: heater voltages.

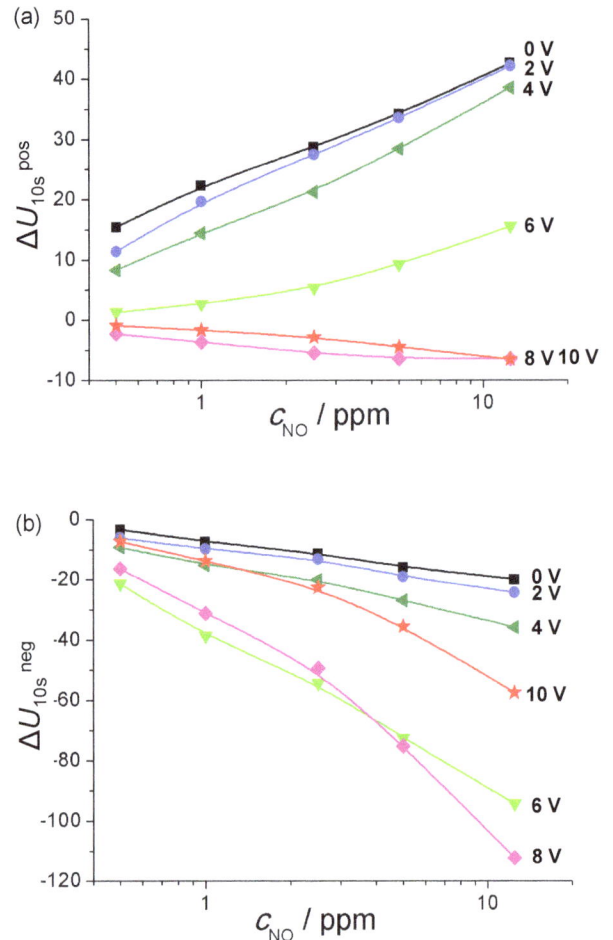

Figure 11. (a) Sensor characteristic towards small NO concentration obtained by evaluating voltage difference $\Delta U_{10\,s}^{pos}$ after positive voltage pulse at various heater voltages and $T_{gas} = 350\,°C$. (b) Sensor characteristic towards small NO concentration obtained by evaluating voltage difference $\Delta U_{10\,s}^{neg}$ after negative voltage pulse at various heater voltages and $T_{gas} = 350\,°C$.

obtained by evaluating the voltage difference after 10 s (right before the next polarization starts) between the curve in the base gas with and without NO. The sensor output in Fig. 11 is defined by $\Delta U_{10\,s} = U_{10\,s}^{NO} - U_{10\,s}^{base\,gas}$.

The characteristics show a semi-logarithmic dependency for all investigated heater voltages at NO concentrations above 1 ppm. The deviations at even lower NO concentrations from the semi-logarithmic behavior may be due to inaccuracies when dosing such small amounts of NO.

As discussed, the NO sensitivity decreases with increasing heater voltage after a positive voltage sign, whereas after the opposite polarization sign, the influence is vice versa. The highest sensitivity of −88.0 mV per decade NO is achieved at a heater voltage of 8 V after negative voltage pulses. Using these polarization parameters, there is almost no response after positive voltage pulses. The NO sensitivities at different heater voltages after both polarization signs are summarized in Table 1.

As illustrated in Fig. 12, the evaluation of ΔU^*, which is described in Sect. 2.2 and illustrated in Fig. 2, can be considered a very suitable sensor signal. Although only the slope of the discharge curve between 2 and 3 s after voltage pulse is determined every 22 s (the cycle time contains 1 s for polarization and 10 s for discharging for each polarization sign), the NO gas dosing program, which consists of 5 min long steps for each concentration, is reflected very clearly, espe-

cially at higher heater voltages as already discussed. The sensor signal ΔU^* at the beginning and the end of the gas dosing process is almost identical, too.

For application in exhausts with variable gas temperatures, a constant NO response in a wide temperature range is required. Therefore, the sensor temperature should be adjusted independently of the outside temperature. The resis-

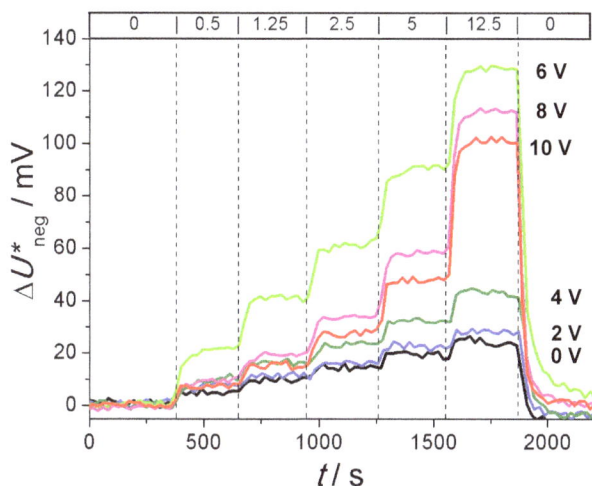

Figure 12. Sensor signal ΔU^*_{neg} (for a definition, see Fig. 2) after negative voltage pulse at various heater voltages and $T_{\text{gas}} = 350\,°\text{C}$. The dosed NO concentrations in the base gas are shown on top of the curve (in ppm).

tance (or also the impedance) between the inner electrode and the outer electrode is an obvious parameter corresponding to this sensor characteristic. According to these results, an impedance control of the lambda probe seems to further improve the applicability of the thimble-type lambda probe for NO sensing using the pulsed polarization technique.

4   Conclusion and outlook

The sensor operation temperature of a thimble-type lambda probe was varied by different gas temperatures as well as by different heater voltages. The sensors were operated in the pulsed polarization technique, and the discharge curves after both polarization signs as well as the sensor responses regarding low NO concentrations (10 ppm range) were evaluated in detail. According to the asymmetric sensor setup that includes a reference atmosphere at the inner electrode, the discharge curves were different for opposite polarization signs. This causes different NO response signals for both polarization signs. Impedance (or resistance) control of the sensor electrodes is suggested to compensate for exhaust gas temperature fluctuations in order to keep the NO sensitivity constant.

The robustness of the sensor in harsh exhaust gas environments offers many opportunities. However, with respect to the reference electrode, one big disadvantage is immanent in the setup: the high NO sensitivity occurs only after one polarization sign, so that the response time is not high enough: 22 s per polarization cycle may be too slow for automotive applications.

A further sensor development should therefore step away from the thimble-type design, but should instead use a planar symmetric design (with electrodes on each side and both electrodes facing the exhaust). Such a setup would not need a reference atmosphere. This symmetric setup should lead to a symmetric polarization and depolarization behavior, which allows the use of positive and negative polarization signs. As a result, the response time should be halved.

References

Badwal, S. P. S.: Zirconia-based solid electrolytes: microstructure, stability and ionic conductivity, Solid State Ionics, 52, 23–32, 1992.

Baunach, T., Schänzlin, K., and Diehl, L.: Sauberes Abgas durch Keramiksensoren, Physik Journal, 5, 33–38, 2006 (in German).

Fergus, J. W.: Materials for high temperature electrochemical NO_x gas sensors, Sensor Actuat. B-Chem., 121, 652–663, 2007.

Fischer, S., Pohle, R., Farber, B., Proch, R., Kaniuk, J., Fleischer, M., and Moos, R.: Method for detection of NOx in exhaust gases by pulsed discharge measurements using standard zirconia-based lambda sensors, Sensor Actuat. B-Chem., 147, 780–785, 2010.

Fischer, S., Schönauer-Kamin, D., Pohle, R., Magori, E., Farber, B., Fleischer, M., and Moos, R.: NO_x-detection by pulsed polarization of lambda probes, 14th International Meeting on Chemical Sensors, IMCS 14, Nuremberg, Germany, 20–23 May 2012, 1050–1053, doi:10.5162/IMCS2012/P1.6.4, 2012a.

Fischer, S., Pohle, R., Magori, E., Schönauer-Kamin, D., Fleischer, M., and Moos, R.: Pulsed Polarization of Platinum Electrodes on YSZ, Solid State Ionics, 225, 371–375, 2012b.

Fischer, S., Schönauer-Kamin, D., Pohle, R., Fleischer, M., and Moos, R.: NO Detection by Pulsed Polarization of Lambda Probes – Influence of the Reference Atmosphere, Sensors, 13, 16051–16064, 2013.

Guth, U. and Zosel, J.: Electrochemical solid electrolyte gas sensors – hydrocarbon and NO_x analysis in exhaust gases, Ionics, 10, 366–377, 2004.

Kato, N., Nakagaki, K., and Ina, N.: Thick film ZrO_2 NO_x sensor, SAE Paper 960334, doi:10.4271/960334, 1996.

Kim, Y. W. and van Nieuwstadt, M.: Threshold monitoring of urea SCR systems, SAE Paper 2006-01-3548, doi:10.4271/2006-01-3548, 2006.

Moos, R.: A Brief Overview on Automotive Exhaust Gas Sensors Based on Electroceramics, Int. J. Appl. Ceram. Tec., 2, 401–413, 2005.

Moos, R.: Automotive Exhaust Gas Sensors, in: Encyclopedia of Sensors, edited by: Grimes, C. A., Dickey, E. C., and Pishko, M. V., 1, 295–312, American Scientific Publishers, Valencia, CA, USA, 2006.

Riegel, J., Neumann, H., and Wiedenmann, H.-M.: Exhaust gas sensors for automotive emission control, Solid State Ionics, 152–153, 783–800, 2002.

Siegberg, D. and Kilinc, M.: Thermal and Chemical Robustness of the Smart NOx-Sensor, 12th CTI International Conference: Ex-

haust Systems – Euro VI and Beyond – Focus on CO_2 Reduction,
Mainz, Germany, 28–29 January 2014.

Zhuiykov, S. and Miura, N.: Development of zirconia-based poten-
tiometric NO_x sensors for automotive and energy industries in
the early 21st century: What are the prospects for sensors?, Sensor Actuat. B-Chem., 121, 639–651, 2007.

3

Magnetic noise contribution of the ferromagnetic core of induction magnetometers

C. Coillot[1], M. El Moussalim[2], E. Brun[3], A. Rhouni[2], R. Lebourgeois[3], G. Sou[4], and M. Mansour[2]

[1]BioNanoNMRI-group, Laboratoire Charles Coulomb (L2C), Universite de Montpellier, Place Eugene Bataillon,
34095 Montpellier, France
[2]Laboratoire de Physique des Plasmas (LPP), Ecole Polytechnique, Route de Saclay, 91128 Palaiseau, France
[3]Thales Research and Technology, Palaiseau, France
[4]Laboratoire d'Electronique et d'Electromagnetisme (L2E), Université Pierre et Marie Curie, Paris, France

Correspondence to: C. Coillot (christophe.coillot@univ-montp2.fr)

Abstract. The performance of induction magnetometers, in terms of resolution, depends both on the induction sensor and the electronic circuit. To investigate accurately the sensor noise sources, an induction sensor, made of a ferrite ferromagnetic core, is combined with a dedicated low voltage and current noise preamplifier, designed in CMOS $0.35\,\mu$m technology. A modelling of the contribution of the ferromagnetic core to the noise through the complex permeability formalism is performed. Its comparison with experimental measurements highlight another possible source for the dominating noise near the resonance.

1 Introduction

Induction magnetometers are used in a wide range of applications (Ripka, 2000; Coillot, 2013) to measure extremely weak magnetic fields over a wide frequency range (from mHz up to GHz). At 1 Hz, for magnetotelluric waves observation purposes, noise equivalent magnetic induction about $0.2\,\mathrm{pT\,Hz^{-1}}$ is reported in Bin (2013). The context of this work concerns the study and the design of an induction magnetometer in the very low–low frequency (VLF–VF) range to investigate plasma waves in space in Jupiter's environment for an ESA mission. For this purpose, the goal of electromagnetic wave measurement, given in terms of noise equivalent magnetic induction (NEMI in $\mathrm{T\,Hz^{-1}}$), is challenging. An ability to reach NEMI lower than $10\,\mathrm{fT\,Hz^{-1}}$ at 10 kHz is mandatory. Due to the severe radiation environment, it has been considered to locate the preamplifier either inside the hollow ferromagnetic core of the induction sensor (Grosz, 2010) or inside the mechanical tri-axis structure to take advantage of an efficient radiation shielding provided by the sensor itself. An ASIC preamplifier designed in $0.35\,\mu$m CMOS technology (Rhouni, 2012; Ozaki, 2014)

offers a possibility of achieving very efficient induction magnetometers. In the context of this work we designed an ASIC low noise preamplifier (called MAGIC2) which offers especially low noise parameters which make possible the investigation of the noise source of the induction magnetometer. This work aims to extend the induction magnetometer modelling presented in Coillot and Leroy (2012) by introducing the noise source arising from the ferromagnetic core based on physical modelling of the complex permeability. Usually this noise source, which is frequency dependent, can be hidden by other dominating noise sources (especially the equivalent input current noise from the preamplifier). However, the use of a low input current noise preamplifier permits one to enhance the noise source from the sensor itself near the resonance. For this purpose, a comparison between the modelling and the measurement of the NEMI is performed on a 12 cm length sensor using a commercial Mn–Zn ferrite core (3C95 from Ferroxcube) of diabolo shape.

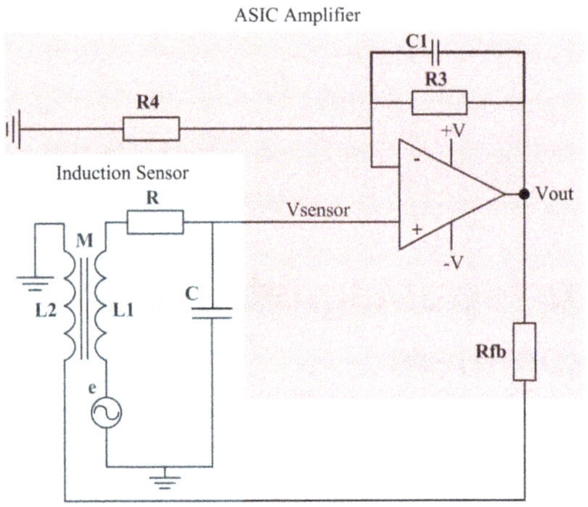

Figure 1. Feedback flux principle.

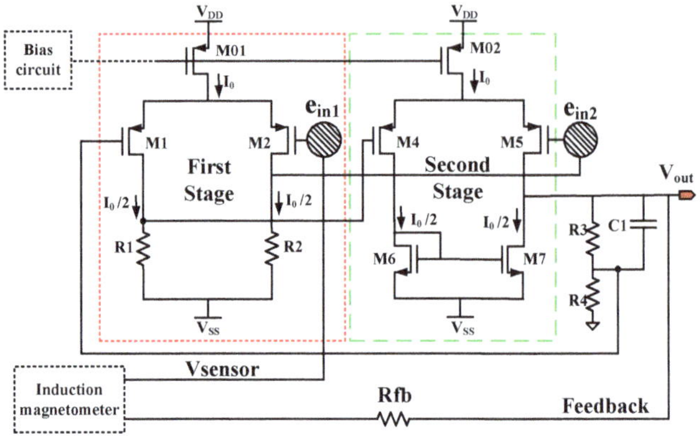

Figure 2. Schematic of the ASIC amplifier design.

2 Induction magnetometer using feedback flux: generalities

In this section we briefly remind the reader of the basis of an induction magnetometer using feedback flux. Induction sensors are basically built with an N turns coil of section S. When the coil is wound around a ferromagnetic core, the induced voltage is multiplied by a factor μ_{app} known as apparent permeability (described in Sect. 4.1). In harmonic regime at angular frequency ω, the induction voltage is written as

$$e = -j\omega NS\mu_{\mathrm{app}}B, \tag{1}$$

where $j^2 = -1$ is the imaginary unit and B is the magnetic flux density to be measured. The electrokinetic modelling assumes that the induced voltage e is in series with the resistance R and the inductance L, while the accessible voltage (V_{out}) is got at the capacitance C terminals. The transmittance of the induction sensor exhibits a resonance at angular frequency $\omega_0 = 1/\sqrt{(LC)}$. In order to remove the resonance, two kinds of electronic conditioning are classically implemented: a feedback flux amplifier or a transimpedance amplifier (Tumanski, 2007). In this work, we will focus only on the feedback flux amplifier schematically presented in Fig. 1.

The transmittance of the feedback flux amplifier is expressed as

$$T(j\omega) = \frac{V_{\mathrm{OUT}}}{B} = \frac{-jNSG\mu_{\mathrm{app}}\omega}{(1 - LC\omega^2) + j\omega(RC + \frac{GM}{R_{\mathrm{fb}}})}, \tag{2}$$

where j is the unity imaginary number, G is the voltage gain of the amplifier, M is the mutual inductance between the measurement winding and the feedback one and R_{fb} is the feedback resistance. In the following section we will focus on the ASIC amplifier design and its noise parameters.

3 Low voltage and current noises ASIC amplifier design for feedback flux induction magnetometers

To preserve the sensor noise performances in terms of NEMI, the equivalent input voltage noise (e_{PA}) and the input current noise (i_{PA}) of the amplifier must be as low as possible with a special awareness of $1/f$ noise. The requirement of the ASIC amplifier design is to satisfy $3\,\mathrm{nV\,Hz^{-1}}$ of equivalent input voltage noise and a few tens of $\mathrm{fA\,Hz^{-1}}$ of equivalent input current noise on a frequency range from $10\,\mathrm{kHz}$ up to $1\,\mathrm{MHz}$. The gain needs to be about $50\,\mathrm{dB}$ to be suited to the 16 bit ADC and the power consumption should be lower than $30\,\mathrm{mW}$. In this context, CMOS technology, which is mainly composed of MOSFET transistors, is an adequate solution. In the following section, design steps, voltage noise modelling and some measurement results of the low noise ASIC preamplifier are given.

3.1 Open loop noise considerations

It is detailed in Rhouni (2012) that, for the same gate size (W/L), the $1/f$ noise of a PMOS transistor is lower than a NMOS one. To achieve a very low noise performance and a high gain, the amplifier is composed of two stages (Fig. 2): the first stage represents the main contribution to the total output noise, while the second one aims to increase the gain of the open loop amplifier. The first stage is a simple PMOS differential pair with resistive charges (R_1 and R_2). In this configuration the input transistor (M_1 and M_2) design is related to the low input voltage noise performance, while the combination of the drain resistance (R_1) and the transconductance $g_{\mathrm{m_1}}$ of the input transistors M_1 and M_2 is used to set the gain (A) of the differential pair:

$$A = g_{\mathrm{m_1}}R_1. \tag{3}$$

By considering the thermal noise of the input pair transistor, the low-frequency noise from the input pair transistor and the thermal noise arising from the drain resistance, the power spectrum density of the equivalent input noise (e_{in1}^2) of the preamplifier's first stage can be obtained:

$$e_{in1}^2 = 2\left(\frac{8kT}{3g_{m_1}} + \frac{KFI_{d1}^{AF}}{C_{ox}L_1W_1fg_{m_1}^2} + \frac{4kT}{g_{m_1}^2R_1}\right), \quad (4)$$

where k is the Boltzmann constant, T is the temperature in K, $I_{d1} = I_0/2$ is the drain resistance, W_1 is the channel width, and L_1 is the gate length of the input transistors. AF ($= 1.4$) and KF ($= 1.8e-26$) are PMOS noise parameters. Knowing that $g_{m_1} = \sqrt{2I_{d1}K'_P\frac{W_1}{L_1}}$ and $I_0 = 2I_{d1}$, we get

$$e_{in1}^2 = \frac{16kT}{3\sqrt{I_0K'_P\frac{W_1}{L_1}}} + \frac{KF/2^{AF}}{K'_PC_{ox}W_1^2f}I_0^{AF-1} + \frac{8kT}{I_0K'_P\frac{W_1}{L_1}R_1}, \quad (5)$$

where I_0 is the bias current and $K'_P = C_{ox}\mu_p$.

As shown in the equation, e_{in1} depends on the transistor gate size (W_1/L_1) and the bias current I_0. A trade-off between the size and the power consumption of the final circuit must be specified considering the noise objective. Increasing the transistor size allows one to decrease its $1/f$ noise contribution. The current is also set to help in reducing the noise regarding the power consumption and the gain. In order to achieve $3\,\text{nV}\,\text{Hz}^{-1}$ at $10\,\text{kHz}$ of equivalent input noise and a minimum gain $A_1 = 30\,\text{dB}$, I_0 was set to $2\,\text{mA}$, $W_1 = W_2 = 5000\,\mu\text{m}$, $L_1 = L_2 = 1.2\,\mu\text{m}$ and $R_1 = R_2 = 3\,\text{k}\Omega$.

The second stage, which is a PMOS differential pair (M4–M5) with a NMOS load (M6–M7), will allow one to enhance the open loop gain to achieve the closed loop gain specification ($G_{dB} = 50\,\text{dB}$) over the desired frequency bandwidth ($> 50\,\text{kHz}$). This stage contains a minimum number of transistors to save power consumption, silicon area and specifically the noise performance of the first stage. The power spectrum density of the voltage noise of this stage referred to as M4–M5 input is written as

$$e_{in2}^2 = \frac{2B_PI_0^{AF-1}}{W_5^2f}\left(1 + \frac{K'_NB_N}{K'_PB_P}\frac{L_5W_5}{L_7^2}\right) \quad (6)$$
$$+ \frac{16kT}{3\sqrt{I_0K'_P\frac{W_5}{L_5}}}\left(1 + \sqrt{\frac{K'_NW_7L_5}{K'_PW_5L_7}}\right),$$

with $B_N = KF/(2K'_NC_{ox})$ and $B_P = KF/(2K'_PC_{ox})$.

The second stage provides a gain $A_2 = 55\,\text{dB}$ if $W_4 = W_5 = 5000\,\mu\text{m}$ and $W_6 = W_7 = 50\,\mu\text{m}$.

The total equivalent input noise voltage e_{PA} is finally calculated using the PSDs of each stage e_{in1}^2 and e_{in2}^2 and their open loop gains A_1 and A_2:

$$e_{PA} = \sqrt{\frac{e_{out}^2}{A_1^2A_2^2}} = \sqrt{\frac{2A_1^2A_2^2e_{in1}^2 + A_2^2e_{in2}^2}{A_1^2A_2^2}}. \quad (7)$$

This last equation can be used to get the noise objective for the second stage, ensuring that e_{PA} is equal to $3\,\text{nV}\,\text{Hz}^{-1}$ at $10\,\text{kHz}$.

Figure 3. Photographs of the low noise ASIC amplifier named MAGIC2.

3.2 Closed loop noise considerations

To make the gain of the amplifier weakly sensitive to temperature variation, the amplifier is used in a closed-loop configuration. The closed-loop gain is set by the R_3 to R_4 ratio ($G = 1 + R_3/R_4$), while the C_1 capacitance is needed to ensure the phase margin. Since this capacitance does not impact the noise analysis, it will not be considered in the rest of the article. As decribed in Sobering (1999), in the context of operational amplifier noise analysis, our noise analysis can be summarized as three contributions: the equivalent voltage input noise at the non-inverting input (e_{PA}), the Johnson noise in R_4 at the inverting input and the Johnson noise in R_3. According to the Sobering (1999) analysis, two gains should be considered: the inverting (A_{v-inv}) and non-inverting ones ($A_{v-non-inv}$). However, in the case of an ideal op amp (which is a correct hypothesis in our design since the open loop gain is $A_1 + A_2 = 85\,\text{dB}$), these two gains are written $A_{v-inv} = R_3/R_4$ and $A_{v-non-inv} = 1 + R_3/R_4$. Moreover, in the case of high closed loop gain (i.e. $R_3/R_4 \gg 1$), these two gains can be considered to be identical. That allows us to consider the equivalent op amp input noise to be

$$e_{OpAmp}^2 = e_{PA}^2 + 4kTR4. \quad (8)$$

Lastly, the input referred noise contribution coming from gain resistance of the preamplifier (R_4) is neglected, since its value is small (in our design we got $R_4 = 28\,\Omega$, which leads to an equivalent voltage noise contribution of about one-tenth of e_{PA}).

3.3 Measurement results and performances

The amplifier was fabricated in a standard $0.35\,\mu\text{m}$ four-metal bulk CMOS process. The manufactured circuit has a $1.21\,\text{mm}^2$ area. The silicon chip contains one amplifier. Its microphotograph is shown in Fig. 3.

The gain transfer function and the equivalent input noise have been characterized. A high pass filter, with cut-off frequency at $1\,\text{Hz}$, is inserted to remove DC offset, while the low pass filtering cut-off frequency is due to the combination R_3 and C_1. Figure 4 shows that the gain is about $50.7\,\text{dB}$ from a few Hz (a high pass filtering is inserted to remove

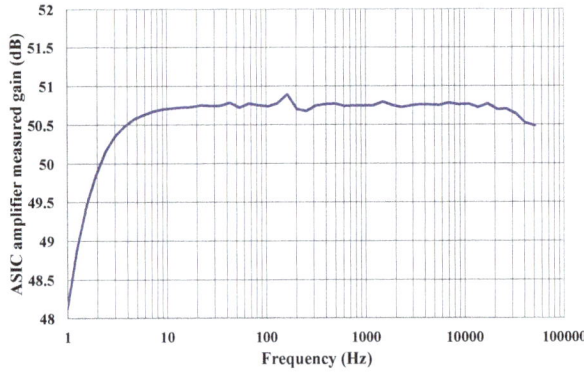

Figure 4. Amplifier transfer function (in dB) of MAGIC2.

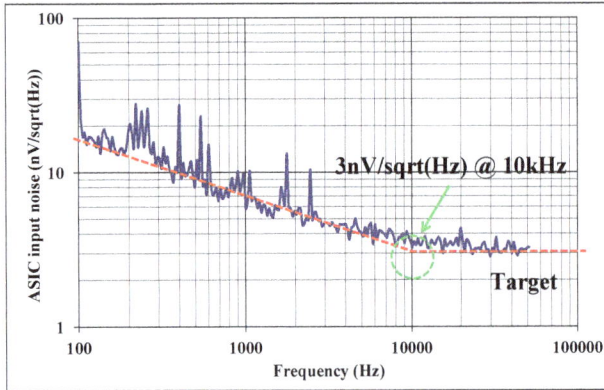

Figure 5. Equivalent input referred noise (in nV $\sqrt{(Hz)}^{-1}$) measured using a 50 Ω input resistor.

the offset) up to 50 kHz. The measured gain value is consistent with the $1 + R_3/R_4$ ratio ($R_3 = 10k\Omega$ and $R_4 = 28\Omega$). Figure 5 demonstrates a measured input voltage noise about 3 nV Hz^{-1} at 10 kHz when connecting a 50 Ω resistor at the input of the amplifier.

The induction sensor has a very high input impedance. It implies that it is essential to minimize the input noise current of the amplifier since it will lead to a high contribution to the output noise voltage. In our design, the input current noise contribution is less than 20 fA $\sqrt{Hz}^{-1}$, which is achieved thanks to the CMOS technology. It can be concluded, for this part, that the combination of low power consumption, low input voltage noise and low current noise, which are essential to induction sensors, can be achieved using CMOS technology at the price of a significant design work.

4 Modelling of the ferromagnetic core noise source contribution

The NEMI reaches its minimum value in the decade around the resonance frequency. The usual modelling of the NEMI will underestimate its value in this frequency range. In rare works, to our best knowledge, noise sources related to the

ferromagnetic material are evoked either through an empirical correlation (Seran and Fergeau, 2005) or a set of quality factors (Korepanov, 2010) to take into account the NEMI increase near to the resonance. In the first quoted paper, the coefficient of the correlation is determined experimentally for a given core size, which does not allow one to take it into account in a preliminary design stage, while, in the second paper, the quality factor values are given from a tentative estimation. At low ambient field ($<$ mT), the noise in the ferromagnetic core comes either from an eddy current or from magnetization mechanisms like domain wall relaxation and magnetization rotation. At a high magnetic field (typ. mT), Barkhausen noise, related to domain wall jumps, will occur. The usual domain of application of an induction magnetometer is related to a quiet electromagnetic environment; thus, Barkhausen noise will not be considered in this study.

4.1 Complex permeability of the Mn–Zn ferrite

The mentioned noise source can be modelled through the concept of complex permeability (Tsutaoka, 2003) where the imaginary part of the permeability is related to the ferromagnetic noise source. For high permeability Mn–Zn sintered ferrite, we use the complex susceptibility of resonance type given in Dosoudil (2004). The first fraction in the susceptibility relation (Eq. 9) corresponds to the frequency dispersion of domain wall motion contribution, while the second fraction represents the magnetic moment rotation contribution.

$$\mu = 1 + \frac{\omega_d^2 \chi_{d0}}{\omega_d^2 - \omega^2 + i\omega\beta} + \frac{(\omega_s + j\omega\alpha)\omega_s\chi_{s0}}{(\omega_s + i\omega\alpha)^2 - \omega^2}, \tag{9}$$

where χ_{d0} and χ_{s0} are the static susceptibilities for domain wall motion and magnetic moment rotation, $\omega_d (= 2\pi f_d)$ and $\omega_s (= 2\pi f_s)$ are resonance frequencies of domain wall motion and magnetic moment rotation, β and α are the damping factors, and $f = \omega/(2\pi)$ is the operating frequency.

The apparent permeability can be written as

$$\mu = \mu' - j\mu''. \tag{10}$$

So, the real and imaginary parts, deduced from Eq. (9), are expressed as follows:

$$\mu' = 1 + \frac{\omega_d^2 \chi_{d0} \left(\omega_d^2 - \omega^2\right)}{\left(\omega_d^2 - \omega^2\right)^2 + (\omega\beta)^2} \tag{11}$$
$$+ \frac{\chi_{s0}\omega_s^2 \left(\omega_s^2 - \omega^2 + \alpha^2\omega^2\right)}{\left(\omega_s^2 - \omega^2(1 + \alpha^2)\right)^2 + (2\omega\omega_s\alpha)^2},$$

$$\mu'' = \frac{\chi_{d0}\omega\beta\omega_d^2}{\left(\omega_d^2 - \omega^2\right)^2 + (\omega\beta)^2} \tag{12}$$
$$+ \frac{\chi_{s0}\omega_s\omega\alpha \left(\omega_s^2 + \omega^2(1 + \alpha^2)\right)}{\left(\omega_s^2 - \omega^2(1 + \alpha^2)\right)^2 + (2\omega\omega_s\alpha)^2}.$$

Table 1. Susceptibility dispersion parameters for spin and domain wall resonance of Ferroxcube 3C95 Mn–Zn ferrite.

χ_{d0}	f_d (MHz)	β	χ_{s0}	f_s (MHz)	α
1400	1.4×10^6	7.5×10^6	900	8×10^6	5

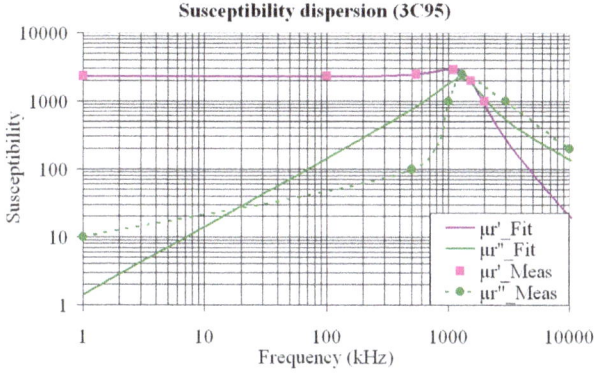

Figure 6. Measured ($\mu'_r\_$Meas and $\mu''_r\_$Meas) and fitted ($\mu'_r\_$Fit and $\mu''_r\_$Fit) susceptibility dispersions of 3C95 Mn–Zn ferrite.

The expressions of real and imaginary parts of susceptibilities are quite similar to the one given by Tsutaoka (2003) at a sign near in the numerator of the real component of the susceptibility. We will now consider Mn–Zn ferrite from Ferroxcube of 3C95 type, which appears to be a good candidate for designing an induction sensor thanks to its high relative permeability ($\mu_r > 2000$), its availability in different shapes and its stability over a wide temperature range (from -100 up to $+200\,°C$). For this material, we have determined on a toroidal core sample the values of the complex susceptibility model parameters (ω_d, ω_s, χ_{d0}, χ_{d0}, and ω_r). These parameters are summarized in Table 1, while the measured and fitted susceptibility dispersions (real and imaginary parts) are plotted in Fig. 6. The obtained values are in the same magnitude range as those reported in Tsutaoka (2003).

4.2 Complex apparent permeability

The magnetic gain produced by the ferromagnetic core, known as apparent permeability (Bozorth and Chapin, 1942), allows one to increase the induced voltage. This one results in the combination of the relative permeability of the material (μ_r) and its shape, through the demagnetizing coefficient ($N_{x,y,z}$) in a given direction (x, y or z). For a long cylinder core of length to diameter ratio m, the approximation of the ellipsoid demagnetizing coefficient, given in Osborn (1945), is repeated here:

$$N_z(m) = \frac{1}{m^2}(\ln(2m) - 1). \tag{13}$$

In the current study, a diabolo core shape (shown in Fig. 7) is used, whose apparent permeability (μ_{app} given in Coillot

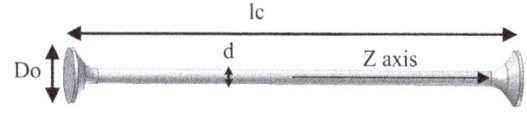

Figure 7. Diabolo core induction sensor.

et al., 2007) is expressed as

$$\mu_{app} = \frac{\mu_r}{1 + N_z(m)\frac{d^2}{D_O^2}(\mu_r - 1)}, \tag{14}$$

where $N_z(m = L_C/D_O)$ is the demagnetizing coefficient in the z direction for a cylinder of length L_C and diameter D_O.

Assuming that apparent permeability owns real and imaginary parts, it can be written under the following form:

$$\mu_{app} = \mu'_{app} - j\mu''_{app}. \tag{15}$$

By substituting, in apparent permeability (Eq. 14), the equation of complex permeability derived for ferrites (Eq. 9), and by identifying it with Eq. (15), we deduce the real and imaginary parts of the apparent complex permeability, respectively Eqs. (16) and (17).

$$\mu'_{app} = \frac{\mu'\left(1 + N_z(m)\frac{d^2}{D_O^2}(\mu' - 1)\right) + N_z(m)\frac{d^2}{D_O^2}\mu''^2}{\left(1 + N_z(m)\frac{d^2}{D_O^2}(\mu' - 1)\right)^2 + \left(N_z(m)\frac{d^2}{D_O^2}\mu''\right)^2} \tag{16}$$

$$\mu''_{app} = \frac{\mu''\left(1 - N_z(m)\frac{d^2}{D_O^2}\right)}{\left(1 + N_z(m)\frac{d^2}{D_O^2}(\mu' - 1)\right)^2 + \left(N_z(m)\frac{d^2}{D_O^2}\mu''\right)^2} \tag{17}$$

In the case of a ferromagnetic core induction sensor, the inductance equation (Tumanski, 2007) is

$$Ł = \lambda N^2 \frac{\mu_0 \mu_{app} S}{L_C}, \tag{18}$$

where (S) is the ferromagnetic core section, μ_0 is the vacuum permeability and $\lambda = (L_C/L_w)^{2/5}$ is a correction factor. Thus, the inductance will also have a real part (L') and an imaginary part (L''):

$$Ł = L' - jL'', \tag{19}$$

which are written as follows:

$$Ł' = \lambda N^2 \mu_0 \frac{\mu'_{app} S}{L_C}, \tag{20}$$

$$Ł'' = \lambda N^2 \mu_0 \frac{(\mu''_{app}) S}{L_C}. \tag{21}$$

Finally, the noise source contribution arising from the ferromagnetic core will look like a Johnson noise whose power spectrum density can be written as

$$\text{PSD}_L = 4kT\Re(jL\omega), \tag{22}$$

which becomes

$$\text{PSD}_L = 4kTL''\omega. \tag{23}$$

In the same way, the mutual inductance will exhibit real and imaginary parts; however, since the mutual inductance is much smaller than the self-inductance, its imaginary part will be neglected and the mutual inductance will be assumed to be a real number.

5 Modelling and experimental results comparison

5.1 The noise equivalent magnetic induction

The block diagram of Fig. 8 is used to facilitate the computation of the output noise contribution for each noise source. The transmittance of the feedback flux amplifier, given by Eq. (2), is modified to take into account the contribution of the complex inductance:

$$T(j\omega) = \frac{V_{\text{OUT}}}{B} = \frac{-jNSG\mu_{\text{app}}\omega}{(1 - LC\omega^2) + j\omega((R + L''\omega)C + \frac{GM}{R_{\text{fb}}})}. \tag{24}$$

In this block diagram, the noise source coming from the ferromagnetic core is directly added to the thermal noise of the coil resistance. Since this block diagram is dedicated to noise analysis, it is assumed that measured flux (φ) is null. For the reasons given in Sect. 3.2, the noise contribution coming from the input resistance of the preamplifier (R_4) is neglected.

The block diagram permits one to determine the transfer function between the output noise contribution (referred to as the V_{OUT} node) and each of the noise sources. The method is the following: the block diagram is drawn for a given source, while the other noise sources are cancelled thanks to the superposition theorem (for instance, see the block diagram for the feedback resistance noise source shown in Fig. 9).

Then, the closed loop transfer function seen by the R_{fb} noise is obtained:

$$T(j\omega)_{R_{\text{fb}}} = \frac{\frac{j\omega MG}{R_{\text{fb}}}}{1 - L'C\omega^2 + j(R + L''\omega)C\omega + \frac{j\omega MG}{R_{\text{fb}}}}. \tag{25}$$

Using the general relation between input and output PSD (namely, $\text{PSD}_{\text{OUT}} = |T(j\omega)|^2\text{PSD}_{\text{IN}}$), we deduce the output noise contribution of the feedback resistance:

$$\text{PSD}_{R_{\text{fb}}} = 4kTR_{\text{fb}}\frac{\left(\frac{\omega MG}{R_{\text{fb}}}\right)^2}{(1 - L'C\omega^2)^2 + \left((R + L''\omega)C\omega + \frac{\omega MG}{R_{\text{fb}}}\right)^2}.$$

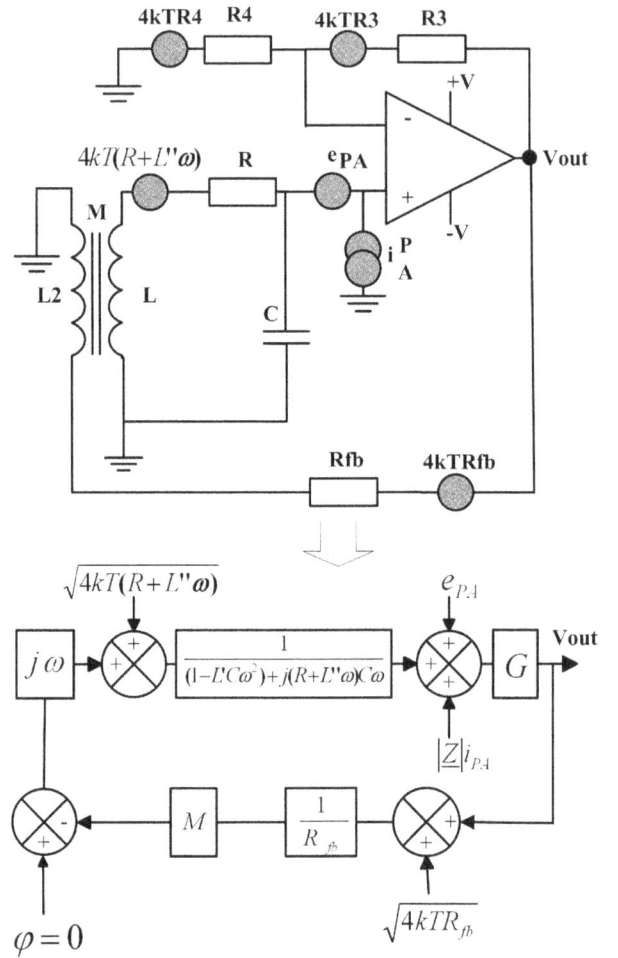

Figure 8. Noise sources in the feedback flux induction configuration and block diagram representation.

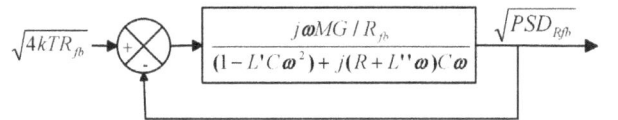

Figure 9. Block diagram representation of the feedback resistance noise source.

$$\tag{26}$$

This latter expression can be simplified in the frequency range where the feedback flux operates:

$$\text{PSD}_{R_{\text{fb}}} \simeq 4kTR_{\text{fb}}. \tag{27}$$

In a similar manner, the noise source contribution from the coil's resistance is derived:

$$\text{PSD}_R = 4kT\frac{G^2(R + L''\omega)}{(1 - L'C\omega^2)^2 + \left((R + L''\omega)C\omega + \frac{GM\omega}{R_{\text{fb}}}\right)^2}. \tag{28}$$

The $1/f$ noise contribution of the preamplifier input voltage noise being neglected, the noise source contribution of the preamplifier input voltage noise is

$$\mathrm{PSD}_{e_{\mathrm{PA}}} = e_{\mathrm{PA}}^2 \frac{G^2 \left((1 - L'C\omega^2)^2 + (C\omega(R + L''\omega))^2 \right)}{(1 - L'C\omega^2)^2 + \left((R + L''\omega)C\omega + \frac{GM\omega}{R_{\mathrm{fb}}} \right)^2}. \quad (29)$$

Similarly, the noise source contribution of the preamplifier input current noise is obtained:

$$\mathrm{PSD}_{i_{\mathrm{PA}}} = (|Z| \, i_{\mathrm{PA}})^2 \frac{G^2 \left((1 - L'C\omega^2)^2 + (C\omega(R + L''\omega))^2 \right)}{(1 - L'C\omega^2)^2 + \left((R + L''\omega)C\omega + \frac{GM\omega}{R_{\mathrm{fb}}} \right)^2}, \quad (30)$$

where $|Z|$ is the equivalent impedance modulus of the induction sensor seen at the positive input of the amplifier, which is expressed (after some computations) as

$$|Z| = \sqrt{\frac{\left((R + L''\omega + \frac{(M\omega)^2}{R_{\mathrm{fb}}})^2 + (L'\omega)^2 \right)}{(1 - L'C\omega^2)^2 + \left((R + L''\omega + \frac{M^2\omega^2}{R_{\mathrm{fb}}})C\omega + \frac{GM\omega}{R_{\mathrm{fb}}} \right)^2}}. \quad (31)$$

Finally, the total output noise contribution ($\mathrm{PSD}_{\mathrm{out}}$) is computed by adding the individual power spectral density contribution of each noise source (under the hypothesis of uncorrelated noise):

$$\mathrm{PSD}_{\mathrm{out}} = \mathrm{PSD}_Z + \mathrm{PSD}_{e_{\mathrm{PA}}} + \mathrm{PSD}_{i_{\mathrm{PA}}} + \mathrm{PSD}_{R_{\mathrm{fb}}}. \quad (32)$$

Finally, the noise equivalent magnetic induction (NEMI), which is the square root of the power spectrum density of the total output noise ($\mathrm{PSD}_{\mathrm{OUT}}$) related to the transfer function modulus of the induction magnetometer ($T(j\omega)$ given by Eq. (24) for feedback flux magnetometer) can be determined.

5.2 Experimental results and discussion

A single axis induction magnetometer has been built with an induction sensor using a diabolo core shape made of 3C95 Mn–Zn ferrite from Ferroxcube. The sensor has been combined with the MAGIC2 ASIC amplifier. The parameters of the induction sensor design and the preamplifier are summarized in Table 2.

The parameters of the sensor lead to the following value of the electrokinetic modelling: $R = 48\,\Omega$ (copper wire operating at $300\,\mathrm{K}$ is assumed, i.e. $\rho = 1.7 \times 10^{-8}\,\Omega m$), $L = 0.306\,\mathrm{H}$ (assuming λ equal to 1), $M = 3\,\mathrm{mH}$, $C = 150\,\mathrm{pF}$ and $\mu_{\mathrm{app}} = 420$. The sensor weight is lower than $30\,\mathrm{g}$, while the ASIC amplifier power consumption supplied with a $12\,\mathrm{V}$ battery is lower than $30\,\mathrm{mW}$. The noise measurement ($\mathrm{PSD}_{\mathrm{out}}$) of the induction magnetometer (i.e. sensor connected to its

Table 2. Design parameters.

Sensor length	$L_{\mathrm{C}} = 120\,\mathrm{mm}$
Winding length	$L_{\mathrm{w}} = 100\,\mathrm{mm}$
Sensor diameter	$d = 4\,\mathrm{mm}$
Diabolo ends diameter	$D_{\mathrm{O}} = 14\,\mathrm{mm}$
Turns number	$N = 2350$
Feedback coil turns number	$N_2 = 24$
Copper wire diameter	$d_{\mathrm{w}} = 0.12\,\mathrm{mm}$
Insulator thickness	$t = 25\,\mu m$
Layer number	$n_l = 4$
Feedback resistance	$R_{\mathrm{fb}} = 10\,k\Omega$
Amplifier gain	$G_{\mathrm{dB}} = 50.7\,\mathrm{dB}$
Voltage noise	$e_{\mathrm{PA}} = 3.3\,\mathrm{nV}\,\sqrt{(\mathrm{Hz})^{-1}}$
Current noise	$i_{\mathrm{PA}} = 20\,\mathrm{fA}\,\sqrt{(\mathrm{Hz})^{-1}}$

preamplifier) has been performed inside a shielded box consisting of three layers of mu-metal materials and one of conductive material (connected to the preamplifier ground in a way that minimizes the current loop via ground connection. The thickness of each layer is 1 mm and the inner box side length is 40 cm. Each layer of the shielding box is separated by 1 cm air gaps (the size of the magnetic shielded box should be much wider than the one of the sensor). The transfer function ($T(j\omega)$) of the induction magnetometer has been measured in gain and phase in a large diameter Helmholtz coil (1 m) mounted on a wood structure to ensure a homogeneous magnetic field at the scale of the sensor. The accuracy of the facility was verified using a small air-core coil whose theoretical transfer function is fully known. For both measurements (transfer function and noise), an Agilent 35670 spectrum analyser was used. The sensor was equipped with a very thin electrostatic shielding to be insensitive to the electric field component of the electromagnetic waves. The electrostatic shielding was designed to minimize the additional noise from induced current (Ozaki, 2015). The measurements have been done in two configurations: one with the electrostatic shielding and the other one without (in this case, the shielded box plays the role). The simulated NEMI curve, using the modelling of the complex permeability, is compared to the measured one (in the configuration without electrostatic shielding) in Fig. 10.

The result is that the theoretical NEMI (computed for both real and complex permeability) leads to an extremely low minimum NEMI value ($< 2\,\mathrm{fT}\,\sqrt{(\mathrm{Hz})^{-1}}$), while practical measurement leads to a higher NEMI value ($\sim 4\,\mathrm{fT}\,\sqrt{(\mathrm{Hz})^{-1}}$) in the same frequency range (namely, 20 and 30 kHz). The contribution of the complex permeability increases weakly the NEMI (at least in the frequency range between 10 and 100 kHz). A significant difference between the measured and computed NEMI remains, which suggests that a noise source other than ferromagnetic core contribution dominates and limits the NEMI value in the frequency range where the feedback operates (namely, from 10 to 100 kHz

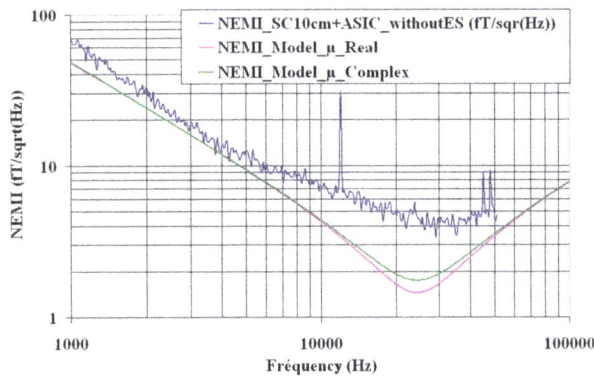

Figure 10. NEMI curves comparison: NEMI with real permeability (pink), NEMI with complex permeability (green) and NEMI measured on a prototype (blue).

in this design). Since the coil was wound directly on the ferromagnetic core, magnetostriction has been suspected of modifying the complex permeability dispersion and thus the higher NEMI measurement. In this aim, a sensor wound on an epoxi tube was realized and a ferrite core with comparable apparent permeability was compared to the sensor reference, but no significant differences have been noticed.

Next, the occurence of an extra noise coming from the coil AC resistance (Butterworth, 1925) is suspected of increasing the Johnson noise contribution coming from the coil resistance (namely, PSD_R). The AC resistance increase of the coil comes from the skin effect enhanced by the proximity effect. This effect is taken into account by designers of transformers (Dowell, 1966). In these devices, the AC resistance increase causes extra losses and thus temperature elevation of the transformer. The model proposed by Dowell (1966) is mono-dimensional and assumes a skin depth depending on the distance between wires. However, a lateral skin effect occurring at the end of the winding is also expected (Butterworth, 1925; Belevitch, 1971), making Dowell's model unusable. The contribution of the skin effect enhanced by proximity and the lateral skin effect is also well known to increase strongly the AC resistance and thus to reduce the signal-to-noise ratio of induction sensors for nuclear magnetic resonance (Hoult, 1976). Consequently the skin effect enhanced by the proximity and lateral effect in the case of a multi-layer winding is one of the possible causes which could explain a part of the difference between measurement and modelling.

6 Conclusions

While fitting methods usually assume that the extra noise from an induction magnetometer comes from the ferromagnetic core, we have undertaken a modelling attempt of the noise source contribution from a high-permeability Mn–Zn ferrite core. The way to take it into account has been achieved by modelling the apparent complex permeability through the susceptibility frequency dispersion of the domain wall motion and magnetization rotation. We have assumed that the machining of the core did not modify the complex permeability. The comparison of the NEMI measurement on a prototype with the NEMI modelling has shown a significant difference around the frequency resonance. Thus, the ferromagnetic core noise seems too weak to explain the difference between the model and the measured NEMI. Thus, the occurrence of an extra noise due to the AC resistance increase is suspected of playing a role. The ferromagnetic core noise contribution (through the apparent complex permeability modelling) should be studied on other ferromagnetic core materials (especially Ni–Fe alloy ferromagnetic material). The accurate and rigorous modelling of the NEMI around the resonance frequency remains an issue for $fT \sqrt{Hz}^{-1}$ induction magnetometer design.

Acknowledgements. The authors would like to thank the reviewers for the time they spent to perform the review and for the stimulating discussions. The authors would also like to thank A. Grosz for the fruitful exchange concerning the delicate NEMI measurements.

References

Belevitch, V.: The lateral skin effect in a flat conductor, Philips Tech Rev., 32, 221–231, 1971.

Bin, Y.: An Optimization Method for Induction Magnetometer of 0.1 mHz to 1 kHz, IEEE Trans. on Mag., 49, 5294–5300, 2013.

Bozorth, R. M. and Chapin, D.: Demagnetizing factors of rods, J. Appl. Phys., 13, 320–327, 1942.

Butterworth, S.: On the alternating current resistance of solenoidal coils, Proc. R. Soc. Lon. Ser.-A, 107, 693–715, 1925.

Coillot, C., Moutoussamy, J., Leroy, P., Chanteur, G., and Roux, A.: Improvements on the design of search coil magnetometer for space experiments, Sens. Lett., 5, 167–170, 2007.

Coillot, C. and Leroy, P.: Induction Magnetometers: Principle, Modeling and Ways of Improvement, in: Magnetic Sensors – Principles and Applications, edited by: Kuang, K., InTech, ISBN: 978-953-51-0232-8, 2012.

Coillot, C., Moutoussamy, J., Boda, M., and Leroy, P.: New ferromagnetic core shapes for induction sensors, J. Sens. Sens. Syst., 3, 1–8, doi:10.5194/jsss-3-1-2014, 2014.

Dosoudil, R., Usakova, M., and Slama, J.: Permeability dispersion in Ni-Zn-Cu ferrite and its composite material, J. Phys., 54, D675–D678, 2004.

Dowell, P. L.: Effects of eddy currents in transformer windings, Proc. the IEEE, 113, 1387–1394, 1966.

Grosz, A., Paperno, E., Amrusi, S., and Liverts, E.: Integration of the electronics and batteries inside the hollow core of a search coil, J. App. Phys., 107, 09E703-1–09E703-3, 2010.

Hoult, D. I. and Richards, R. E.: The signal-to-noise ratio of the nuclear magnetic resonance experiment, J. Magn. Reson., 24, 71–85, 1976.

Korepanov, V. and Pronenko, V.: Induction Magnetometers: Design Peculiarities, Sens. Transducers J., 120, 92–106, 2010.

Osborn, J. A.: Demagnetizing factors of the general ellipsoids, 67, 351–357, 1945.

Ozaki, M., Yagitani, S., Takahashi, K., and Nagano, I.: Dual-Resonant Search Coil for Natural Electromagnetic Waves in the Near-Earth Environment, IEEE Sens. J., 13, 644–650, 2013.

Ozaki, M., Yagitani, S., Kojima, H., Takahashi, K., and Kitagawa, A.: Current-Sensitive CMOS Preamplifier for Investigating Space Plasma Waves by Magnetic Search Coils, IEEE Sensors Journal, 2, 421–429, 2014.

Ozaki, M., Yagitani, S., Takahashi, K., Imachi, T., Koji, H., and Higashi, R.: Equivalent Circuit Model for the Electric Field Sensitivity of a Magnetic Search Coil of Space Plasma, IEEE Sensors J., 15, 1680–1689, 2015.

Prance, R. J., Clarck, T. D., and Prance, H.: Ultra low noise induction magnetometer for variable temperature operation, Sensor. Actuator., 85, 361–364, 2000.

Rhouni, A., Sou, G., Leroy, P., and Coillot, C.: A Very Low 1/f Noise and Radiation-Hardened CMOS Preamplifier for High Sensitivity Search Coil Magnetometers, IEEE Sens. J., 13, 159–166, 2012.

Ripka, P. (Ed.): Magnetic sensors and magnetometers, Artech House Publ., London, UK, 2000.

Seran, H. C. and Fergeau, P.: An optimized low frequency three axis search coil for space research, Rev. Sci. Instrum., 76, 044502–0044502-9, 2005.

Sobering, T.: Op Amp Noise Analysis, Technnote 5, available at: http://www.k-state.edu/ksuedl/publications/Technote5-OpampNoiseAnalysis.pdf, 2005.

Tsutaoka, T.: Frequency dispersion of complex permeability in Mn-Zn and Ni-Zn spinel ferrites and their composite materials, J. Appl. Phys., 93, 2789–2796, 2003.

Tumanski, S.: Induction coil sensors – A review, Meas Sci. Technol., 18, R31–R46, 2007.

4

Ammonia storage studies on H-ZSM-5 zeolites by microwave cavity perturbation: correlation of dielectric properties with ammonia storage

M. Dietrich[1], D. Rauch[1], U. Simon[2], A. Porch[3], and R. Moos[1]

[1]Bayreuth Engine Research Center (BERC), Zentrum für Energietechnik (ZET), Department of Functional Materials, University of Bayreuth, 95440 Bayreuth, Germany
[2]Institute of Inorganic Chemistry (IAC), RWTH Aachen University, 52074 Aachen, Germany
[3]School of Engineering, Cardiff University, Cardiff CF24 3AA, Wales, UK

Correspondence to: M. Dietrich (functional.materials@uni-bayreuth.de)

Abstract. To meet today's emission standards, the ammonia-based selective catalytic reduction (SCR) has become the major NO_x control strategy for light and heavy diesel engines. Before NO_x reduction can proceed, adsorption of ammonia on the acidic sites of the catalyst is necessary. For improvements in efficiency and control of the exhaust gas aftertreatment, a better understanding of the ammonia storage on the acidic sites of zeolite-based SCR catalysts is needed. Thereby, the correlation of dielectric properties of the catalyst material itself with the ammonia storage is a promising approach. Recently, a laboratory setup using microwave cavity perturbation to measure the dielectric properties of catalyst material has been described. This study shows the first experimental data on zeolite-based SCR materials in their H-form. The SCR powder samples are monitored by microwave cavity perturbation while storing and depleting ammonia, both with and without admixed NO_x at different temperatures. Its complex dielectric permittivity is found to correlate closely with the stored mass of ammonia. The influence of the temperature and the Si / Al ratio of the zeolite to the ammonia storage behavior are also examined. These measurements disclose different temperature dependencies and differing sensitivities to ammonia storage for both real and imaginary parts of the complex permittivity. The apparent constant sensitivity of the real part can be related to the polarity of the adsorbed ammonia molecules, whereas the imaginary part depends on the Si / Al ratio and is related to the conductivity mechanisms of the zeolite material by proton hopping. It provides information about the zeolite structure and the number of (and the distance between) acidic storage sites, in addition to the stored ammonia mass.

1 Introduction

The stringent regulations for emissions of nitrogen oxides (NO_x) from combustion engines are a continuous factor in forcing automotive manufacturers to improve the efficiency of their exhaust gas aftertreatment systems. This is especially true for light- and heavy-duty diesel engines, which are operated leanly, where the ammonia-based selective catalytic reduction (SCR) has become the major NO_x control strategy to meet emission standards like the upcoming Euro 6 (Johnson, 2009). Metal oxides like V_2O_5–WO_3–TiO_2 (VWT) were established as SCR catalysts, but as a consequence of their clas-

sification as being toxic and harmful for the environment, zeolites with active components like iron (Fe) and copper (Cu) have received more attention in the past years (Rahkamma-Tolonen et al., 2005; DiIorio et al., 2015). In automotive applications, the ammonia-based SCR uses a non-toxic, aqueous urea solution (AdBlue) as a reducing agent. The injected solution decomposes thermally to ammonia (NH_3) in the hot exhaust. An essential precondition for SCR reactions is a previous NH_3 adsorption on the acidic sites on the zeolite surface. This NH_3 storage mechanism also buffers against changes of flow and temperature in order to secure a permanent NO_x conversion. The catalyst reduces NO_x selectively to nitrogen (N_2) and water (H_2O). The two main SCR reactions are shown in the following: the standard SCR Reaction (R1) and the fast SCR Reaction (R2) with equimolar amounts of NO and NO_2 (Koebel et al., 2000):

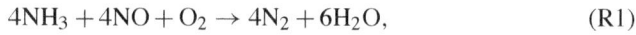

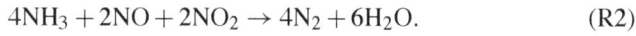

$$4NH_3 + 4NO + O_2 \rightarrow 4N_2 + 6H_2O, \quad (R1)$$
$$4NH_3 + 2NO + 2NO_2 \rightarrow 4N_2 + 6H_2O. \quad (R2)$$

NH_3 storage capacity and catalytic activity of zeolite SCR catalysts depend on the number and strength of their acid sites (Lewis and Brønsted sites). NH_3 can adsorb strongly on Brønsted acid sites and weakly on top of each other on these sites via hydrogen bonds or on Lewis sites (Rodriguez-Gonzalez et al., 2008; Giodanino et al., 2014). Both acidic sides can be determined by temperature-programmed desorption (TPD) of NH_3, but a direct differentiation between Lewis and Brønsted sites is still not possible (Niwa and Katada, 2013). Therefore, the ability to measure dielectric properties under reaction conditions during NH_3 storage, and to correlate them with the catalytic behavior of the materials in situ, offers new opportunities to analyze and identify acidic sites and to optimize the catalyst material in general. Similarly, the NH_3 storage on SCR active materials has been investigated under defined gas atmospheres by impedance spectroscopy in a frequency range from $0.1\,Hz$ to $1\,MHz$ (Franke and Simon, 2004; Rodríguez-González et al., 2008). NH_3 loading and TPD experiments were performed with zeolite powders deposited on interdigitated capacitor chips to evaluate mechanistic models of proton transport in zeolites. In comparison with such impedance methods, the cavity perturbation method, which uses microwaves in the GHz range and a metal cavity resonator, holds promise for additional information and applications, not least because it is non-invasive (apart from interaction with a low power microwave field) and contactless.

Recently, a similar approach has been suggested and serial-type catalyst devices (as applied in automotive exhaust gas aftertreatment systems) have been examined (Moos et al., 2013). The catalytic converter had a volume of about 1.5 to 2 L. Since the sample occupied most of the cavity volume (metal canning), these systems are suitable for real-world applications but not to characterize material properties owing to their very large perturbation of the sample on the cavity space, making the inversion analysis for extracting complex permittivity very difficult. Instead, they are intended to detect the status of full-sized exhaust gas aftertreatment devices during operation on the road. Typical applications are the determination of the oxygen loading of three-way catalytic converters (Moos et al., 2008, 2013; Beulertz et al., 2013; Reiß et al., 2011a), or the soot loading (Sappok et al., 2010; Feulner et al., 2013) or ash loading (Kulkarni et al., 2013) of full-sized diesel particulate filters. The storage degree of NO in lean NO_x traps (Fremerey et al., 2011; Moos et al., 2009) and the NH_3 loading on SCR catalyst devices have also been successfully monitored (Reiß et al., 2011b; Rauch et al., 2014, 2015) using the cavity perturbation method.

In order to determine the dielectric properties of a catalyst material in operando, we have developed a laboratory test setup for catalyst powder characterization under reaction conditions by microwave cavity perturbation (introduced in Dietrich et al., 2014). It enables direct measurement of the complex permittivity of catalytic powder samples undergoing gas storage and catalytic reactions in a defined gas atmosphere, with gas analyzers upstream and downstream of the catalyst sample. In its first version, it operates within a temperature range from room temperature (where usually no reactions occur) to $300\,°C$.

2 Microwave cavity perturbation

The microwave cavity perturbation technique uses electromagnetic standing waves (resonances) inside a defined, hollow metal canning. The presence of a small sample inside the cavity resonator leads to a perturbation of the electromagnetic field distribution. For a sample placed within a region of maximum electric field (and zero magnetic field), the resulting decrease of the resonance frequency and the increase of the $3\,dB$ bandwidth (i.e., decrease of the quality factor Q) of the resonance curve are related to real and imaginary parts, respectively, of the complex dielectric permittivity $\varepsilon = \varepsilon_1 - j\varepsilon_2$ of the sample: the real part ε_1 (or, more properly, $\varepsilon_1 - 1$) quantifies the polarization of the material and the imaginary part ε_2 quantifies the dielectric loss. The setup, fully described in Dietrich et al. (2014), is designed to use the TM_{010} mode of a cylindrical cavity resonator because of its uniform electric field along its axis, where the sample is positioned. The resonant frequency of the TM_{010} mode is set to around $1.2\,GHz$ by suitable choice of the inner cavity radius. On inserting the sample, the real part ε_1 of its permittivity leads to a change of the resonance frequency from f_0 to f_s and its imaginary part ε_2 to a change of the unloaded quality factor from Q_0 to Q_s. With the volume of the sample V_s and the mode volume of the resonator V_{eff} (which is the effective volume occupied by the electric field energy and depends on the particular cavity mode), the complex dielectric permittivity can be calculated using Eqs. (1) and (2) (Porch et al., 2012).

$$\frac{f_0 - f_s}{f_0} \approx (\varepsilon_1 - 1)\frac{V_s}{2V_{eff}} \quad (1)$$

$$\frac{1}{Q_s} - \frac{1}{Q_0} = \Delta\left(\frac{1}{Q}\right) \approx \varepsilon_2 \frac{V_s}{V_{eff}} \quad (2)$$

The mode volume V_{eff} for the TM_{010} mode is $26.9\,\%$ of the enclosed volume of the cylinder. It should be noted that the fundamental interaction of the material with the microwave electric field via the material's polarization is exactly the same as that in microwave heating applications, but in cavity perturbation the electric field levels are so low (with input power typically 1 mW or 0 dBm) that there is negligible sample heating.

3 Experimental

The samples under investigation are H-ZSM-5 zeolites with different Si / Al ratios (27, 90, 200 and 800). The powders were kindly provided by Clariant International Ltd. For zeolites in their H-form, the charge compensating cations are simply protons and no active metal ions are present (Franke and Simon, 1999). The number of acidic sites of the zeolite is directly dependent to the Al content. All samples were weighed and their skeletal volumes were determined using a helium gas pycnometer. The powders were placed on a porous frit in an almost lossless quartz glass tube and were heated independently from the cavity by a flow of hot air. The TM_{010} cavity mode is excited inductively by means of two loop-terminated coaxial lines around the perimeter of the cavity, with suitably oriented loops. The cavity measurement has been improved by using active water cooling to reduce measurement errors as a result of a non-uniform cavity temperature.

The sequence of the NH_3 storage experiments was identical for all samples and is shown for an H-ZSM-5 zeolite with a Si / Al ratio of 27 at 250 °C in Figs. 1a and 2. Figure 1a represents raw data for one measurement at 250 °C with the resonance peak of the TM_{010} mode shown without sample (black), with sample (red) and with sample loaded with NH_3 (blue). These measurements are taken with a vector analyzer (Anritsu VNA Master MS2028B) and here we plot the transmitted voltage amplitude $|S_{21}|$ as a function of frequency, which is a measure of resonator impedance. The insertion of the sample into the cavity leads to a reduction in resonance frequency, but not to an increased bandwidth (or, equivalently, a reduction in transmitted power or Q factor). This demonstrates that the unloaded zeolites have inherent low microwave loss, i.e., small values of ε_2. On loading with NH_3, the resonant frequency shifts further downwards and the bandwidth increases due to the high inherent electric dipole moment of the ammonia molecule. The parameters considered for microwave analysis are shown in Fig. 1b: the resonance frequency with the sample f_s, the 3 dB (or "half-power") bandwidth BW and the peak height $|S_{21,\mathrm{max}}|$. To calculate ε_1 from Eq. (1), the resonance frequency shift is used, determined simply from a change in frequency of the peak's maximum. For the calculation of ε_2 by Eq. (2), the unloaded quality factor Q is required, from which the effects of cavity coupling have been removed. Our cavity is designed (and measured) to have symmetric coupling, i.e., equal inductive coupling strength at each of its two ports, so the coupling unloading process can be calculated using Eq. (3) (Porch et al., 2012).

$$Q = \frac{f}{\mathrm{BW}}\left(1 - 10^{-|S_{21,\mathrm{max}}|/20}\right), \qquad (3)$$

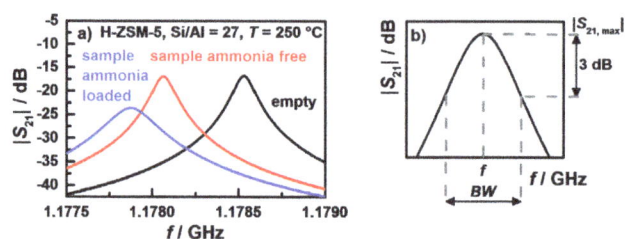

Figure 1. Example of a resonance curve of an H-ZSM-5 zeolite with Si / Al ratio of 27 at 250 °C: **(a)** the decrease of the resonance frequency from an empty sample tube (black) over a filled one without ammonia loading (red) to an ammonia loaded one (blue) can be clearly evaluated. The additional increase in bandwidth by ammonia loading is also visible. **(b)**: the resonance peak in more detail, showing the resonant frequency f, 3 dB bandwidth BW and maximum peak height $|S_{21,\mathrm{max}}|$.

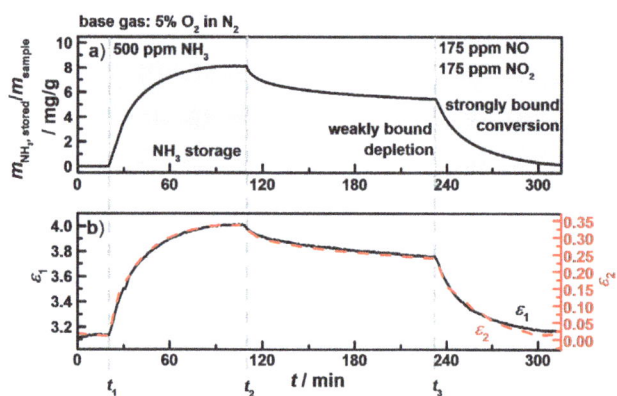

Figure 2. Experimental run for an H-ZSM-5 zeolite with a Si / Al ratio of 27 at 250 °C: **(a)** the stored amount of ammonia in the sample, and **(b)** the measured complex dielectric permittivity.

where the value of resonant frequency is f_0 for calculation of Q_0 (without the sample), or f_s for calculation of Q_s (with the sample).

Figure 2a shows the stored amount of NH_3 in the sample related to the sample mass, calculated from the difference between the measured upstream and downstream NH_3 concentrations. The results of the microwave measurements are displayed in Fig. 2b. The response of the material is immediate, but the response time of the measurement system is reduced by the acquisition time of the network analyzer, or more precisely, the time a frequency sweep requires. In the measurements shown in this paper, one frequency sweep took 75 s, but can be speeded up with different settings of the network analyzer. Initially, the samples were heated up to the measurement temperature in nitrogen with 5 % O_2 added with a total volume flow rate of 500 mL min^{-1}, leading to a mass-specific space velocity (WHSV) between 30 and 80 l h^{-1}. At time t_1, 500 ppm NH_3 is admixed. The feed is stopped when the downstream gas analysis detected the inlet concentration (t_2), meaning that the catalyst powder is

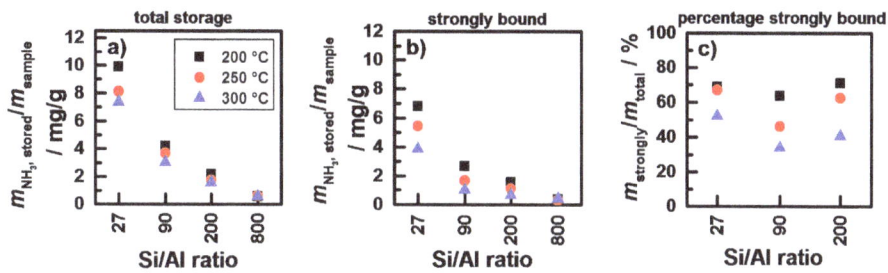

Figure 3. Ammonia storage behavior as a function of the Si / Al ratio at 200, 250 and 300 °C: (**a**) total storage, (**b**) strongly bound ammonia, and (**c**) percentage of strongly bound ammonia in relation to total storage.

saturated with NH_3 and the weakly bound NH_3 can desorb in the base gas. When no more NH_3 desorption can be detected downstream of the powder (t_3), the feed gas composition is changed to equimolar amounts of 175 ppm NO and 175 ppm NO_2 to convert the strongly bound NH_3 in the fast SCR Reaction (R2). At the end of each measurement run, the samples are again completely free of NH_3, as can be seen in the NH_3 balance and the microwave signal. It is clearly demonstrated in Fig. 2b that both the real part ε_1 (black) and imaginary part ε_2 (red dashed) of the complex dielectric permittivity are very strongly correlated to the mass of stored NH_3. This experiment was performed for each sample at temperatures of 200, 250 and 300 °C. In all experimental runs, no hints are found on sample degradation or deposition of unwanted reaction products, which could have either been detected by the gas analyzer, or would have led to visible changes of the powder samples.

4   Results and discussion

Before discussing the microwave measurement data, we first take a look at the storage behavior of the samples at different temperatures. Figure 3a shows the NH_3 mass at saturation (t_2 in Fig. 2) and Fig. 3b the strongly bound NH_3 after the free desorption (t_3 in Fig. 2) as a function of the Si / Al ratio at temperatures of 200, 250 and 300 °C. For comparison, the NH_3 mass is normalized to the mass of the examined sample powder. As expected, the total stored mass of NH_3 decreases with increasing temperature for all Si / Al ratios because of the temperature-dependent adsorption (Niwa and Katada, 2013). At 200 °C, the H-ZSM-5 zeolite with a Si / Al ratio of 27 is found to store 10 mg per gram of SCR sample, and with a Si / Al ratio of 800 less than 0.6 mg per gram of sample. The strongly bound NH_3 shows the same dependency on the Si / Al ratio and for a ratio of 27 the maximum storage of 7 mg/g is attained at a temperature of 200 °C. The calculated percentage of the strongly bound NH_3 is displayed in Fig. 3c for Si / Al ratios of 27, 90 and 200. The NH_3 mass for a ratio of 800 is too small for an accurate consideration. At 200 °C, the percentage of strongly bound NH_3 for the three considered samples is around 70 %, and at

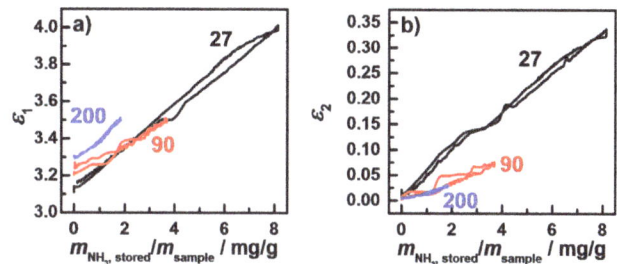

Figure 4. Measurement for an H-ZSM-5 zeolite with Si / Al ratios of 27 (black), 90 (red) and 200 (blue) at 250 °C: (**a**) ε_1 and (**b**) ε_2 as a function of the stored amount of ammonia in the samples.

300 °C between 35 and 50 %. These results indicate that the percentage of strongly bound NH_3 for HZSM-5 zeolites appears to be independent of the Si / Al ratio but dependent on temperature, as expected. Of course, the NH_3 mass for total and strongly bound storage is still a function of the number of storage sites and therefore a function of the Si / Al ratio.

Figure 4 shows the complex dielectric permittivity as a function of normalized stored NH_3 for Si / Al ratios of 27 (black), 90 (red) and 200 (blue) at 250 °C: (a) the real part ε_1 and (b) the imaginary part ε_2. In both plots, the entire measurement runs, including NH_3 loading, free desorption and SCR reaction, are displayed. It is clearly visible that a linear relation between the NH_3 loading and the complex permittivity occurs, regardless of whether the samples are storing or releasing NH_3. This basic behavior is observed in all measurements and indicates that both real and imaginary parts of the complex permittivity are suitable to detect NH_3 within zeolite-based SCR catalysts.

However, for some measurement runs the determined complex permittivity appear very noisy for the higher Si / Al ratios, generally as a result of lower NH_3 storage and higher sensitivity to small temperature changes. For an easier direct comparison of the observed samples, the following discussion is based on the permittivity values at three steady-state points of each measurement run (according to Fig. 2): the NH_3 free sample at the beginning of the measurement (t_1), total storage at saturation (t_2) and the remaining strongly bound NH_3 after free desorption (t_3). The results for Si / Al

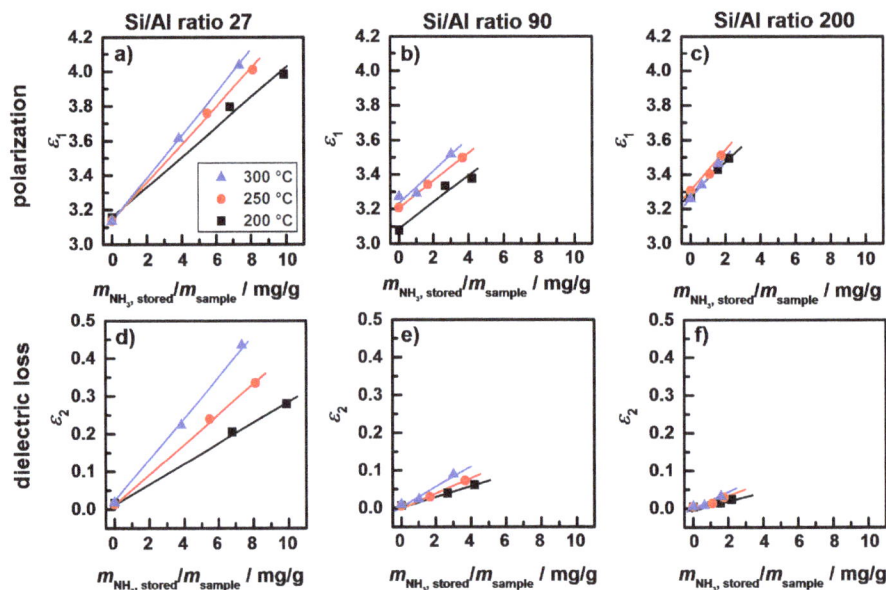

Figure 5. Dielectric permittivity of H-ZSM-5 zeolites as a function of the stored amount of ammonia normalized to the sample mass for different temperatures and Si / Al ratios (27, 90 and 200): (**a**) to (**c**) ε_1, and (**d**) to (**f**) ε_2.

ratios of 27, 90 and 200 are displayed in Fig. 5. The ratio of 800 is excluded since no NH_3 storage could be measured (Fig. 3a), so that the resulting measurement signal is poor. Figure 5a–c show ε_1 and (d) to (f) ε_2 as a function of the normalized stored NH_3 mass at 200, 250 and 300 °C.

The response of ε_1 to NH_3 appears similar for all samples. Completely free of NH_3, ε_1 takes values between 3.1 and 3.3. The variation in these values can be explained by the uncertainties in volume and mass determination in the samples' preparation. For all experiments, ε_1 increases linearly with the stored mass of NH_3 and the rate of increase is similar for all samples. The maximum permittivity value of 4.0 is obtained for the sample with the lowest Si / Al ratio of 27 at the highest temperature of 300 °C. The influence of temperature is visible in increasing sensitivity, especially for the Si / Al ratio of 27. The reason for this is the thermal activation of the NH_4^+ ions, as they become more mobile at higher temperatures. This sensitivity change becomes smaller with increasing Si / Al ratio.

ε_2 increases linearly with the NH_3 content as well, but shows a different dependence on the Si / Al ratio. Without NH_3, ε_2 is almost zero for all samples at all temperatures, as zeolites are low-loss materials. Increasing temperature affects the measured ε_2 significantly, with a higher increase in sensitivity. With an increasing Si / Al ratio, the sensitivity of ε_2 to NH_3 decreases. However, the sensitivity is strongly related to the Si / Al ratio. The highest value for ε_2 of 0.45 was determined for the lowest Si / Al ratio of 27 at 300 °C.

For a closer look at the sensitivity of ε_1 and ε_2 to NH_3, Fig. 6 shows the sensitivities calculated by Eq. (4) as a function of the Al content (corresponding to the given Si / Al ra-

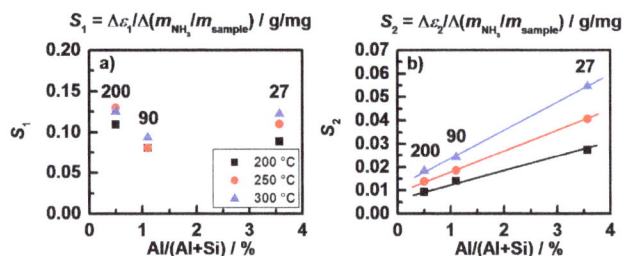

Figure 6. Sensitivities of the dielectric permittivity of H-ZSM-5 zeolites to ammonia: (**a**) S_1 for ε_1, and (**b**) S_2 for ε_2, for Si / Al ratios of 27, 90 and 200 at temperatures of 200, 250 and 300 °C.

tio) as calculated from the rates shown in Fig. 5. The sensitivity S_1 of ε_1 to NH_3 in Fig. 6a appears to be independent of the Si / Al ratio. The small deviations, especially for the measurements of the Si / Al ratio of 90, result from generally less NH_3 storage with increasing Si / Al ratio and from the consequently higher impact of temperature inconstancies to the small changes of the resonance frequency. The sensitivity S_2 of ε_2 to NH_3 in Fig. 6b increases linearly with the Al content for each temperature and increases with increasing temperature.

$$S_i = \frac{\Delta \varepsilon_i}{\Delta (m_{NH_3} / m_{sample})} \quad i = 1, 2 \quad (4)$$

A possible explanation for these observed dependencies of the sensitivity is that the change in ε_1, as a measure for polarization, is mostly dependent on the stored amount of NH_3, as the NH_3 molecules are highly polar. With higher temperature, the polarity of the NH_3 molecule increases, which is most visible for the Si / Al ratio of 27 in Fig. 6a. The change in ε_2, which represents the dielectric loss, is strongly dependent on the observed material. It can include conductivity mechanisms (ionic and/or electronic) for conducting samples. With decreasing Al content of the zeolite, the number of the Brønsted acid sites decreases and the distance between them increases. Consequently, the proton mobility due to proton hopping between neighboring Brønsted sites (Rodriguez-Gonzalez et al., 2008) decreases with increasing storage site distance. The part of the dielectric loss related to the proton mobility of the adsorbed NH_3 molecules increases with temperature, resulting in lower activation energies (thermal activation) for proton hopping, corresponding to the observed measurement results.

5 Conclusions and outlook

In this study, initial measurements with a recently introduced measurement setup using microwave cavity perturbation for catalyst powder samples are performed on H-ZSM-5 zeolites with different Si / Al ratios under reaction conditions. The observed temperature range is 200 to 300 °C and all measurements are performed without the influence of humidity. The amount of stored NH_3 is mirrored by both the real ε_1 and the imaginary parts ε_2 of the complex dielectric permittivity. From this we conclude that both values are suitable for NH_3 detection. Through comparison of different Si / Al ratios, ε_1 shows a similar sensitivity to NH_3 for all samples with the same temperature dependence. The sensitivity of ε_2 to NH_3 has a strong dependence on the Si / Al ratio and on temperature. A possible explanation is that the change of ε_1 represents only the polarity of the present NH_3 molecules, and the change of ε_2 is additionally related to the conductivity mechanisms of the zeolite material, for example, by proton hopping.

In future work, the influence of humidity will be observed and ion-exchanged samples will be investigated. Another focus is to further enhance the setup to access a higher temperature range, to incorporate the ability to perform simultaneous temperature-programmed desorption experiments and to increase measurement accuracy. Additionally, frequency-dependent measurements of the complex permittivity by analyzing several cavity modes (from 1.1 to 4.2 GHz) are planned.

Author contributions. M. Dietrich, D. Rauch, A. Porch, and R. Moos conceived and designed the test setup. M. Dietrich performed the experiments and analyzed the data. All authors evaluated the results and wrote the paper.

Acknowledgements. R. Moos is indebted to the German Research Foundation (DFG) for supporting this work under grant MO 1060/19-1.

U. Simon acknowledges financial supported by the German Research Foundation (DFG), contract No: Si609/14-1, and by the Exploratory Research Space of RWTH Aachen University within the Center for Automotive Catalytic Systems Aachen (ACA).

A. Porch acknowledges the support of Merck GKaA.

References

Beulertz, G., Herbst, F., Hagen, G., Fritsch, M., Gieshoff, J., and Moos, R.: Microwave Cavity Perturbation as a Tool for Laboratory In Situ Measurements of the Oxidation State of Three Way Catalysts, Top. Catal., 56, 405–409, doi:10.1007/s11244-013-9987-3, 2013.

Di Iorio, J. R., Ribeiro, F. H., Bates, S. A., Verma, A. A., Miller, J. T., and Gounder, R.: The Dynamic Nature of Brønsted Acid Sites in Cu–Zeolites During NOx Selective Catalytic Reduction: Quantification by Gas-Phase Ammonia Titration, Top. Catal., 58, 424–434, doi:10.1007/s11244-015-0387-8, 2015.

Dietrich, M., Rauch, D., Porch, A., and Moos, R.: A laboratory test setup for in situ measurements of the dielectric properties of catalyst powder samples under reaction conditions by microwave cavity perturbation: set up and initial tests, Sensors, 14, 16856–16868, doi:10.3390/s140916856, 2014.

Feulner, M., Hagen, G., Moos, R., Piontkowski, A., Müller, A., Fischerauer, G., and Brüggemann, D.: In-Operation Monitoring of the Soot Load of Diesel Particulate Filters: Initial Tests, Top. Catal., 56, 483–488, doi:10.1007/s11244-013-0002-9, 2013.

Franke, M. and Simon, U.: Proton mobility in H-ZSM5 studied by impedance spectroscopy, Solid State Ionics, 118, 311–316, doi:10.1016/S0167-2738(98)00436-6, 1999.

Franke, M. and Simon, U.: Solvate-Supported Proton Transport in Zeolites, Chem. Phys. Chem., 5, 465–472, doi:10.1002/cphc.200301011, 2004.

Fremerey, P., Reiß, S., Geupel, A., Fischerauer, G., and Moos, R.: Determination of the NOx Loading of an Automotive Lean NOx Trap by Directly Monitoring the Electrical Properties of the Catalyst Material Itself, Sensors, 11, 8261–8280, doi:10.3390/s110908261, 2011.

Giodanino, F., Borfecchia, E., Lomachenko, K., Lazzarini, A., Agostini, G., Gallo, E., Soldatov, A. V., Beato, P., Bordiga, S., and Lamberti, C.: Interaction of NH_3 with Cu-SSZ-13 Catalyst: A Complementary FTIR, XANES, and XES Study, Journal of Physical Chemistry Letters, 5, 1552–1559, 2014, doi:10.1021/jz500241m, 2014.

Johnson, T. V.: Review of diesel emission and control, Int. J. Engine Res., 10, 275–285, doi:10.1243/14680874jer04009, 2009.

Koebel, M., Elsener, M., and Kleemann, M.: Urea-SCR: a promising technique to reduce NOx emissions from automotive diesel engines, Catal. Today, 59, 335–345, doi:10.1016/s0920-5861(00)00299-6, 2000.

Kulkarni, V. P., Leustek, M. E., Michels, S. K, Nair, R. N., Snopko, M. A., and Knitt, A. A.: Ash Detection in Diesel Particulate Filter, U.S. Patent 8, 470, 070 B2, 25 June 2013.

Moos, R., Spörl, M., Hagen, G., Gollwitzer, A., Wedemann, M., and Fischerauer, G.: TWC: lambda control and OBD without lambda probe – an initial approach, SAE paper 2008-01-0916, 2008, doi:10.4271/2008-01-0916, 2008.

Moos, R., Wedemann, M., Spörl, M., Reiß, S., and Fischerauer, G.: Direct Catalyst Monitoring by Electrical Means: An Overview on Promising Novel Principles, Top. Catal., 52, 2035–2040, doi:10.1007/s11244-009-9399-6, 2009.

Moos, R., Beulertz, G., Reiß, S., Hagen, G., Votsmeier, M., Fischerauer, G., and Gieshoff, J.: Overview of the Microwave-Based Automotive Catalyst State Diagnosis, Top. Catal., 56, 358–364, doi:10.1007/s11244-013-9980-x, 2013.

Niwa, M. and Katada, N.: New Method for the Temperature-Programmed Desorption (TPD) of Ammonia Experiment for Characterization of Zeolite Acidity: A Review, Chemical Record, 5, 432–455, doi:10.1002/tcr.201300009, 2013.

Porch, A., Slocombe, D., Beutler, J., Edwards, P., Aldawsari, A., Xiao, T., Kuznetsov, V., Almegren, H., Aldrees, S., and Almaqati, N.: Microwave treatment in oil refining, Applied Petrochemical Research, 2, 37 44, doi:10.1007/s13203-012-0016-4, 2012.

Rahkamma-Tolonen, K., Maunula, T., Lomma, M., Huuhtanen, M., and Keiski, R. L.: The effect of NO_2 on the activity of fresh and aged zeolite catalysts in the NH3-SCR reaction, Catal. Today, 100, 217-222, doi:10.1016/j.cattod.2004.09.056, 2005.

Rauch, D., Kubinski, D., Simon, U., and Moos, R.: Detection of the ammonia loading of a Cu Chabazite SCR catalyst by a radio frequency-based method, Sensor Actuat. B-Chem., 205, 88–93, doi:10.1016/j.snb.2014.08.019, 2014.

Rauch, D., Kubinski, D., Cavataio, G., Upadhyay, D., and Moos, R.: Ammonia Loading Detection of Zeolite SCR Catalysts using a Radio Frequency based Method, SAE Technical Paper 2015-01-0986, doi:10.4271/2015-01-0986, 2015.

Reiß, S., Wedemann, M., Spörl, M., Fischerauer, G., and Moos, R.: Effects of H_2O, CO_2, CO, and flow rates on the RF-based monitoring of three-way catalysts, Sensor Letters 2011, 9, 316–320, doi:10.1166/sl.2011.1472, 2011a.

Reiß, S., Schönauer, D., Hagen, G., Fischerauer, G., and Moos, R.: Monitoring the Ammonia Loading of Zeolite-Based Ammonia SCR Catalysts by a Microwave Method, Chemical Engineering Technology, 34, 791–796, doi:10.1002/ceat.201000546, 2011b.

Rodriguez-Gonzalez, L., Rodriguez-Castellon, E., Jimenez-Lopez, A., and Simon, U.: Correlation of TPD and impedance measurements on the desorption of NH3 from zeolite H-ZSM-5, Solid State Ionics 179, 1968–1973, doi:10.1016/j.ssi.2008.06.007, 2008.

Sappok, A., Parks, J., and Prikhodko, V.: Loading and Regeneration Analysis of a Diesel Particulate Filter with a Radio Frequency-Based Sensor, SAE Technical Paper 2010-01-2126, doi:10.4271/2010-01-2126, 2010.

5

A new low-cost hydrogen sensor build with a thermopile IR detector adapted to measure thermal conductivity

M. Liess

RheinMain University of Applied Sciences, Department of Engineering, Am Brückweg 26,
65428 Rüsselsheim, Germany

Correspondence to: M. Liess (martin.liess@hs-rm.de)

Abstract. It is demonstrated how a commercially available MEMS thermopile infrared radiation sensor can be used as thermal conductivity gas detector (TCD). Since a TCD requires a heater while IR-thermopile sensors have no integrated heater, the thermopile itself is used as heater and temperature sensor at the same time. It is exposed to the measured gas environment in its housing. It is shown that, by using a simple driving circuitry, a mass-produced low-cost IR sensor can be used for hydrogen detection in applications such as hydrogen safety and smart gas metering. The sensor was tested to measure hydrogen in nitrogen with concentration of 0–100 % with a noise equivalent concentration of 3.7 ppm.

1 Introduction

In this paper, a new very low-cost hydrogen sensor with high levels of reliability will be introduced. The sensor is based on a thermal conductivity measurement performed with a commercial thermopile IR sensor device. The sensor's performance makes it suitable for applications such as hydrogen measurement in smart gas metering and hydrogen technology safety.

1.1 Motivation

Hydrogen fuel cell systems are of growing interest in the area of sustainable transportation as well as for stationary electric power in remote areas, distributed electric energy generation, in space and other closed environment and auxiliary power systems (Appleby and Foulkes, 1989).

Since a mixture of hydrogen and air is highly explosive in concentrations between 4 and 75 % hydrogen (Kenneth Barbalace, 1995–2015), leakage monitoring is necessary for safety reasons. In particular, the high pressure in hydrogen pressure tanks can lead to significant safety issues (Larminie and Dicks, 2003).

Another use for this kind of sensor is measuring the hydrogen content in natural gas systems. Hydrogen that is produced with electric energy from excess wind and solar energy by electrolysis can be added to existing gas systems (Gahleitner, 2013) up to a content of 5 %. The GERG (European Gas Research Group; Winkler-Goldstein and Rastetter, 2013) sees the potential to add an amount of up to 20 %. In this case, it is necessary to monitor the hydrogen content of the gas at the consumer side to ensure optimization of combustion and smart metering, since the gross heat of combustion of hydrogen ($286\,kJ\,mol^{-1}$) and methane ($889\,kJ\,mol^{-1}$) are significantly different (Burgess, 2011).

Therefore, reliable and low-cost hydrogen sensors are necessary for leakage monitoring and smart metering of combustible gas.

There are different known sensor principles that allow for hydrogen measurement and detection in a matrix of other gases:

1. The thermal conductivity detector (TCD) or katharometer (Daynes, 1920) detects and distinguishes different gases based on their thermal conductivity and thermal capacity. Since hydrogen has a very high thermal conductivity, the principle of TCD is very suitable for its detection. This principle is treated in more detail in Sect. 1.3.

2. Gas chromatography in combination with TCD, PDD (pulsed discharge ionization detector) BID (barrier discharge ionization detector) as well as mass spectrome-

try are also used to determine hydrogen and other gases in complex mixtures. However such methods are rather slow, expensive and more suitable for scientific analysis, rather than safety and consumer purposes.

3. Catalytic sensors or catalytic bead gas sensors measure the combustion heat generated by the chemical reaction of the gas with oxygen on a heated catalytic surface in comparison to a reference surface. Usually catalytic gas sensors are relatively unspecific and react to all combustible gases. However, depending on the catalyst, they might have an enhanced sensitivity to the specific gas concerned (such as hydrogen). A MEMS-catalytic gas sensor has the advantage of miniaturization (Lee et al., 2011). The disadvantage of all catalytic sensors (in addition to their unspecific response) is that they are prone to catalytic poisoning.

4. Palladium-based sensors react to hydrogen due to the high solubility of hydrogen in palladium and subsequent changes to the palladium conductivity, Fermi energy level or work function (Lewis, 1967). Also hydrogen sensors that are based on a Schottky contact between palladium and a semiconductor have been demonstrated (Hudeish and Abdul Aziz, 2006; Song et al., 2005). The disadvantage of all sensors based on Palladium is the fact that this metal is affected by poisoning, for example by sulfur or lead-containing compounds.

Modern palladium-based sensor designs suitable for low-cost hydrogen measurements exhibit dynamic ranges from 0.025 to 2 % and response times above 1.8 min (Hong et al., 2015), which – for the applications mentioned above – are less suitable than the sensor presented here.

5. Surface acoustic wave (SAW) sensors are based on piezoelectric and, in most cases, additional sensitive materials. In the case of hydrogen detection, the influence of hydrogen on the elasticity of a thin film of WO_3 changes the speed of the surface acoustic waves (Ippolito et al., 2003) and can thus be detected. As with all indirect principles that do not directly react to a physical property of the detected gas, effects of material ageing and poisoning can lead to measurement errors and sensor degeneration.

The sensor presented here is based on a thermal conductivity measurement and a "creative" and new use of a standard, commercially available thermopile IR sensor. Thermopile sensors are mostly used for infrared radiation measurements. They consist of a sensor element that is sealed in a housing and quantifies the energy carried by the radiation to be detected by converting it to heat. In this study, however, an unsealed thermopile sensor is used. The thermopile sensor element not only measures a temperature difference but is also heated by an electric alternating current (AC). At the

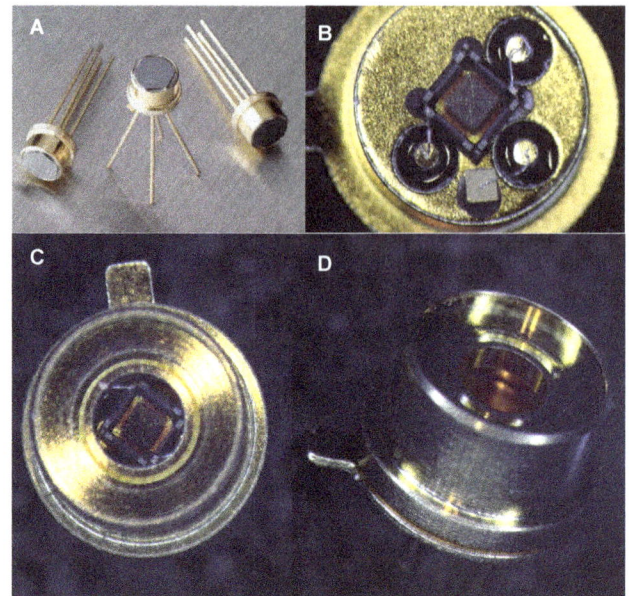

Figure 1. TPS 23B Thermopile sensors. These sensors are a mass product and normally used for IR radiation measurement for example in ear thermometers. (**a**) The image on the upper left (courtesy of Excelitas Technologies GmbH & Co. KG.) shows the sensor as it is used for broadband IR detection in applications such as ear thermometers. (**b**) The image on the upper right shows the interior of the device comprising the sensor element with its thermoelements, the (partially transparent) membrane and the IR-absorber patch. A separate thermistor for ambient temperature compensation is located on the bottom of the device. The parts (**c**) and (**d**) of the figure show sensors that were manufactured without an IR transparent window. Through the "open window", the thermopile sensor element is exposed to the gas environment. In this study they are used as thermal conductivity gas sensors.

same time it is cooled by the gas environment which the thermopile is exposed to and measures. Due to the high thermal conductivity of hydrogen, this gas can be detected.

1.2 Thermopile IR-radiation sensors

MEMS thermopile sensors (Fig. 1) consist of a thin thermally insulating membrane made of silicon oxide and/or silicon nitride. The membrane is surrounded by a silicon rim/periphery with high thermal conductivity. A large number of thermocouples, forming a thermopile, are placed on the thermally insulating membrane and the silicon periphery in such a way that a temperature difference between membrane and periphery results in a thermoelectric voltage that is multiplied by the number of elements. In the case of the sensor presented, those thermocouples are made of n-doped and p-doped polysilicon.

Thermopile sensors are usually applied in infrared (IR) radiation measurements. In such applications, the IR radiation to be measured heats the so-called "hot contacts" located on the thermally insulating membrane, while the "cold contacts" are on the cold heat-conducting silicon periphery and are not warmed significantly by the incoming IR radiation. Since its thermal behaviour depends on the thermal conductivity of the ambient gas, the IR-thermopile sensor element is usually packaged hermetically with an inert gas filling. IR radiation can enter the packaging through an IR transparent window. In some cases, the packaging is filled with a gas with low thermal conductivity gas. This leads to better isolation of the heated membrane and thus to a higher sensitivity of the device (Graf et al., 2007; Liess, 2012).

Thermopile IR sensors are classically used for contactless temperature measurements through the IR radiation emitted by the object to be measured. They are also used in non-dispersive infrared (NDIR) gas sensors to detect gases through their specific IR absorption lines (Graf et al., 2007). The method presented in this paper is not related to the well-known method of NDIR or any kind of IR detection.

Thermopile sensors are very sensitive to electric currents driven into their output terminals since this leads to resistive heating of the thermocouple structure and thus of the membrane of the sensor element. The resulting thermal gradient generates an error signal. This can be of relevance since, for example, a chopper amplifier (Wu et al., 2013) can capacitively couple alternating currents unintentionally into the sensor. This AC error itself is filtered by the low-pass properties of the amplifier and signal conditioning electronics and is therefore of no relevance. However it leads to an ohmic heating effect within the thermopile, which in turn generates a DC error that has the same properties as the measured IR signal and cannot be filtered.

Thermopile IR sensors are thus cross-sensitive to input average currents and to a leakage of the filling gas. Exactly these properties are used in this study to build a gas sensor.

1.3 Stability of the presented sensor

The modified CMOS production process of thermopile sensors involves high temperatures and harsh processes such as potassium hydroxide etching and photoresist removal. Thus the sensor element can be expected to be very stable against thermal and chemical damage caused by environmental influences.

In automotive applications (Liess et al., 2004), including racing sports, the sensor design has demonstrated a quasi-unlimited lifetime over the years and high mechanical robustness as an IR-radiation sensor.

The physical measurement principle is based on heat conduction and does not involve the chemical interaction of the measured environment with any part of the sensor. The sensor surface is passivated by a layer of silicon nitride so that the sensing elements are not exposed to any materials that could give rise to poisoning or a change to the doping of the polysilicon. Furthermore, the films are heavily doped with doses of 5×10^{15}–15×10^{15} atoms cm^{-3} and dopant diffusion is expected to be low at operation temperatures, which

Table 1. Examples of the thermal conductivity of different gases. Data from Young and Sears (1992).

Gas	Thermal conductivity at 20 °C [W (m · K)$^{-1}$]
Nitrogen	0.0234
Oxygen	0.0238
Hydrogen	0.172
Air (at 0 °C)	0.024
Helium	0.138

are significantly below the diffusion temperature of around 1000 °C. Therefore, sensitivity changes or any kind of poisoning of the sensitive parts of the device cannot be expected due to the chemical effects of the measured environment.

1.4 Thermal conductivity gas sensors

Commercially available TCDs are based on four identical platinum resistors that are arranged in a bridge configuration and are electrically heated. The voltage drop on each platinum resistor indicates its resistance and is thus a measure of its temperature. Since two resistors are exposed to the gas to be measured and the two other resistors are exposed to a reference gas, differences in the heat conductivity between the measured gas and reference gas lead to a bridge voltage. As can be seen in Table 1, hydrogen has a 7 times higher thermal conductivity than air. Hydrogen's high thermal conductivity is paralleled only by helium, which is not present in the typical environment where hydrogen needs to be detected.

De Graaf and Wolffenbuttel (2012) developed a micromachined thermal conductivity gas sensor, based on a thermopile for temperature measurement and a dedicated heater for heating the hot contacts of the thermopile. The sensor has, in addition, a dedicated sample chamber, made by a surface micromachining processes. To obtain a stable signal, heat modulation and a lock-in technique is used.

1.5 Operating principle of the sensor presented here

The sensor used in this work (Fig. 1) is a commercially available thermopile IR sensor that is available for applications such as ear thermometers on the mass market. The sensor itself has two output contacts. Usually they output a DC voltage that is proportional to the IR radiation, which the sensor element receives. However, in this application, the same two pins are used to input an AC heating voltage into the device and to output, at the same time, the DC signal voltage, which measures the hydrogen concentration. Also the thermopile structure itself is now used for two purposes: its resistive properties are used for heating (by an AC voltage) and its thermoelectric properties are used for measuring a temperature difference (by a DC voltage). In order to enable this mode of operation the sensor is connected through a high-pass circuit to the AC heating supply and gives out its DC signal through a low-pass circuit. The gas to be measured is simply supplied to the surface of the sensor and contained by the standard sensor housing (Fig. 2).

2 Experimental section

2.1 A MEMS thermopile used as a thermal gas sensor

A thermopile sensor can be used as a hydrogen sensor because of

- the high thermal conductivity of hydrogen,

- the cross-sensitivity of thermopile sensors toward their gas environment,

- the error sensitivity toward AC input currents into the sensor's output terminals.

For gas sensor operation, commercially available thermopile IR-sensor devices have been used that are manufactured without an IR window (Fig. 1) so that the sensor element can be directly exposed to the gas environment. The membrane of the sensor element is heated by an AC that is applied to the input terminals of the thermopile sensor (Liess, 2014). The thermopile's DC output voltage is then measured.

2.2 Electrical setup

An Agilent Technologies Function/Arbitrary-Waveform generator 80 MHz 33250A frequency generator was used to supply the AC driving voltage to the sensor. Using a capacitor, the frequency generator is decoupled from DC voltages generated in the circuit. The AC voltage is applied to a thermopile sensor. The DC voltage generated by the thermopile is filtered by a low-pass filter comprising a resistor and a capacitor and measured by a Fluke 77 IV digital multimeter. The electrical setup is shown in Fig. 2. Measurement curves over extended periods of time were recorded using the capabilities of a modified gas analyser manufactured by Emerson Process Management GmbH & Co. OHG.

2.3 Driving a thermopile with AC voltage

The heat P_{heat} generated inside the sensor element by an AC voltage U_{AC} applied to the output terminals of the thermopile sensor follows the equation

$$P_{\text{heat}} = \frac{U_{\text{AC}}^2}{R_{\text{sensor}}}. \tag{1}$$

One can assume that the thermal contact between the periphery and the ambient is so good that the temperatures of the periphery and the ambient environment are identical. They are equal to T_{amb}. Since a fraction A of the heat P_{heat} is generated on the membrane, its temperature T_{mem} depends

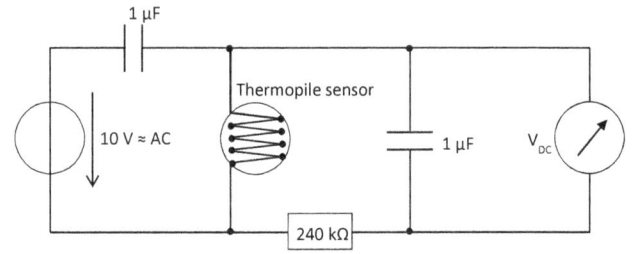

Figure 2. Electrical circuit for driving the thermopile as gas sensor. Inner resistance of the sensor is $120\,\text{k}\Omega$.

on the thermal contact between the membrane and the periphery λ_{mem} and the gas λ_{gas} as

$$T_{\text{mem}} = \frac{A \cdot P_{\text{heat}}}{\lambda_{\text{mem}} + \lambda_{\text{gas}}} + T_{\text{amb}}. \tag{2}$$

Here, the thermal contact λ is defined as (thermal conductivity) · (contact area)/(length of the contact). The generated thermopile voltage U is proportional to the temperature difference ΔT between the membrane and the periphery

$$U_{\text{DC}} \propto \Delta T = \left(T_{\text{mem}} - T_{\text{per}}\right) \approx \frac{A \cdot P_{\text{heat}}}{\lambda_{\text{mem}} + \lambda_{\text{gas}}}. \tag{3}$$

Thus,

$$U_{\text{DC}} \propto \frac{U_{\text{AC}}^2}{\lambda_{\text{mem}} + \lambda_{\text{gas}}}. \tag{4}$$

2.4 Mechanical setup

Two thermopile sensors (Excelitas Technologies, Wiesbaden, Germany) type TPS 23B without an IR window (Fig. 1) were connected to the electrical setup. They were exposed to nitrogen, hydrogen, or any mixture of these gases at different temperatures using a climate chamber. The gas mixtures were produced by a DIGAMIX KM301 Wösthoff gas mixing pump and measured with a variable area flow meter. Different flow rates of nitrogen were generated using a needle valve and a variable area flow meter. Figure 3 shows the mechanical setup. The sensors are exposed to the gas flow using T-junctions and are sealed with O-rings. By mounting the sensor in the slipstream (Fig. 3) about 1 cm off of the main stream, the sensor is exposed to the gas but not to its direct flow; 6 mm diameter Swagelok stainless steel tubing and materials were used.

3 Results and discussion

3.1 Response of the thermopile sensor to an AC voltage

In the first experiments, the functionality of the electronic setup was verified. Figure 4 demonstrates that the idea of applying an AC voltage to the sensor through a high-pass filter,

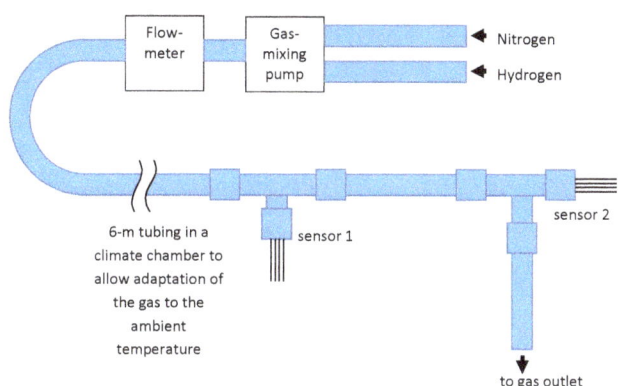

Figure 3. Drawing of the gas tubing with the thermopiles. The sensors are fitted in T-junctions within stainless steel tubing inside a climate chamber.

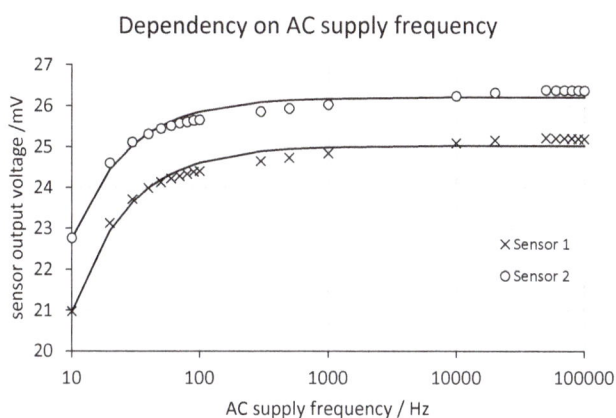

Figure 4. Frequency dependence of the sensor signal in the ambient environment. The lines represent the behaviour of first-order high-pass filters with cut-off frequencies of 1.9 and 2.4 Hz.

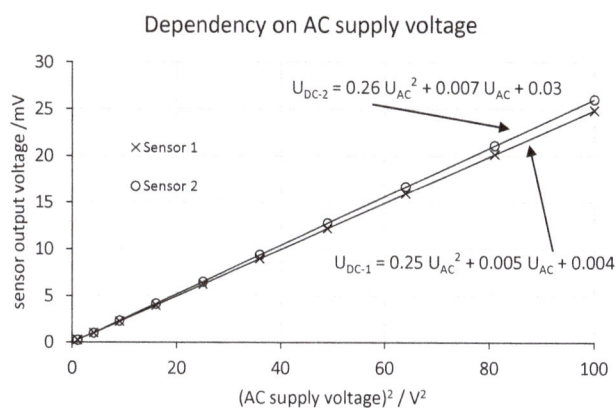

Figure 5. AC supply voltage dependence of the sensor signal in a room environment. The lines represent polynomial fits of the third order with no constraints. It can be seen that, within the range of supply voltages, the quadratic term is 370–500 times greater than any other term, indicating that the measured signal represents a DC thermovoltage generated by ohmic heating of alternating currents through the thermopile. The power applied to the sensor calculates as $P = U^2/R$. Since $R = 120\,\mathrm{k\Omega}$, the maximum power is 0.83 mW.

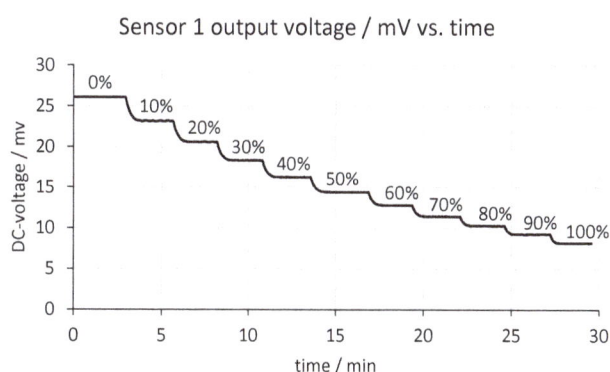

Figure 6. Measurements of the sensor 1 signal with different H_2 concentrations. The measurements were performed at 5 °C in a climate-controlled room. If not indicated differently, all measurements were performed at flow rates of 0.5 L min^{-1}.

constituted by a capacitor and the sensor's inner resistance, and measuring the output DC voltage through a low-pass filter works well. It can be seen that the AC input to the thermopile sensor generates a DC output.

Figure 5 shows a simple quadratic dependence between input AC and output DC voltage with practically no zeroth- and first-order terms, indicating that the input AC voltage is converted to ohmic heat, which is – within a reasonable approximation – the only source of a temperature gradient that leads to the signal from the thermopile sensor. A significant first-order term would have indicated a prominent Peltier effect. This was not observed, which is in agreement with the fact that, firstly, the quadratic effect of ohmic heating dominates over Peltier heat transport at high driving voltages and, secondly, AC voltage was used for driving the sensor. A significant zeroth-order term (offset) would have indicated a strong external heat source or a high temperature difference between the gas and the sensor. This was not observed since

the sensor was operated in equilibrium with the measured gas.

3.2 Response of the DC sensor output voltage to the H_2 concentration

To test the effect of the gas environment, the sensors were exposed to mixtures of hydrogen in nitrogen in steps of 10 or 0.5 % of a few minutes in duration, while the sensor's output voltage was measured (Figs. 6 and 7). It can be seen that, despite the simple measurement setup, the signal can easily be distinguished from the noise.

Due to the transfer of inertia and energy between the different atoms, the heat conductivity of mixtures between hydrogen and nitrogen is not a simple linear function of

Sensor 1 output voltage / mV vs. time

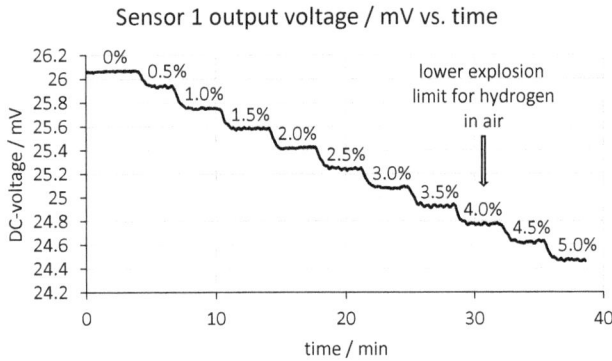

Figure 7. Measurements of the sensor 1 signal with different low H_2 concentrations. The measurements were performed at $5\,^{\circ}C$ in a climate-controlled room. The lower detection limit of 4 % hydrogen can easily be distinguished by a sensor output voltage drop of 4.8 % as compared to the output voltage in the air.

Sensor 1 signal vs. hydrogen concentration

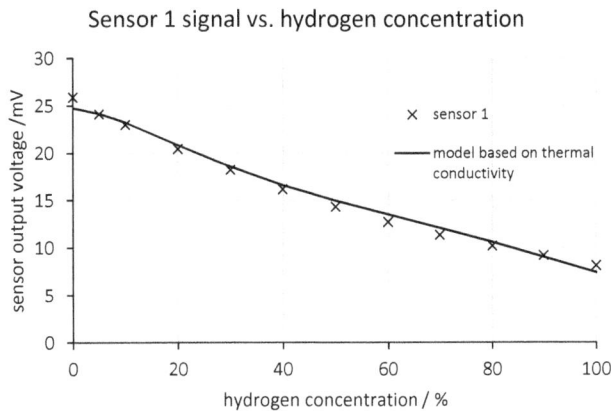

Figure 8. Comparison of the measured data with sensor 1 to a model based on Eq. (1) and literature heat conductivity values of mixtures between hydrogen and nitrogen (Mason and Saxena, 1958).

their concentration. Experimental heat conductivity data of mixtures between hydrogen and nitrogen at $0\,^{\circ}C$ (Mason and Saxena, 1958) were used in Eq. (1) together with the heat conductivity of the membrane that was fitted using the Levenberg–Marquardt algorithm to match the sensor signal (Fig. 8). A good agreement was obtained. The heat conductivity of a hydrogen environment can be calculated as 4.38 times higher than the heat conductivity of the membrane, while the heat conductivity of a nitrogen environment is lower than the heat conductivity of the membrane by a factor of 0.61.

3.3 Basic sensor specifications

The sensitivity (Fig. 9), the dynamic response ($\tau = 2.5\,s$) and the measurement's rms noise level ($3.7\,\mu V$) were calculated from the data of the output voltage changing H_2 concentration (Fig. 6). It can be seen that the dynamic response

Sensitivity (sensor 1 and 2 - avarage)

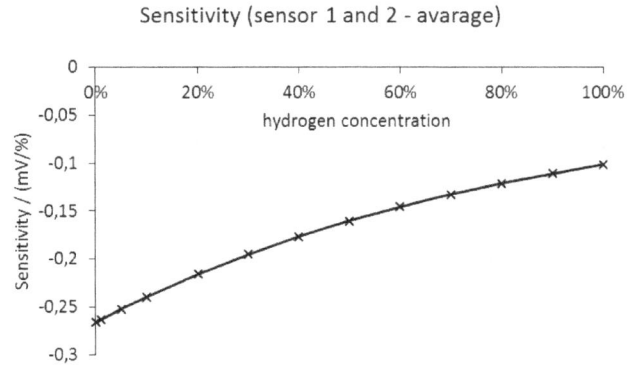

Figure 9. Sensitivity at different hydrogen concentrations.

Temperature dependence of sensor signal

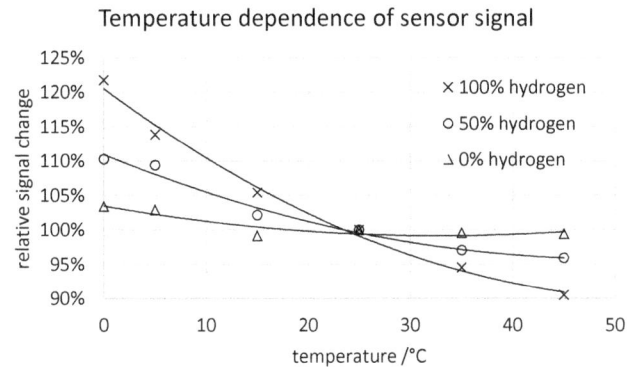

Figure 10. Temperature dependence of the sensor output voltage at different hydrogen concentrations.

is significantly smaller than the thermal response (15 ms) of the device as indicated by the sensor's manufacturer. This indicates that the dynamic response is limited by the diffusion of the gas to the sensor that is mounted in the slipstream of the dead end of the T-junction (Fig. 3) and not by the sensor properties themselves.

The noise level of the sensor can be calculated from the thermal or Johnson noise due to its inner resistance ($120\,k\Omega$). It can be seen that the noise level of the measurement is not due to the sensor itself but due to the driving and signal recording electronic circuitry. The basic specifications of the sensor and the measurement are compiled in Table 2.

3.4 Temperature and flow rate cross-sensitivity measurements

Even though the relative temperature dependence of the heat conductivity of hydrogen and air (Davies, 2006) remains similar in the relevant temperature range, the temperature dependence of the sensor reading increases with larger hydrogen concentrations (Fig. 10). This can be attributed to the fact that, with rising hydrogen concentration and thus rising thermal conductivity of the gas environment, its temperature-dependent contribution rises. In contrast, the thermal conduc-

Table 2. Compilation of basic sensor and measurement specifications.

	Value/unit	Comment
Sensitivity	-27 to -10 mV %$^{-1}$	At concentrations of 0–100 % hydrogen (compare Fig. 9)
Intrinsic thermal time constant	15 ms	From manufacturer's data sheet
Time constant limited by gas diffusion	2.5 s	As observed in the measurements shown in Fig. 6
Sensor chip inner resistance	120 kΩ	From manufacturer's data sheet
Rms noise level of the measurement	3.7 μV	Calculated during the first 3 min of the measurement shown in Fig. 6
Rms noise level of the sensor during 3 min of measurement with measurement rate of 1.75 Hz	0.06 μV	Calculated from the thermal or Johnson noise at 20 °C for a 120 kΩ sensor
Noise equivalent concentration of the measurement	1.4–3.7 ppm (H$_2$)	Calculated from measured rms noise and the measured sensitivity at 0–100 % hydrogen concentration
Cross-sensitivity towards the flow velocity	max. 15 ppm (H$_2$)/(m s^{-1})	Calculated from the data of Fig. 11 and the device's minimum sensitivity at 100 % %H$_2$ concentration (Fig. 9)
Cross-sensitivity towards the temperature	-9 to -30 ppm (H$_2$) K^{-1}	Calculated from the data of Fig. 10 and the sensitivity shown in Fig. 9 for 0–100 % %H$_2$

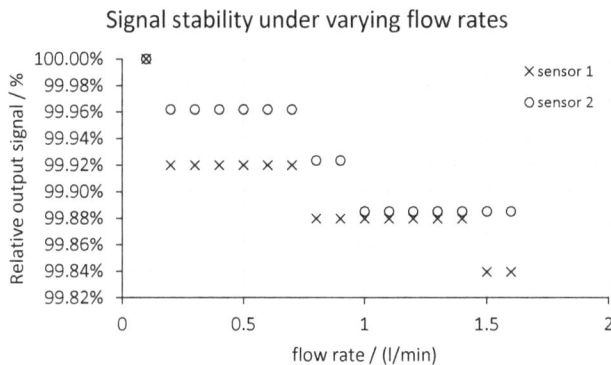

Figure 11. Flow rate dependence of the sensor output voltage for flow velocities up to 1 m s^{-1}. At the flow rate of 1 L min^{-1} the flow velocity is 0.59 m s^{-1}.

tivity of the sensor membrane is constant with temperature and dominates the behaviour of the sensor at low hydrogen concentrations. As with other hydrogen sensors, such as classical TCDs, the temperature of the sensor must be stabilized to allow for precise measurements, or the effect of the ambient temperature must be compensated after the measurement. Since a thermistor temperature probe is part of most thermopile sensor devices, temperature measurement and compensation is possible without any additional effort.

Within the commercial temperature range from 0 to 85 °C, the maximum measurement error caused by temperature effects can be calculated. It is smaller than 0.26 % (H$_2$) within the full measurement range, 0.08 % (H$_2$) for hydrogen concentrations of below 5 % and smaller than 0.1 % (H$_2$) for hydrogen concentrations of less than 20 % (Fig. 10). Therefore, for hydrogen safety and smart metering, no temperature compensation of the sensor signal is necessary.

Since for hydrogen concentrations up to 20 % the maximum measurement error caused by gas flow velocities of up to 1 m s^{-1} is smaller than 7.3 ppm (H$_2$) (Fig. 11), no flow rate compensation must be applied for use of these sensors in hydrogen safety and smart metering.

4 Conclusions and outlook

It was demonstrated that a thermopile sensor with suitable simple driving circuitry, which is exposed to a mixture of hydrogen and a heavier gas (for example nitrogen or air), can be used as a low-cost thermal hydrogen gas sensor. It behaves according to the theoretical expectations.

The sensor is suitable for applications such as hydrogen technology safety and smart gas metering without compensation of ambient temperature and gas flow velocity. If mounted in the slipstream of an approximately 1 cm long dead-end tube, the sensor exhibits basically no cross-sensitivity toward the flow rate and reacts sufficiently fast (2.5 s).

The idea of using a thermopile as a heater and temperature difference sensor at the same time can also be applied in other sensor principles. This allows for simplification of the design or even for use of the same MEMS sensor element for different purposes (like IR sensing, gas sensing or flow sensing) depending on the driving circuitry, housing and exposure to the measured magnitudes.

It was also demonstrated how sensor shortcomings (of a MEMS thermopile IR-radiation sensor) can be used to create a new sensor for a different purpose (when the sensor is used as a gas sensor). Thus also here Edward W Ng's famous quote "One man's noise is another man's signal" (Blackslee, 1990) applies well.

Acknowledgements. I thank M.-K. Winter and H. Krause of Emerson Process Management in Hasselroth for their support in facilitating the measurements and for financial support. I also express my gratitude to the team of Excelitas Technologies GmbH & Co. KG, who manufactured the custom thermopile sensor devices without IR windows for these experiments.

References

Appleby, A. J. and Foulkes, F. R.: Fuel cell handbook, Van Nostrand Reinhold, New York, USA, XXVI, 762 pp., 1989.

Blackslee, S.: Lost on Earth: Wealth of Data Found in Space: "In science, one man's noise is another man's signal", New York Times, available at: http://www.nytimes.com/1990/03/20/science/lost-on-earth-wealth-of-data-found-in-space.html (last access: 12 May 2015), 1990.

Burgess, D. R.: Thermochemical Data, in: NIST Chemistry WebBook: NIST Standard Reference Database Number 69, edited by: Linstrom, P. J. and Mallard, W. G., Gaithersburg, MD, USA, available at: http://webbook.nist.gov (last access: 30 April 2015), 2011.

Davies, T. W.: Thermal Conductivity Values, in: A-to-Z Guide to Thermodynamics, Heat and Mass Transfer, and Fluids Engineering, Begellhouse, Redding, Connecticut, USA, 2006.

Daynes, H. A.: The Theory of the Katharometer (with an Introductory Note by G. A. Shakespear), Proc. Roy. Soc. London A, 273–286, 1920.

de Graaf, G. and Wolffenbuttel, R. F.: Surface-micromachined thermal conductivity detectors for gas sensing, in: 2012 IEEE International Instrumentation and Measurement Technology Conference (I2MTC), 13–16 May 2012, Graz, Austria, 1861–1864, 2012.

Gahleitner, G.: Hydrogen from renewable electricity: An international review of power-to-gas pilot plants for stationary applications, Int. J. Hydrogen Energ., 38, 2039–2061, 2013.

Graf, A., Arndt, M., Sauer, M., and Gerlach, G.: Review of micromachined thermopiles for infrared detection, Meas. Sci. Technol., 18, R59–R75, 2007.

Hong, J., Lee, S., Seo, J., Pyo, S., Kim, J., and Lee, T.: A highly sensitive hydrogen sensor with gas selectivity using a PMMA membrane-coated Pd nanoparticle/single-layer graphene hybrid, ACS Appl. Mater. Interfaces, 7, 3554–3561, doi:10.1021/am5073645, 2015.

Hudeish, A. Y. and Abdul Aziz, A.: A hydrogen sensitive Pd/GaN schottky diode sensor, J. Phys. Sci., 17, 161–167, 2006.

Ippolito, S. J., Kandasamy, S., Kalantar-Zadeh, K., Trinchi, A., and Wlodarski, W.: A Layered Surface Acoustic Wave ZnO/LiTaO$_3$ Structure with a WO$_3$ Selective Layer for Hydrogen Sensing, Sen. Lett., 1, 33–36, doi:10.1166/sl.2003.007, 2003.

Kenneth Barbalace: Periodic Table of Elements – Hydrogen – H, EnvironmentalChemistry.com, available at:http://EnvironmentalChemistry.com/yogi/periodic/H.html (last access: 31 August 2015), 1995–2015.

Larminie, J. and Dicks, A.: Fuel cell systems explained, SAE International and John Wiley & Sons, Ltd., 428 pp., 2003.

Lee, E.-B., Hwang, I.-S., Cha, J.-H., Lee, H.-J., Lee, W.-B., Pak, J. J., Lee, J.-H., and Ju, B.-K.: Micromachined catalytic combustible hydrogen gas sensor, Sensor. Actuat. B-Chem., 153, 392–397, doi:10.1016/j.snb.2010.11.004, 2011.

Lewis, F. A.: The Palladium Hydrogen System, Academic Press, London, UK, New York, USA, 1967.

Liess, M.: 14 Fotoelektrische Sensoren, in: Sensoren in Wissenschaft und Technik: Funktionsweise und Einsatzgebiete, edited by: Hering, E., Vieweg+Teubner Verlag, Wiesbaden, Germany, 609–628, 2012.

Liess, M.: A new principle for low-cost hydrogen sensors for fuel cell technology safety, in: 4th International Conference on Mathmatics and Natural Sciences (ICMNS 2012): Science for Health, Food and Sustainable Energy, Bandung, Indonesia, 8–9 November 2012, AIP Conference Proceedings, AIP Publishing LLC, 522–525, 2014.

Liess, M., Hausner, M., Schilz, J., Lauck, G., Karagoezoglu, H., and Ernst, H.: Temperature radiation sensors for automotive climate control, in: IEEE Sensors, 2004, Vienna, Austria, 24–27 October 2004, 5–7, 2004.

Mason, E. A. and Saxena, S. C.: Approximate Formula for the Thermal Conductivity of Gas Mixtures, Phys. Fluids, 1, 361, doi:10.1063/1.1724352, 1958.

Song, J., Lu, W., Flynn, J. S., and Brandes, G. R.: AlGaN/GaN Schottky diode hydrogen sensor performance at high temperatures with different catalytic metals, Solid State Electron., 49, 1330–1334, doi:10.1016/j.sse.2005.05.013, 2005.

Winkler-Goldstein, R. and Rastetter, A.: Power to Gas: The Final Breakthrough for the Hydrogen Economy?, Green, 3, 69–78, doi:10.1515/green-2013-0001, 2013.

Wu, R., Huijsing, J. H., and Makinwa, K. A. A.: Precision Instrumentation Amplifiers and Read-Out Integrated Circuits, Springer New York, New York, USA, Online-Ressource, 196 pp., 2013.

Young, H. D. and Sears, F. W.: University physics, 8th ed., Addison-Wesley Pub. Co., Boston, USA, xx, 1132 pp., 1992.

6

Calibration of uncooled thermal infrared cameras

H. Budzier and G. Gerlach

Technische Universität Dresden, Electrical and Computer Engineering Department, Solid-State Electronics Laboratory, Dresden, Germany

Correspondence to: H. Budzier (helmut.budzier@tu-dresden.de)

Abstract. The calibration of uncooled thermal infrared (IR) cameras to absolute temperature measurement is a time-consuming, complicated process that significantly influences the cost of an IR camera. Temperature-measuring IR cameras display a temperature value for each pixel in the thermal image. Calibration is used to calculate a temperature-proportional output signal (IR or thermal image) from the measurement signal (raw image) taking into account all technical and physical properties of the IR camera. The paper will discuss the mathematical and physical principles of calibration, which are based on radiometric camera models. The individual stages of calibration will be presented. After start-up of the IR camera, the non-uniformity of the pixels is first corrected. This is done with a simple two-point correction. If the microbolometer array is not temperature-stabilized, then, in the next step the temperature dependence of the sensor parameters must be corrected. Ambient temperature changes are compensated for by the shutter correction. The final stage involves radiometric calibration, which establishes the relationship between pixel signal and target object temperature. Not all pixels of a microbolometer array are functional. There are also a number of defective, so-called "dead" pixels. The discovery of defective pixels is a multistep process that is carried out after each stage of the calibration process.

1 Introduction

In recent years, thermography has had a dramatic development with annual growth rates of over 20 % (Mounier, 2011). This development will have an even more dynamic impact in the future. Enabling this huge market success are infrared (IR) image sensors based on microbolometer arrays, which have excellent thermal and spatial resolution (Kruse, 1997; Vollmer and Möllmann, 2010). Also, because no cooling is required, they have low power consumption and have a low entry-level price.

Mainly due to the rapid development of micro- and nanotechnology, microbolometers have become significantly cheaper and more efficient. While early in 2000 the maximum image size was 320×240 pixels, nowadays commercial sensor arrays are available with a maximum size of 1024×768 pixels. IR image sensors with full HDTV resolution (Black et al., 2011) are already being advertised. The development trend is towards even smaller pixel grids ($< 17\,\mu$m), to lower power loss and compact ceramic packages. While up until a few years ago, the stabilization of the temperature sensor was a requirement, nowadays current arrays manage without this. The use of microbolometers without temperature stabilization allows for compact, energy-efficient camera designs.

An uncooled IR camera consists of the following main components (Fig. 1) (Budzier and Gerlach, 2011):

 – IR optics

 – sensor arrays

 – processor-based camera electronics.

Both these components and the calibration process play an especially crucial role in ensuring the quality of an IR camera. The calibration is implemented both in the hardware and the software (firmware). For the two device concepts

 – vision display device and

 – temperature-measuring image device

there are different calibration concepts (Budzier, 2014).

Vision devices display the measured radiation distribution of the scene qualitatively. They serve primarily as a night

Figure 1. Structure of an uncooled IR camera with microbolometers.

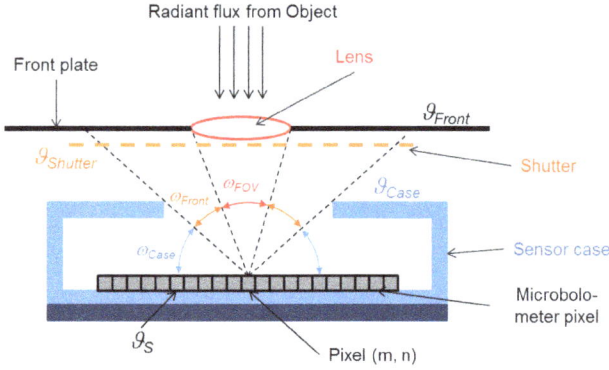

Figure 2. Optical channel in an uncooled IR camera. ω_x: reduced solid angles; ϑ_x: temperatures.

vision device. This class of devices is produced in very large quantities. They are used in military applications, security technology and increasingly in automotive technology (night driving aid). The aim of calibration is to produce as closely as possible an optically flawless image. This process is referred to generally as "smooth out".

Temperature-measuring IR cameras also display a temperature value for each pixel. Here, after the "smooth out", a radiometric calibration must also be carried out. Practically every pixel of an IR camera is a separate pyrometer. The main problem is that the calibration parameters of microbolometers depend on both the ambient temperature and the camera temperature (Budzier and Gerlach, 2011).

In the following text the calibration of temperature-measuring, uncooled IR cameras will be described for microbolometer arrays[1]. This is based on a radiometric camera model, which is described in Sect. 2. The individual steps of the calibration process will be discussed in the following sections.

2   Radiometric camera model

The theoretical basis for the calibration is based on a radiometric model of the thermal uncooled IR camera. In this case, the sensor array including the sum of radiant fluxes from the object and from inside the camera will be considered (Fig. 2).

A pixel "sees" inside the camera essentially the edge and the bracket (front panel) of the optics and its own sensor

---

[1]This article is a summarized presentation of the habilitation thesis of Budzier (2014).

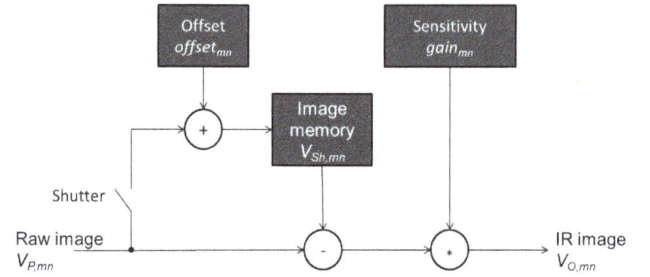

Figure 3. Camera model for the calculation of the IR image (explanation in text).

housing. In addition, the pixel gives out emissions in the entire half-space. The irradiance E_P of the pixel (m, n) is

$$E_{P,mn} = L_{O,mn}\omega_{FOV,mn} + L_{C,mn}\omega_{C,mn} - L_S\pi, \qquad (1)$$

with the radiance $L_{O,mn}$ of the object, $L_{C,mn}$ of the camera interior and L_S of the pixel and the reduced solid angle of the object $\omega_{FOV,mn}$ and the camera interior $\omega_{C,mn}$ ($= \omega_{case,mn} + \omega_{Front,mn}$).

The reduced solid angle $\omega_{FOV,mn}$ of the object depends on the position of the pixel in the microbolometer and the f-number k of the optics used. For the central pixel and an f-number $k = 1$

$$\omega_{FOV} = \frac{1}{5}\pi \qquad (2)$$

applies (Budzier and Gerlach, 2011).

Since the pixel receives radiation from the entire half-space, the reduced solid angle ω_C for the central pixel of the camera is

$$\omega_C = \pi - \omega_{FOV} = \frac{4}{5}\pi. \qquad (3)$$

Thus, the centre pixel receives 4 times as much radiation from the camera interior as from the object! In off-centre pixels, the ratio is even worse (Budzier, 2014).

In order to determine the radiance L_C of the object from Eq. (1), the radiance L_C of the camera interior must also be known. Since each pixel sees different elements in the camera interior, the radiance L_C must be measured with an optical shutter. For this purpose, the shutter, which has the camera temperature ϑ_C, is closed. The irradiance E_{Sh} with closed shutter now is

$$E_{Sh,mn} = \left(L_{Sh,mn} - L_S\right)\pi. \qquad (4)$$

Assuming that the camera shutter represents the interior camera space (same mean temperature and simplified: $L_C = L_{Sh}$), then from Eqs. (4) and (1) the irradiation intensity of the pixel through the object is obtained for the deviation between the shutter image and the measured object:

$$E_{O,mn} = \left(L_{O,mn} - L_{Sh,mn}\right)\omega_{FOV,mn}. \qquad (5)$$

For calculating the object temperature the sensor temperature is not required. Equation (5) is valid as long as the camera internal temperature and the sensor temperature are constant. If there is a change in any one of these temperatures then the shutter must be activated again. The raw image $V_{P,mn}$ is then obtained by multiplying the irradiance $E_{O,mn}$ with the voltage sensitivity R_V and the pixel area A_P:

$$V_{P,mn} = R_V E_{O,mn} A_P. \qquad (6)$$

In the above considerations, a single DC (direct current) bias (offset$_{mn}$) and a uniform sensitivity (gain$_{mn}$) were assumed for all pixels. Due to the complex manufacturing technology, that is not the case. While the sensitivity can vary by up to $\pm 20\,\%$, the deviation of the offset lies in the range of the signal. Due to the large offset differences between the pixels it is often the case that no signal is visible in the raw image. These differences are eliminated by a conventional two-point correction and apply to the IR image $V_{O,mn}$:

$$V_{O,mn} = \left(V_{P,mn} - V_{Sh,mn} - \text{offset}_{mn}\right)^{*} \text{gain}_{mn}, \qquad (7)$$

with the pixel-specific variables offset$_{mn}$ and the two-point correction gain$_{mn}$. For this purpose, when the shutter is closed, an image is stored in an image memory and then subtracted from the current online IR image (Fig. 3). In order to save an arithmetic operation per pixel in real time, the offset and shutter correction are thereby summarized.

Modern microbolometer arrays do not contain Peltier elements in a vacuum housing. These are called TEC-less microbolometers (TEC: thermo-electric cooler). The microbolometer is no longer stabilized to a constant temperature, i.e. the sensor temperature ϑ_S varies with the temperature ϑ_K of the camera. However, since the sensitivity and the operating point (offset) of a pixel are dependent on the sensor temperature, this must therefore be measured and taken into account in the camera model, as opposed to temperature-stabilized microbolometers.

The dependence of the offset and the sensitivity of a microbolometer array cannot be derived from the physical properties of a bolometer resistance without information concerning the signal processing. Since the internal signal processing of a microbolometer array is not known in detail for reasons of company in-house security, the array must be regarded as a black box. In general, the following polynomials can be assumed for the temperature dependence of the offset O_V and the sensitivity G_V:

$$O_V(\vartheta_S) = o_3 \vartheta_S^3 + o_2 \vartheta_S^2 + o_1 \vartheta_S + o_0, \qquad (8)$$

$$G_V(\vartheta_C) = g_2 \vartheta_S^2 + g_1 \vartheta_S + g_0. \qquad (9)$$

The sensitivity and the offset of the pixel must therefore be corrected using the temperature sensor ϑ_S (Fig. 4):

$$U_{O,mn} = \left[U_{P,mn} - U_{Sh,mn} - \text{offset}_{mn}\right]^{*} \text{gain}_{mn}/G_V(\vartheta_S) - O_V(\vartheta_S).$$

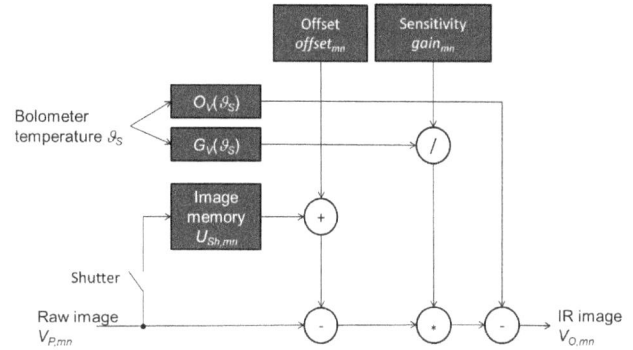

Figure 4. Extension of the camera model from Fig. 3 for the calculation of the IR image for TEC-less microbolometers.

$$(10)$$

3 Calibration

A radiometer, where each individual pixel of an IR camera can be determined, measures the radiant flux of the object and generates an output signal which, as a result of the calibration, is proportional to the temperature of the object (DeWitt and Nutter, 1989). A radiometric IR camera displays as accurately as possible the true temperature of a black body. The calibration is used here to calculate a temperature-proportional output signal (IR image) from the measurement signal (raw image) taking into account all technical and physical properties of the IR camera. The steps necessary for this are summarized in Fig. 5.

In the following section, the non-uniformity correction (Sect. 3.1), the temperature-dependent correction (Sect. 3.2), the defective pixel correction (Sect. 3.3), the shutter correction (Sect. 3.4) and the radiometric calibration (Sect. 3.5) will be presented in detail. There will be no further discussion of the operating point setting which depends significantly on the microbolometer used and would correspond to the example provided by the manufacturer's procedure.

3.1 Non-uniformity correction

Because of the technology, the individual pixels of a microbolometer have uniquely different operating points (DC bias values) and sensitivities and, thus, differing characteristics. During the correction of this non-uniformity all pixels are converted onto a single characteristic curve, the so-called standard characteristic curve. This process is referred to as "smooth out" because the IR image now with uniform illumination has no structure and so is smooth. According to this characteristic adjustment, all pixels behave the same and subsequent calibration steps can be exemplarily performed on any pixel or on any group of pixels.

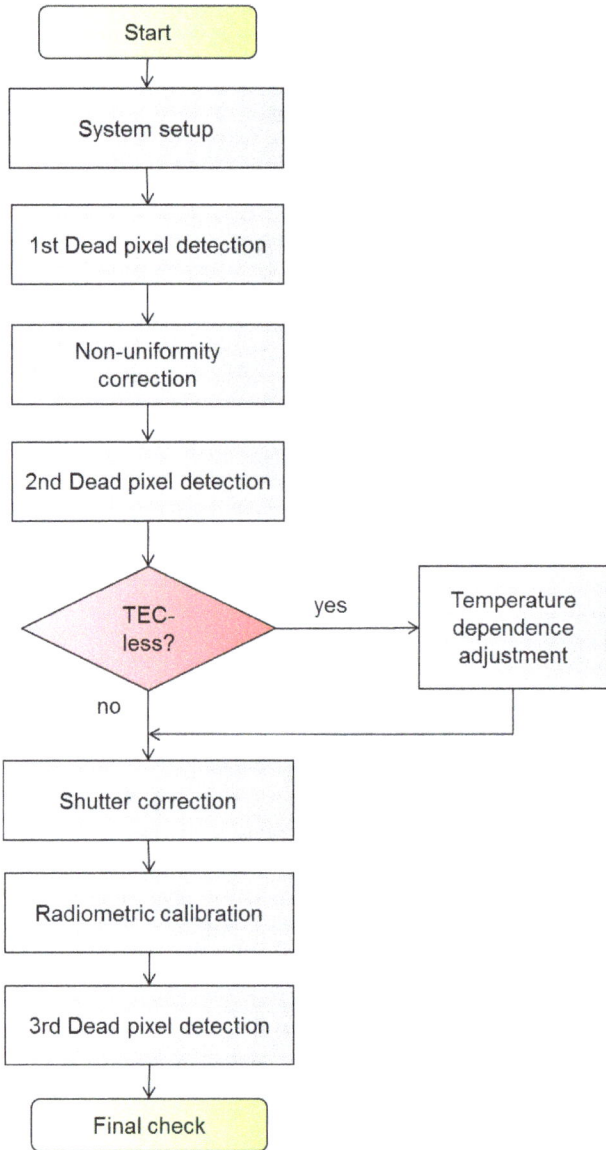

Figure 5. Flow chart of radiometric calibration.

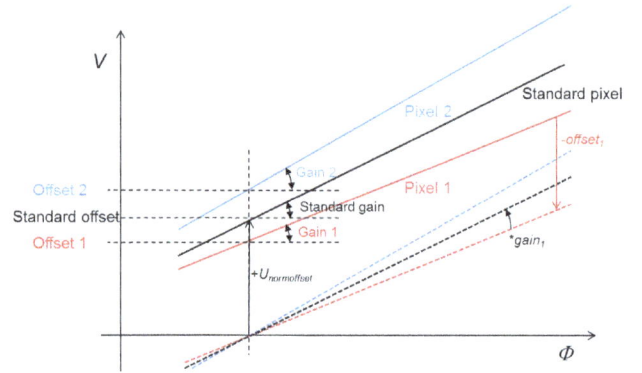

Figure 6. Schematic of the two-point correction procedure, using the example of two pixels. Dashed curves: the value "offset" is shifted parallel to the standard curve.

The description of the pixel characteristic curve is often a function of the object temperature ϑ_O. The function $U_{mn}(\vartheta_O)$ of the pixel at location (m,n) is not linear. Therefore, the characteristic curve is generally described with a second-order polynomial (Schulz and Caldwell, 1995). For radiometric IR cameras this regression is not sufficient. Here, an exponential regression (Horny, 2003) is provided, which will be described in Sect. 3.5.

The characteristic curve corrections described in the literature (Schulz and Caldwell, 1995, and Wallrabe, 2001) also refer to photon sensors, whose function $U_{mn}(\Phi_O)$ is often not linear, as well as IR vision equipment without radiometric adjustment. In contrast, the relationship between the radiant flux Φ and the pixel voltage U_{mn} of a microbolometer is linear and can be used for describing a characteris-

tic curve. A significant simplification of the correction algorithm is achieved by the subsequently described separation of non-uniformity correction and radiometric calibration. This means that the smooth out of the image is performed first and then subsequently the temperature of connectivity of the now common pixel characteristic curve. This approach also leads to a reduction of the computational effort and correction allows for ease of data processing in real time.

Firstly, it is assumed that both the sensor temperature and the ambient temperature are constant.

The linear relationship between the voltage of the pixel U_{mn} and the radiant flux Φ can be described with a linear equation:

$$U_{mn} = a_{mn}\Phi + b_{mn}, \tag{11}$$

with the slope of the straight line represented by a_{mn} and the intercept by b_{mn}. Slope and intercept behave pixel-specific and must be calculated so that a standard characteristic curve is obtained for all pixels:

$$U_{norm} = a_{norm}\Phi + b_{norm}, \tag{12}$$

where a_{norm} is the slope and b_{norm} is the intercept. For each pixel (m,n), a constant pair offset$_{mn}$ and gain$_{mn}$ must be determined so that

$$U_{korr,mn} = (U_{mn} - \text{offset}_{mn})^* \text{gain}_{mn} + U_{normoffset} \tag{13}$$

applies. The voltage $U_{korr,mn}$ is then corrected, i.e. on the standard curve traced back to the voltage value of the pixel. Figure 6 shows the principle of this two-point correction procedure. First, the pixel graph is shifted in parallel ($-$offset$_{mn}$) and then the slope is corrected (*gain$_{mn}$). Finally, for all pixels a valid constant voltage can be added. The standard curve is thus shifted back ($+U_{normoffset}$). This last step is not necessary in every case but guarantees that the corrected pixel voltages are in the same range of values as the measured pixel values. This is important, for example, if the correction is implemented in hardware (16 bit fixed-point arithmetic) and the range of values is limited.

In order to reduce the calculation during real-time correction, Eq. (13),

$$U_{\text{korr},mn} = U_{mn} \cdot \text{gain}_{mn} + \text{offset}^*_{mn}, \qquad (14)$$

can be simplified to

$$\text{offset}^*_{mn} = -\text{offset}_{mn} \cdot \text{gain}_{mn} + U_{\text{normoffset}}. \qquad (15)$$

To determine the standard curve and the pixel-related coefficients, all pixel values U_{mn} must be measured with $T_2 > T_1$ on two radiant fluxes: $\Phi(T_1)$ and $\Phi(T_2)$ (Eq. 12). Then the coefficients of the standard curve (Eq. 12) may be initially calculated:

$$a_{\text{norm}} = \frac{\langle U_2(\Phi_2) \rangle - \langle U_1(\Phi_1) \rangle}{\Phi_2 - \Phi_1}, \qquad (16)$$

$$b_{\text{norm}} = \frac{\langle U_1(\Phi_1) \rangle \Phi_2 - \langle U_2(\Phi_2) \rangle \Phi_1}{\Phi_2 - \Phi_1}, \qquad (17)$$

with the mean pixel voltage U_k ($k = 1, 2$) as the average value for all pixels J:

$$\langle U_k(\Phi_k) \rangle = \frac{1}{J} \sum_{j=1}^{J} U_j(\Phi_k). \qquad (18)$$

The standard curve is now the mean curve for all pixel characteristics. Each pixel now deviates at the measurement point k with $\Delta U_{mn,k}$ from the mean:

$$\Delta U_{mn,k} = U_{mn,k} - \langle U_k \rangle. \qquad (19)$$

Using the method of least squares, a function,

$$\Delta U_{mn} = c_{mn} U_{mn} + d_{mn}, \qquad (20)$$

can now be determined so that all voltage differences $\Delta U_{mn,k}$ are minimal. Using a simple regression the pixel-specific coefficients from Eq. (20) can now be obtained:

$$c_{mn} = \frac{\Delta U_{mn,1} - \Delta U_{mn,2}}{\langle U_1 \rangle - \langle U_2 \rangle}, \qquad (21)$$

$$d_{mn} = \frac{\langle U_1 \rangle \Delta U_{mn,2} - \langle U_2 \rangle \Delta U_{mn,1}}{\langle U_1 \rangle - \langle U_2 \rangle}. \qquad (22)$$

The corrected pixel voltage $U_{\text{korr},mn}$ is then calculated from the current measured value U_{mn} of the pixel and the differential voltage ΔU_{mn}:

$$U_{\text{korr},mn} = U_{mn} - \Delta U_{mn}. \qquad (23)$$

The values for Eq. (15) result from a comparison of the following coefficients:

$$\text{gain}_{mn} = (1 - c_{mn}), \qquad (24)$$

$$\text{offset}^*_{mn} = -d_{mn}. \qquad (25)$$

Due to various measurement errors, in particular the temporal noise of the pixel voltage, the determination of the coefficients is faulty. This shows that a small deviation remains between the pixels in the corrected image. This can be interpreted as spatial noise. Now, in order to assess the quality of the correction, the corrigibility parameter C was introduced by Gross et al. (1999) and Horny (2003):

$$C = \sqrt{\frac{U_{\text{SN}}^2}{U_{\text{TN}}^2}}, \qquad (26)$$

with the noise voltages U_{TN} and U_{SN} for the temporal noise and spatial noise, respectively. The corrigibility C is ideal for correcting the characteristics of 0. It is equal to 1 if the spatial noise has the same value as the temporal noise. If the corrigibility is greater than 1, then the spatial noise dominates and the non-uniformity is clearly visible in the IR image.

3.2 Temperature dependence correction

If the temperature of the microbolometer is not constant, as is the case for TEC-less microbolometers, the sensor sensitivity and offset parameters change with temperature (Eqs. 8–10). To correct this temperature dependence, the polynomials of Eqs. (8) and (9) must be determined. For this purpose, the pixel voltages U_P must be measured at various ambient temperatures ϑ_A. The pixel voltage is the sum of the pixel offset U_S, the signal from the camera interior U_C and the constant object voltage U_O:

$$U_P(\vartheta_A) = U_O + U_C(\vartheta_C) + U_S(\vartheta_S). \qquad (27)$$

It is important that both the camera temperature ϑ_C and the sensor temperature ϑ_S can be distinguished from the ambient temperature ϑ_A. They are always higher by a few kelvin. If the ambient temperature changes, then the camera and sensor temperatures change, in fact, with different time constants. This makes them distinguishable in the output signal. The temperature dependence of the pixel offsets $U_S(\vartheta_S)$ is given by Eq. (8). The signal voltage U_C, resulting from the temperature ϑ_C of the camera interior, is calculated with a quadratic polynomial:

$$U_C(\vartheta_C) = a_C \vartheta_C^2 + b_C \vartheta_C + c_C, \qquad (28)$$

where the polynomial coefficients a_C, b_C and c_C are initially unknown. If a measurement is made at the point in time t_i with the temperatures ϑ_C and $\vartheta_{S\,i}$, then, by superposition of Eqs. (27) and (28),

$$U_i = U_O + o_3 \vartheta_{S,i}^3 + o_2 \vartheta_{S,i}^2 + o_1 \vartheta_{S,i} + a_C \vartheta_{C,i}^2 + b_C \vartheta_{C,i} + c \quad (29)$$

applies using the combined constant

$$c = o_0 + c_C. \qquad (30)$$

By making a series of measurements with $i = 1, 2, \ldots$, i.e. measurements under varying ambient temperatures, then the following vectors are obtained:

$$U_i = \begin{bmatrix} U_O & \vartheta_{C,i}^3 & \vartheta_{C,i}^2 & \vartheta_{C,i} & \vartheta_{K,i}^2 & \vartheta_{K,i} & 1 \end{bmatrix} \begin{bmatrix} 1 \\ o_3 \\ o_2 \\ o_1 \\ a_k \\ b_k \\ c \end{bmatrix}. \tag{31}$$

This series of measurements is taken with three different object temperatures, but they are constant in each case. In the IR image three emitters are thus shown with different temperatures $\vartheta_{O1}, \vartheta_{O2}$ and ϑ_{O3}:

$$U_n = \mathbf{A}_n X, \tag{32}$$

with $n = 1, 2, 3$, and the vectors

$$\mathbf{A}_n = \begin{bmatrix} U_O & \vartheta_{C,i}^3 & \vartheta_{C,i}^2 & \vartheta_{C,i} & \vartheta_{K,i}^2 & \vartheta_{K,i} & 1 \end{bmatrix}, \tag{33}$$

$$X = \begin{bmatrix} 1 \\ o_3 \\ o_2 \\ o_1 \\ a_k \\ b_k \\ c \end{bmatrix}. \tag{34}$$

The series of measurements can now be represented in matrix notation:

$$\mathbf{U} = \mathbf{A}X, \tag{35}$$

with

$$\mathbf{U} = \begin{bmatrix} U_1 \\ U_2 \\ U_3 \end{bmatrix}, \tag{36}$$

$$\mathbf{A} = \begin{bmatrix} \mathbf{A}_1 \\ \mathbf{A}_2 \\ \mathbf{A}_3 \end{bmatrix}. \tag{37}$$

The vector X includes the desired polynomial coefficients. The solution to this over-determined system of equations with $I \gg 3$ can be calculated with the Gaussian standard equation:

$$X = \left(\mathbf{A}^T \mathbf{A} \right)^{-1} \mathbf{A}^T \mathbf{U}. \tag{38}$$

So now the desired polynomial coefficients are known, in particular the sensor offsets o_3, o_2 and o_1. The missing constant o_0 can be chosen arbitrarily. When the calculated value is subtracted from the sensitivity corrected signal according to Eq. (8),

$$U_{\text{Pixel}} = \left[U_{\text{korr}} / G_V (\vartheta_C) \right] - \left(o_3 \vartheta_C^3 + o_2 \vartheta_C^2 + o_1 \vartheta_C + o_0 \right), \tag{39}$$

the camera will behave like a camera with temperature-stabilized microbolometers.

3.3 Defective pixel correction

Due to the difficult manufacturing process for microbolometer arrays, all pixels have different parameters such as operating points, characteristic curves and noise. Pixels that either do not work or whose parameters vary greatly from the mean are defined as non-functional or defective. Defective pixels are generally referred to as "dead" pixels.

Pixel defects manifest themselves as defective pixels in the IR image. Their actual value can only be estimated with the help of neighbouring pixels. The measured value at this point of the IR image is not reconstructable. Therefore, the number of dead pixels is an important quality characteristic of microbolometers. Normally not more than a maximum of 1 % of all pixels should be defective.

A pixel is considered defective if any of the following conditions is met.

– The operating point is outside of the previously defined voltage range ΔU_{AP} of the offset value dispersion.

– The sensitivity differs more than $\pm 10\,\%$ from the mean value.

– The noise voltage is 1.5 times greater than the average noise voltage of the array.

In addition, a group of defective pixels exist, which, although they do not meet the above criteria, behave differently and are classified as defective. These are, for example, short circuits between adjacent pixels or non-linear characteristics of individual pixels. Figure 7 shows a raw image with a plurality of defective pixels (black dots).

Defective pixels occur not only individually but also in clusters. A cluster of dead pixels is a group of at least two defective pixels that are adjacent or gather together in a corner. Clusters are characterized by their size, that is, by the number of defective pixels. In the image section in Fig. 7b clusters are clearly visible. Particularly critical are defective rows or columns, because, despite a correction in the IR image, they are always conspicuous. A column or row is usually considered defective if more than 50 % of the pixels do not work.

Since the number of defective pixels of a microbolometer is an important quality attribute, the manufactures will

Table 1. Specification of allowable defective pixels.

Type	Zone	Defective pixel	Example in Fig. 7
Cluster	A	None with more than 2 defective pixels	1 cluster with 2 pixels
	B	Maximum cluster size of 9 pixels	4 clusters with 4 pixels
			22 clusters with 3 pixels
			90 clusters with 2 pixels
Columns and rows	A	0	0
	B	1 row or column	0
Functionality	A	99.5 %	99.7 % (59 pixels)
	B	99.0 %	99.3 % (1867 pixels)

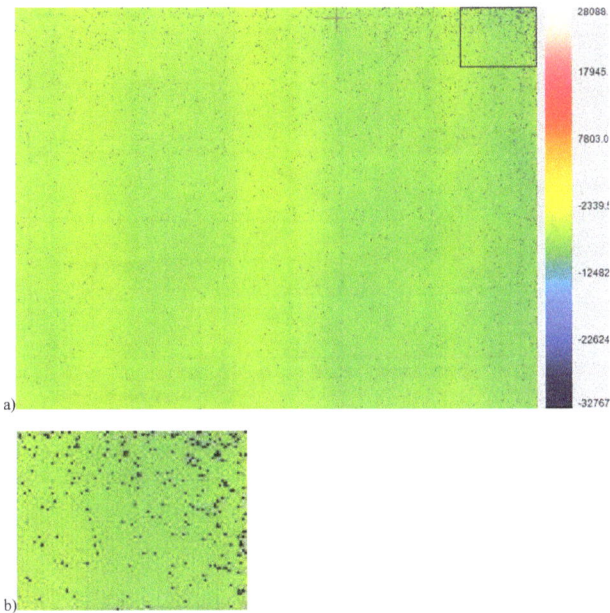

Figure 7. Raw image with 1926 defective pixels (0.6 %): **(a)** total microbolometer array and **(b)** the section marked in the top right-hand corner (90 × 70 pixels). Defective pixels are shown in black. Microbolometer array with 640 × 480 pixels.

always indicate in their specifications the maximum permissible number of defective pixels. In the centre of an IR image defective pixels are particularly noticeable. Therefore, the image area is divided into at least two zones (Fig. 8). In the central region (zone A) higher demands are placed on the functionality of pixels than of those on the edges. Table 1 shows a specification of permissible defective pixels.

The detection of defective pixels proceeds in three steps (flow chart in Fig. 5).

1. The first defective pixel detection must be done before the uniformity correction. Here all defective pixels are detected that are located outside of the previously defined range of variation of the pixel operating points. These are primarily pixels which are outside the con-

trol range and, thus, affect the position of the standard curve. Furthermore, all pixels which are too noisy are eliminated.

2. The second defective pixel detection is performed using the calculated gain and offset values according to the characteristic curve correction. Here defective pixels are identified that have too great a deviation from the standard curve.

3. The third defective pixel detection is carried out at the end of the calibration. In this case, all defective pixels which have not yet been detected are recorded. This is done by considering the IR image with different adjustments and richly contrasting scenes.

While the first two defective pixel detections can be performed computationally, the final detection is carried out manually. This also means to verify the correction of the cluster and, if necessary, to change the correction method.

There is no reading at the location of a defective pixel. This can only be estimated from the surrounding area. The aim of the recalculation of the pixel value is always to produce a high-quality visual image, i.e. so that an observer of the IR image may not notice any defective pixels. The calculation of the pixel value is carried out by methods of image pre-processing, such as with median operators.

3.4 Shutter correction

In order to calculate the radiance and the object temperature from Eq. (5), the radiance L_C of the camera interior must be known. Should the camera temperature ϑ_C vary as a result of a change in the ambient temperature, the radiance value of the camera interior needs to be adjusted accordingly. This process is called shutter correction.

The starting point for consideration is that the radiance of the camera interior space was determined using a known ambient temperature and thus a known camera temperature $\vartheta_{C0}(\vartheta_{C0} = \vartheta_{Sh0})$; therefore, according to Eq. (5) the object radiance and the object temperature can be calculated.

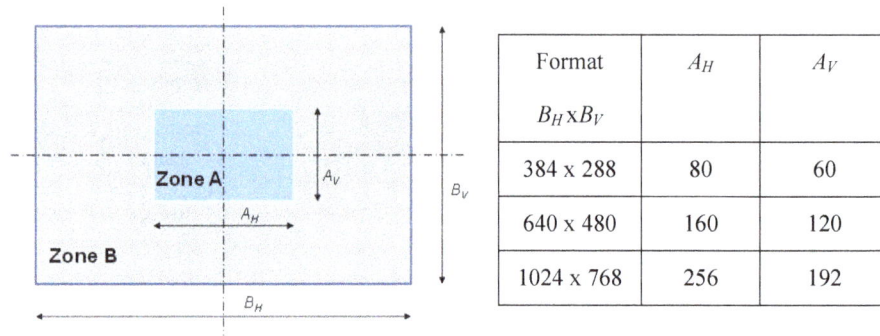

Format $B_H \times B_V$	A_H	A_V
384 x 288	80	60
640 x 480	160	120
1024 x 768	256	192

Figure 8. Typical subdivision of infrared microbolometer arrays in the mid (zone A) and marginal zones (zone B).

When the shutter is open, the pixel voltage is

$$U_{\text{open}}(\vartheta_C) = U_O + \omega_C U_C(\vartheta_C), \tag{40}$$

with the voltage equal to

$$U_C(\vartheta_C) = R_V A_P L_C. \tag{41}$$

When $\omega_C U_C$ is known, then the ambient temperature correction can be carried out:

$$U_O = U_O(\vartheta_C) - \omega_C U_C(\vartheta_C). \tag{42}$$

The voltage U_O in Eq. (42) is now independent of the ambient temperature as well as the camera temperature and is used to calculate the object radiance L_O.

The projected solid angle ω_C is known theoretically. With an f-number $k = 1$, then, for the centre pixel

$$\omega_C = \frac{4}{5}\pi = 0{,}8\pi \tag{43}$$

is obtained.

For each pixel the voltage U_C is determined with the aid of shutter correction. For this purpose, the shutter is closed and the pixel voltages (shutter image U_{Sh}) are measured. When the shutter is closed, the pixel sees the entire half-space ($\omega_c = \pi$):

$$U_{\text{Sh}} = \pi U_C(\vartheta_C). \tag{44}$$

Therefore, the shutter signal U_{Sh} must be multiplied by a factor a_{sh}. This results from the so-called shutter characteristic curve. The shutter characteristic curve is the ratio of the shutter-open signal U_{open} to shutter-off signal U_{Sh}:

$$\frac{U_{\text{open}}}{U_{\text{Sh}}} = \frac{\omega_C U_C(\vartheta_C) + U_O}{\pi U_C(\vartheta_C)}. \tag{45}$$

It follows that

$$U_{\text{open}} = a_{\text{Sh}} U_{\text{Sh}} + U_O, \tag{46}$$

with

$$a_{\text{Sh}} = \frac{\omega_C}{\pi}. \tag{47}$$

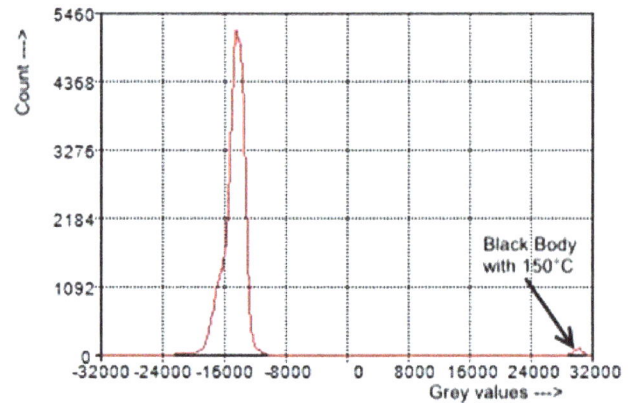

Figure 9. Histogram of the raw image at the operating point showing a black body with a temperature of $150\,°C$ (microbolometer with 384×288 pixels).

The projected solid angle ω_C of the camera interior can be read from Eq. (46):

$$\omega_C = a_{\text{Sh}}\pi, \tag{48}$$

with the increase a_{Sh} in the shutter curve. Now the signal voltage of the object can be calculated independently of the ambient temperature using the shutter image:

$$U_O = U_O(\vartheta_C) - a_{\text{Sh}} U_{\text{Sh}}(\vartheta_C). \tag{49}$$

3.5 Radiometric calibration

The previous corrections in Sect. 3.1 and 3.4 lead to all pixels of the IR image having the same behaviour and the IR image not being dependent on the ambient temperature. Finally, the radiometric calibration calculates the temperature of the object to be measured from the grey values of the pixel. It works with a voltage object temperature characteristic, so for each grey value U_D a temperature value T_O is assigned.

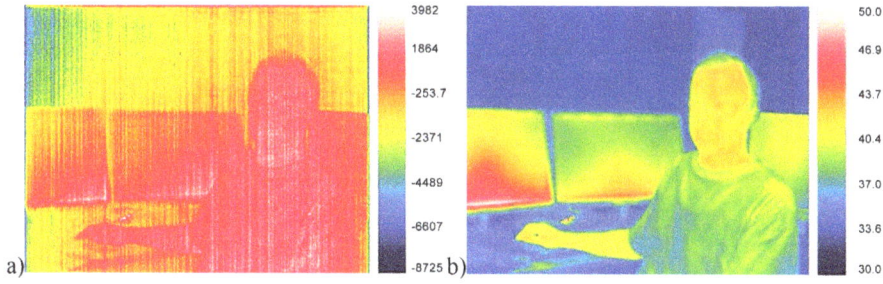

Figure 10. Effect of two-point correction. **(a)** Raw image (colour bar chart in grey values) and **(b)** thermal image (colour bar chart in °C).

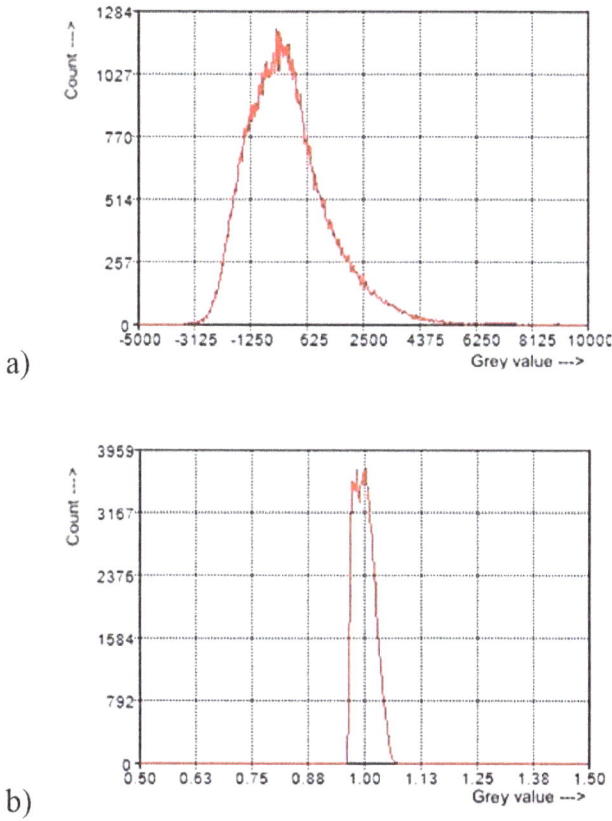

Figure 11. Histograms of **(a)** the offset values and **(b)** the gain values of the microbolometer arrays under consideration.

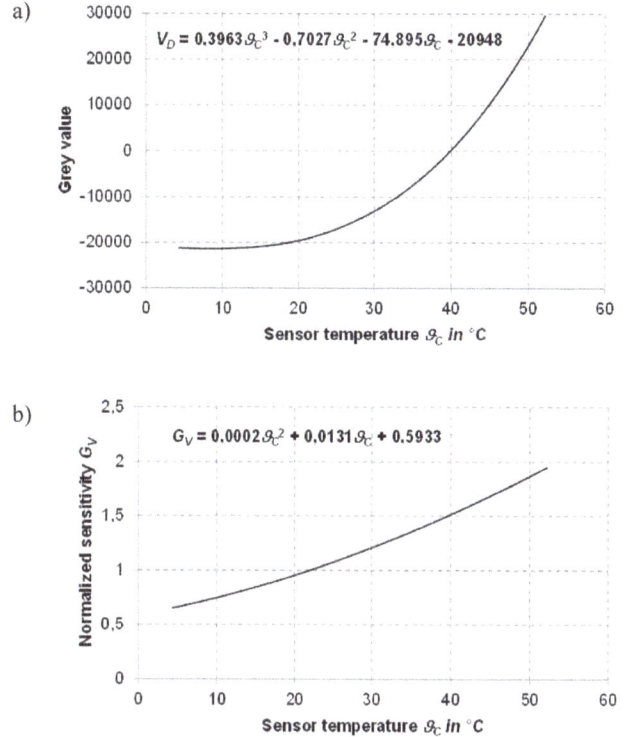

Figure 12. Temperature dependence of **(a)** the offset values and **(b)** the sensitivity of the observed TEC-less microbolometer arrays.

According to Horny (2003) it is possible to approximate the sensor output signal with a Planck curve:

$$U_D = \frac{R}{e^{\frac{B}{T_O}} - F} + O, \qquad (50)$$

where in B, F, O and R are the regression coefficients to be determined. Then the inverse function from Eq. (50) serves to calculate the object temperature T_O:

$$T_O = \frac{B}{\ln\left(\frac{R}{U_D - O} + F\right)}. \qquad (51)$$

The coefficients allow for a physical interpretation. The value of O is a general offset. Using this value, the characteristic curve along the ordinate can be displaced in parallel. The coefficient R represents the system response of the IR camera and is the counterpart to the system sensitivity. Planck's radiation law can be applied for the coefficient B:

$$B = \frac{c_2}{\lambda_B}. \qquad (52)$$

It thus describes the spectral behaviour of the system. The wavelength λ_B is the effective wavelength of the IR camera. The coefficient F allows for an alignment of the non-linearity of the system. The coefficients can be determined using a non-linear regression analysis.

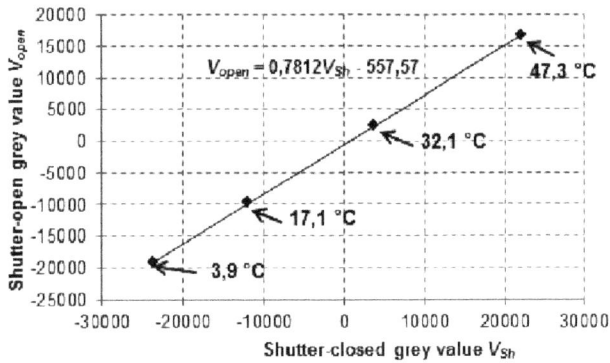

Figure 13. Shutter characteristic curve. Parameter: camera temperature of the respective measurement points.

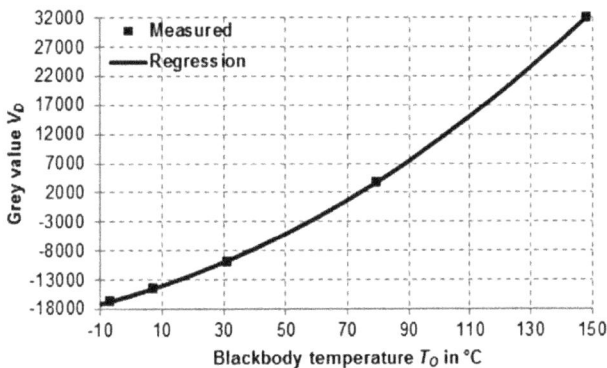

Figure 14. Object temperature ϑ_S vs. signal voltage V_D characteristic. Regression according to Eq. (50) where $B = 1514.3\,\mathrm{K}$; $F = 0.920$; $R = 1\,967\,454$; and $\mathrm{Off} = -23\,392$.

4 Example of application

In the following, the calibration process will be illustrated by an example.

The calibration process begins with the setting of the operating point. Figure 9 shows a histogram of the raw image with optimized operating point (dynamic range of $-32\,000$ to $+32\,000$ grey values). To demonstrate the modulation, a black body with a temperature of $150\,^{\circ}\mathrm{C}$ has additionally been shown.

After the operating point has been fixed, in the first defective pixel detection procedure, all pixels which lie outside of the dynamic range are defined as defective. Figure 10 shows the effect of a two-point correction of the pixel graph, in which the variation of the values of the individual pixels occurring in the raw image (Fig. 10a) is eliminated. The recognizable characteristic stripe structure is formed by the column-wise arrangement of "blind" bolometers. The optical image resulting from the natural vignetting of the optical signal and delivered to the image edge is also always present in raw image where in Fig. 10a it is hardly recognizable by the content of the thermal image.

By analysing the offset and gain values calculated in the two-point correction (Fig. 11), in a second defective pixel detection procedure more abnormal pixels that lie outside a defined scatter band can be sorted out. Fixed limits of $\pm 20\,\%$ have been proven here by the gain values.

For TEC-less microbolometers, the temperature dependence of the offset values and the sensitivity has now to be determined (Fig. 12). The regression analysis includes the known relationships between camera or sensor temperature and the respective measured variables. These are for the sensitivity of a polynomial of the second order (see Eq. 9) and the offset values for a third-order polynomial (see Eq. 8).

After correction of the non-uniformity, the correction of the ambient temperature dependence is carried out. The so-called shutter characteristic curves are recorded, which represent the ratio of the shutter-open signal to shutter-off signal as a function of the temperature of the camera (Fig. 13). For this, the camera temperature (in Fig. 13 from 3.9 to 47.3 $^{\circ}\mathrm{C}$) varies and in each case the signal V_{open} of the object and the shutter V_{sh} is measured in a temperature chamber. Both signals are directly proportional to each other.

Subsequently, the radiometric calibration is carried out by determining the signal voltage V_D object temperature characteristic (Fig. 14). The inclusion of this characteristic is done with a black body as a measuring object. Its temperature T_O will vary within the measurement range, e.g. from -10 to $150\,^{\circ}\mathrm{C}$. With the recorded measurement points, a regression is carried out in accordance with Eq. (50).

Finally, a third defective pixel detection takes place during the final check. Here visually conspicuous pixels are rejected by visual inspection.

5 Summary

The calibration of an uncooled IR camera is a complex and lengthy process, which significantly affects the cost of an IR camera. Depending on the measurement technology used, such as black bodies, references and climatic chambers, the proposed calibration allows for the measurement of absolute temperatures with a maximum measurement uncertainty of about $\pm 1\,\mathrm{K}$. This is only true for steady ambient temperatures, i.e. when the camera is at a constant temperature. When changing the ambient temperature a shutter cycle is always required. With sudden changes of the camera temperature between two shutter cycles, e.g. as a result of a jump in the ambient temperature, strong variations in measurements can appear. The behaviour of the IR camera is not ergodic. An additional measurement of the ambient temperature is not usually possible. In commercial IR cameras, this problem is solved by predictive models which estimate an expected camera internal temperature from the previous temperature change of the camera. Solutions to these problems are not known or published.

For rapid changes in temperature of the camera, the shutter needs to be frequently operated, e.g. several times within 1 min. Since the operation of the shutter always causes an interruption of the measuring process, the user should operate the shutter as little as possible or even completely avoid using it. For the shutterless operation of IR cameras, however, much more complex calibration algorithms are required, which build on those described here. Such an approach is pursued by Tempelhahn et al. (2014).

References

Black, S. H., Sessler, T., Gordon, E., Kraft, R., Kocain, T., Lamb, M., Williams, R., and Yang, T.: Uncooled detector development at Raytheon, Proc. SPIE, 8012, 80121A-1–12, 2011.

Budzier, H.: Radiometrische Kalibrierung ungekühlter Infrarot-Kameras, TUDpress, Dresden, 2014.

Budzier, H. and Gerlach, G.: Thermal Infrared Sensors, John Wiley & Sons, Chichester, 2011.

DeWitt, D. P. and Nutter, G. D.: Theory and Practice of Radiation Thermometry, Wiley, New York, 1989.

Gross, W., Hierl, T., and Schulz, M.: Correctability and long-term stability of infrared focal plane arrays, Opt. Eng., 38, 862–869, 1999.

Horny, N.: FPA camera standardization, Infrared Phys. Technol., 44, 109–119, 2003.

Kruse, P. W.: Uncooled Infrared Imaging Arrays and systems, Academic Press, San Diego, 1997.

Mounier, E.: Technical and market trends for microbolometers for thermography and night vision, Proc. SPIE, 8012, 80121U-1–6, 2011.

Schulz, M. and Caldwell, L.: Nonuniformity correction and correctability of infrared focal plane arrays, Infrared Phys. Technol., 36, 763–777, 1995.

Tempelhahn, A., Budzier, H.; Krause, V., and Gerlach, G.: Development of a shutterless calibration process for microbolometer-based infrared measurement systems, International Conference on Quantitative Infrared Thermography, 7–11 July 2014, Bordeaux, QIRT-2014-060, 2014.

Vollmer, M. and Möllmann, K.-P.: Infrared Thermal Imaging, Wiley-VCH, Weinheim, 2010.

Wallrabe, A.: Nachtsichttechnik, Vieweg, Braunschweig/Wiesbaden, 2001.

7

Correlation of BAW and SAW properties of langasite at elevated temperatures

M. Schulz[1], E. Mayer[2], I. Shrena[2], D. Eisele[2], M. Schmitt[2], L. M. Reindl[2], and H. Fritze[1]

[1]Institute of Energy Research and Physical Technologies, Clausthal University of Technology, Goslar, Germany
[2]Department of Microsystems Engineering, Albert-Ludwigs-Universität Freiburg, Freiburg, Germany

Correspondence to: M. Schulz (michal.schulz@tu-clausthal.de)

Abstract. The full set of electromechanical data of langasite ($La_3Ga_5SiO_{14}$) is determined in the temperature range from 20 to 900 °C using differently oriented bulk acoustic wave resonators. For data evaluation a physical model of vibration is developed and applied. Thereby, special emphasis is taken on mechanical and electrical losses at high temperatures. The resulting data set is used to calculate the properties of surface acoustic waves. Direct comparison with experimental data such as velocity, coupling coefficients and propagation loss measured using surface acoustic wave devices with two different crystal orientations shows good agreement.

1 Introduction

Langasite (LGS, $La_3Ga_5SiO_{14}$) is a single crystalline piezo-electric material suited for high-temperature applications (Fukuda et al., 1998; Fachberger et al., 2001; Fritze and Tuller, 2001; Fritze, 2011b). It belongs to the point group 32 and exhibits the same crystal structure as quartz, but exceeds its operation temperature limit significantly. LGS does not undergo any phase transformation up to its melting point at 1470 °C and may be piezoelectrically excited up to at least 1400 °C provided that stable electrodes are available (Fritze, 2011b). The crystals exhibit mixed electronic and ionic conductivity, which contributes to the loss at high temperatures. The oxygen partial pressure-dependent conductivity impacts the performance of the resonators in a minor way as long as the oxygen partial pressure is kept, e.g., above 10^{-20} bar at 600 °C (Fritze et al., 2006; Fritze, 2006, 2011a).

LGS can be used as a resonant sensor. When operated in the microbalance mode, small mass changes of a layer with affinity to specific gas particles cause a shift of the resonance frequency (Tuller and Fritze, 2002). Anticipated applications of such sensors include fuel cells and gas reformers (Schneider et al., 2005). The stability of LGS in harsh environments such as high temperatures and low oxygen partial pressures makes it suited for wireless sensors based on surface acoustic wave (SAW) devices (Fachberger et al., 2004; Thiele and Pereira da Cunha, 2006; Bardong et al., 2008;

Wall et al., 2015; Behanan et al., 2015). Here, an accurate set of LGS material constants determined for a wide temperature range is strongly required. The availability of such data sets is scarce, since the published materials data are either limited to a fixed temperature or cover only a narrow range around room temperature °C (Bungo et al., 1999; Malocha et al., 2000). Weihnacht et al. (2012) provide the components of elastic compliance tensor in the 25 to 600 °C temperature range, determined using the pulse-echo ultrasonic technique (Weihnacht et al., 2012).

In this work, the full set of elastic, electric and piezoelectric properties of langasite is determined at elevated temperatures using bulk acoustic wave (BAW) resonators of different orientations. Special attention is drawn to the mechanical and electrical losses, which play an important role for high-temperature applications. Subsequently, the full set of electromechanical data is used to predict the wave velocity of SAW resonators for two selected cuts of LGS. Their Euler angles are (0, 138.5, 26.6°) and (0, 30.1, 26.6°). Finally, the calculated values are compared with the corresponding data extracted from the high-temperature measurements of specially designed SAW test structures.

The LGS crystals for BAW resonators are provided by the Institute for Crystal Growth, Berlin Adlershof, Germany. The wafers for SAW measurements are fabricated by FOMOS-

Materials, Russia, and Mitsubishi Materials Corporation, Japan.

2 Model for BAW and SAW

2.1 High-temperature losses

A piezoelectric resonator operated at room temperature can be described by a one-dimensional physical model of vibration, where electrical losses are very low and thus negligible. At elevated temperatures, this description is not accurate due to, e.g., the finite electrical conductivity. In this case the elastic, dielectric and piezoelectric coefficients must be extended by imaginary parts which express the mechanical, electrical and piezoelectric losses.

2.1.1 Mechanical loss

The mechanical loss at elevated temperatures depends on the resonance frequency f and can be described by the viscosity η. The loss of the resonator at the angular frequency ω, with $\omega = 2\pi f$, is related to the imaginary part of the elastic stiffness according to (Ikeda, 1990):

$$\hat{c} = c + j\omega\eta. \tag{1}$$

The elastic compliance s must be extended by a viscous contributions in a similar way. The relation between the elastic stiffness and the viscosity can be expressed using the inverse resonant quality factor, Q^{-1} and the mechanical loss tangent $\tan\delta_{\mathrm{M}}$, by

$$\tan\delta_{\mathrm{M}} = \frac{\Im(\hat{c})}{\Re(\hat{c})} = \frac{\omega\eta}{c} \equiv Q^{-1}. \tag{2}$$

2.1.2 Electrical loss

A perfectly insulating resonator with two parallel electrodes, excited by a harmonic electric field $E = E_0 e^{j\omega t}$, acts as a capacitor with a dielectric material described by the permittivity ε. In case of materials exhibiting dielectric loss, the dielectric tensor must be regarded as a complex property consisting of real and imaginary parts (Ikeda, 1990)

$$\hat{\varepsilon} = \varepsilon - j\frac{\sigma}{\omega}, \tag{3}$$

with the loss tangent $\tan\delta_E$ given by

$$\tan\delta_E = \frac{\Im(\hat{e})}{\Re(\hat{e})}. \tag{4}$$

The origin of dielectric loss can be attributed to the electrical conductivity σ of the material. Calculating the admittance Y for a capacitor with the thickness a, the electrode area A and the dielectric coefficient described by the complex property $\hat{\varepsilon}$ results in (Ikeda, 1990)

$$Y = \frac{A}{a}(\sigma + j\omega\varepsilon). \tag{5}$$

The equation describes the admittance of an electric circuit with capacitor and resistance connected in parallel.

2.1.3 Piezoelectric loss

In analogy to the mechanical and electrical loss, the imaginary part of the piezoelectric coefficient can be introduced. The nature of this "piezoelectric loss" is however not clear. It may be explained by, e.g., jumping of lattice defects or the movement of domain walls in polycrystalline materials (Sherrit and Mukherjee, 1998; Smits, 1976). However, for most materials this term is negligible and assumed to be zero (Ikeda, 1990). The validity of this assumption for langasite is proven in previous studies, where calculations with complex piezoelectric coefficient are used to evaluate the data (Fritze, 2011b). Therefore, in this article the imaginary part of the piezoelectric coefficient is omitted.

2.2 Piezoelectric equations

In case of piezoelectric materials, the relation between mechanical and electrical properties is described by the piezoelectric equation

$$\begin{bmatrix} \hat{c} & e^T \\ e & \hat{\varepsilon} \end{bmatrix} \cdot \begin{bmatrix} S \\ E \end{bmatrix} = \begin{bmatrix} T \\ D \end{bmatrix}, \tag{6}$$

where the complex stiffness $\hat{c}$ and the piezoelectric constant e are 4th and 3rd rank tensors, respectively. The complex dielectric coefficient $\hat{\varepsilon}$, strain S and stress T are 2nd rank tensors. The electric field E and electric displacement D are vectors. In order to differentiate between tensors and tensor components, the former is written in bold.

Due to the tensor symmetry and the crystal symmetry of LGS many tensor components vanish, so that the number of independent coefficients in Eq. (6) is reduced to 10 (Meitzler et al., 1988; Ikeda, 1990):

$$\begin{bmatrix} \hat{c}_{11} & \hat{c}_{12} & \hat{c}_{13} & \hat{c}_{14} & 0 & 0 & e_{11} & 0 & 0 \\ \hat{c}_{12} & \hat{c}_{11} & \hat{c}_{13} & -\hat{c}_{14} & 0 & 0 & -e_{11} & 0 & 0 \\ \hat{c}_{13} & \hat{c}_{13} & \hat{c}_{33} & 0 & 0 & 0 & 0 & 0 & 0 \\ \hat{c}_{14} & -\hat{c}_{14} & 0 & \hat{c}_{44} & 0 & 0 & e_{14} & 0 & 0 \\ 0 & 0 & 0 & 0 & \hat{c}_{44} & \hat{c}_{14} & 0 & -e_{14} & 0 \\ 0 & 0 & 0 & 0 & \hat{c}_{14} & \hat{c}_{66} & 0 & -e_{11} & 0 \\ e_{11} & -e_{11} & 0 & e_{14} & 0 & 0 & \hat{\varepsilon}_{11} & 0 & 0 \\ 0 & 0 & 0 & 0 & -e_{14} & -e_{11} & 0 & \hat{\varepsilon}_{11} & 0 \\ 0 & 0 & 0 & 0 & 0 & 0 & 0 & 0 & \hat{\varepsilon}_{33} \end{bmatrix}, \tag{7}$$

with $\hat{c}_{66} = (\hat{c}_{11} - \hat{c}_{12})/2$. Here, the $\hat{c}_{ij}$, $\hat{e}_{ij}$ and $\hat{\varepsilon}_{ij}$ represent the components of stiffness tensor c, piezoelectric tensor e and dielectric coefficient tensor ε, respectively.

Equation (6) applies, when the strain S and electric field E are independent variables. This is the case where a resonator is operated in the thickness shear mode of vibration. In case of length-extensional mode of vibration, where the elastic stress T and the electric field E are independent variables, an alternative notation applies

$$\begin{bmatrix} \hat{s} & d^T \\ d & \hat{\varepsilon} \end{bmatrix} \cdot \begin{bmatrix} T \\ E \end{bmatrix} = \begin{bmatrix} S \\ D \end{bmatrix}. \tag{8}$$

Here, the complex elastic compliance $\hat{s}$ is used instead of the stiffness $\hat{c}$, and the piezoelectric constant d replaces the coefficient e. The relation between piezoelectric tensors e and d and the relation between elastic compliance and stiffness are shown, e.g., in (Meitzler et al., 1988; Ikeda, 1990). The latter is used to calculate the component $s_{12}(T)$ of the elastic stiffness tensor from the coefficient $c_{66}(T)$ as follows:

$$s_{12}(T) = s_{11}(T) - \frac{1}{2c_{66}(T)} - \frac{2(s_{14}(T))^2}{s_{44}(T)}. \quad (9)$$

The determination of all components of the elastic stiffness and compliance tensors requires different crystal cuts. The shear components of those tensors determine two different modes of vibration. The stiffness $\hat{c}_{66}$ describes the thickness shear (TS) vibration of a partially electroded Y-cut resonator, and the elastic compliance $\hat{s}_{44}$ describes the face shear (FS) vibration of a rectangular Y-cut plate. The remaining four components of the elastic compliance tensor may be determined using differently oriented rods excited in length-extensional (LE) mode of vibration. Here, the effective compliance s^{eff} is a superposition of several components as function of the angle φ between the rod and the Y axis (Ikeda, 1990; Schulz and Fritze, 2008):

$$s^{\text{eff}}(\varphi) = s_{11}\cos^4\varphi + s_{33}\sin^4\varphi - 2s_{14}\cos^3\varphi\sin\varphi$$
$$+ (s_{44} + 2s_{13})\cos^2\varphi\sin^2\varphi. \quad (10)$$

Several angles φ have to be chosen in such a way that for every φ at least one term in Eq. (10) dominates. The effective piezoelectric coefficient d^{eff} is a superposition of two different coefficients as a function of the φ:

$$d^{\text{eff}}(\varphi) = -d_{11}\cos^2\varphi + d_{14}\cos\varphi\sin\varphi. \quad (11)$$

From Eq. (10) it may be seen that four differently oriented rods with angles φ of -30, 0, 30 and $45°$ are required to determine the majority of components of the elastic compliance $\hat{s}$. The coefficient s_{12} remains to be determined and is obtained from the thickness shear mode of vibration as shown in Eq. (9). The sum of coefficients $(s_{44} + 2s_{14})$ is separated using the face shear mode of vibration described by s_{44}. The stiffness tensor $\hat{c}$ is calculated using the relation between stiffness and elastic compliance. Additionally, two rectangular X- and Z-cut plates are required to determine the two independent complex components of the dielectric tensor, $\hat{\varepsilon}_{11}$ and $\hat{\varepsilon}_{33}$. The imaginary part of dielectric tensor provides the information about conductivity σ_1 and σ_3 (commonly denoted as σ_X and σ_Z, respectively) of langasite.

All crystal cuts used in this work are visualized in Fig. 1.

2.3 Models for different crystal orientations

2.3.1 Length-extensional mode

In case of a rod, where the width b, length l and thickness a are chosen to be parallel to the x_3, x_2 and x_1 axes, respectively, and the applied electric field is parallel to the x_1 axis,

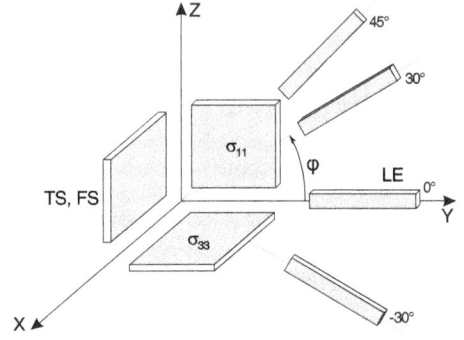

Figure 1. The crystal cuts of langasite used to determine the full set of electromechanical data (Ikeda, 1990). Four differently rotated rods with electrodes perpendicular to X axis vibrate in length extensional mode (LE). The rectangular Y-cut plate vibrates in face shear (FS) and thickness shear (TS) modes. The rectangular X- and Z-cut plates are used for conductivity measurements.

the piezoelectric relation shown in Eq. (8) is reduced significantly. Due to the boundary conditions chosen here, only the stress and strain in x_2 direction are relevant. All other components of the tensors T and S vanish. Similarly, only the electric displacement D_1 and electric field E_1 are taken into account resulting in the following equation:

$$S_2 = \hat{s}_{22}T_2 + d_{12}E_1$$
$$D_1 = d_{12}T_2 + \hat{\varepsilon}_{11}E_1. \quad (12)$$

The strength of the coupling between the mechanical and electrical properties of a piezoelectric medium is expressed as coupling factor $\hat{k}$, defined by

$$\hat{k}_{12}^2 = \frac{d_{12}^2}{\hat{s}_{22}\hat{\varepsilon}_{11}}. \quad (13)$$

With the coupling factor, Eq. (12) can be transformed in

$$T_2 = \frac{1}{\hat{s}_{22}}\frac{\partial u_2}{\partial x_2} - \frac{d_{12}}{\hat{s}_{22}}E_1,$$
$$D_1 = \frac{d_{12}}{\hat{s}_{22}}\frac{\partial u_2}{\partial x_2} + \hat{\varepsilon}_{11}\left(1 - \hat{k}_{12}^2\right)E_1. \quad (14)$$

Solving Newton's equation of motion with T_2 described by Eq. (14) and under the assumption of harmonic time dependence $e^{j\omega t}$ of the electric field E_1, the electric impedance of the resonator is calculated (Schulz and Fritze, 2008)

$$Y = j\omega\frac{lb}{a}\hat{\varepsilon}\left[1 - \hat{k}^2 + \hat{k}^2\frac{2}{lb\sqrt{\rho\hat{s}_{22}}}\tan\left(\frac{lb\sqrt{\rho\hat{s}_{22}}}{2}\right)\right]. \quad (15)$$

The parameter of the tangent function reflects the phase shift and a wave velocity in the bulk of a resonator:

$$\hat{v} = lb\sqrt{\rho\hat{s}_{22}}. \quad (16)$$

The capacity C_0 of the resonator equals

$$C_0 = \hat{\varepsilon}_{11} \frac{lb}{a}. \tag{17}$$

From Eqs. (15), (16) and (17) the admittance Y of a resonator is found to be

$$Y = j\omega C_0 \left[1 - \hat{k}^2 + \hat{k}^2 \frac{2}{\hat{v}} \tan\left(\frac{\hat{v}}{2}\right) \right]. \tag{18}$$

2.3.2 Thickness shear mode

In case of a partially electroded Y-cut resonator with the electric field applied parallel to the x_2 axis, only the strain S_6 and stress T_6 do not vanish. The piezoelectric relation shown in Eq. (6) can be expressed as

$$T_6 = \hat{c}_{66}S_6 + e_{26}E_2,$$
$$D_2 = e_{26}S_6 + \hat{\varepsilon}_{22}E_2. \tag{19}$$

Under the assumption of a harmonic time dependence of an electric field, Eq. (19), Newton's equation of motion may be transformed analogously to the case of length extensional mode of vibration. The resulting equation is found to be (Mason, 1964)

$$\left(\hat{c}_{66} + \frac{e_{26}^2}{\hat{\varepsilon}_{22}} \right) \frac{\partial^2 u_1}{\partial x_2^2} + \omega^2 \rho u_1 = 0. \tag{20}$$

Here, a new property, the piezoelectrically stiffened shear modulus $\bar{c}$ is introduced

$$\bar{c} = \hat{c}_{66} + \frac{e_{26}^2}{\hat{\varepsilon}_{22}} = c_{66} + \frac{e_{26}^2}{\varepsilon_{22} + \sigma^2/\varepsilon_{22}\omega^2}$$
$$+ j\omega \left(\eta + \frac{\sigma_2}{1 + \sigma_2^2/e_{26}^2\omega^2} \right). \tag{21}$$

This shear modulus depends on the piezoelectric and dielectric coefficients, as well as on the electromechanical losses, i.e., electric conductivity and viscosity. The consequences of these relations are discussed elsewhere (Fritze, 2011b).

Using these equations, the electric impedance of the thickness shear resonator is obtained

$$Z = \frac{1}{j\omega C_0} \left[1 - \hat{k}_t^2 \frac{2}{\hat{v}} \tan\left(\frac{\hat{v}}{2}\right) \right], \tag{22}$$

with coupling factor $\hat{k}_t^2 = e_{26}^2/\hat{c}_{66}\hat{\varepsilon}_{22}$ and the reduced wave velocity $\hat{v} = \omega a \sqrt{\rho/\hat{c}}$. Here, a denotes the thickness of the resonator.

2.3.3 Face shear mode

The face shear vibration of a Y-cut resonator is determined by the elastic compliance s_{44}. This two-dimensional motion

cannot be described by a physical model which has a straight-forward analytical solution. Therefore, only an approximate solution for the resonance frequency f_r of such a resonator as presented in (Ikeda, 1990; Bechmann, 1951) is used:

$$f_r = \frac{\kappa \Theta}{\pi l} \sqrt{\frac{1}{\rho s_{44}}}. \tag{23}$$

Here, κ is a solution of transcendental equation $\tan \kappa + \kappa = 0$, and Θ is a correction factor defined as

$$\Theta = 1 - \frac{1}{2\kappa} \left(\frac{\kappa^2 - 2}{\kappa^2 + 2} \right)^{3/2} \sqrt{\frac{s_{11} + s_{33}}{2s_{44}}}. \tag{24}$$

2.4 Parameters of SAW propagation

The parameters of SAW propagation, like velocity v, propagation loss α and coupling coefficient k^2, are calculated from the tensor data, obtained from the impedance measurements of BAW resonators as described in Sect. 2.3.

For this purpose the equations of motion

$$\rho \frac{\partial^2 u_i}{\partial t^2} = \frac{\partial T_{ij}}{\partial x_j},$$
$$\frac{\partial D_i}{\partial x_i} = 0, \tag{25}$$

with $i, j = 1, 2, 3$, have to be solved together with the piezoelectric Eq. (6) for the semi-infinite substrate crystal of the given cut. For this, the Green function method described in (Peach, 1995, 2001; Koskela, 1998) is used. The poles and zeros of the Green function component G_{44} on the complex slowness space s correspond to surface acoustic waves, propagating on a free and on a metallized surface, respectively. A complex pole

$$G_{44}^p = \frac{K}{s - s_0} \tag{26}$$

where the value $s_0 = s_0' + j s_0''$ is associated with a SAW, which propagates on the free surface with the velocity $v_0 = 1/s_0'$. The attenuation is

$$\alpha = 2\pi f \cdot \frac{s''}{s'} \cdot 20 \cdot \lg e [\text{dB}/\mu\text{s}]. \tag{27}$$

Here, f and e are the frequency of the SAW and the Euler number, respectively.

2.5 Parameters of SAW propagation extracted from the SAW measurements

Special test SAW devices are designed for the experimental determination of the surface acoustic wave parameters v, α and k^2. These are four delay lines of two different lengths. Two of them have free propagation paths, the other two propagation paths are metallized as shown in Fig. 2.

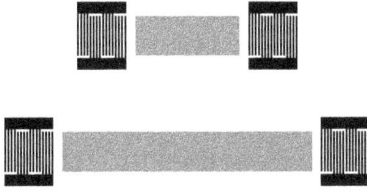

Figure 2. Delay lines of different length.

Figure 3. IDT with eight electrodes per period.

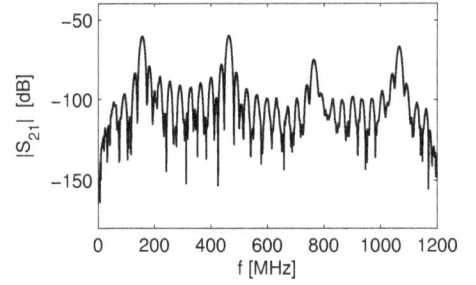

Figure 4. Transfer function S_{21} of a delay line with four harmonics (after signal processing).

and Δt is the difference of delay times of the two delay lines for this harmonic.

3 Experimental

3.1 Sample preparation

3.1.1 Crystal cuts

For the determination of the bulk properties of langasite crystals several orientations as shown in Fig. 1 are used. Two rectangular $10 \times 10 \times 0.5\,\text{mm}^3$ X- and Z-cut plates are used for impedance measurements at low frequencies. The Y-cut sample has the same dimensions and operates in the face shear mode of vibration. A circular Y-cut sample of 0.27 mm in thickness is excited in the thickness shear mode. For determination of most components of the elastic compliance tensor, four differently rotated $10 \times 2 \times 0.5\,\text{mm}^3$ rods with electrodes perpendicular to the X axis are prepared. During the machining of langasite rods, the angle φ could not be adjusted as precise as other dimensions of the sample with typical uncertainty of about $\pm 1°$. In order to overcome this limitation, the samples are characterized at room temperature and the elastic compliance tensor is calculated. The comparison of the resulting tensor with literature data and previous pulse-echo measurements on the same crystals enable precise determination of the cut angles.

For the SAW measurements two cuts with Euler angles (0, 138.5, 26.6°) and (0, 30.1, 26.6°) are selected. They show large k^2 and zero power flow angle (PFA) (Naumenko and Solie, 2001; da Cunha and Fagundes, 1999; Plessky et al., 1998). In addition, the first cut is temperature compensated around room temperature, which makes it attractive for many applications. Because of that, this cut is commercially available from many distributors.

3.1.2 Electrodes

BAW samples

All samples are coated with platinum electrodes by pulsed laser deposition (PLD). The thickness of this electrodes is

Thereby, identical input and output interdigital transducers (IDTs) with eight electrodes per electrical period are used (see Fig. 3).

They can be regarded as multi-electrode transducers with double spatial sampling. As shown in (Engan, 1975), the transfer function S_{21} of such IDT has four pass bands (four harmonics) with central frequencies corresponding to $1/4 f_0$, $3/4 f_0, 5/4 f_0, 7/4 f_0$, where $f_0 = v/(2p)$. The pitch p corresponds to the period of the electrode in the IDT. In our case the transfer function S_{21} shows the 1st, 3rd, 5th and 7th harmonics at about 150, 450, 750 and 1050 MHz (Fig. 4).

During signal processing the difference between time delay of long and short delay lines at each harmonic is obtained very precisely (Shrena et al., 2008). This allows the determination of the group velocity based on the known length difference between both delay lines. The influence of the IDTs is, therefore, automatically eliminated. The results for the group velocity of SAWs on the free surface and on the metallized surfaces are extrapolated to zero frequency. This way, the phase velocity for SAW on the free LGS surface v_0 and on the short-circuited LGS surface v_m (i.e., metallized surface without mass loading by the metal film) is determined. Further, the electromechanical coupling coefficient k^2 is obtained from the phase velocities v_0 and v_m using the Ingebrigtsen relation (Ingebrigtsen, 1972)

$$k^2 = -2 \cdot \frac{v_m - v_0}{v_0}. \tag{28}$$

The propagation losses α at the harmonic frequencies are calculated to be

$$\alpha = \frac{1}{\Delta t} \cdot 20 \lg \frac{|S_{21}^{\text{long}}|}{|S_{21}^{\text{short}}|} [\text{dB}/\mu\text{s}], \tag{29}$$

where S_{21}^{long} and S_{21}^{short} are the transfer function of the long and short delay line at the harmonic frequency, respectively, about 250 nm. A titanium adhesion layer of a few nanometers is applied.

SAW devices

The metallization of the SAW test devices is done by SAW Components Dresden GmbH using the lift-off technique. Thin Pt layers of 45 and 75 nm are deposited. A Zr film of 4 nm in thickness is taken as adhesion layer. Dewetting of the thin Pt film above that temperature determines a temperature limit for the measurements. In contrast to the BAW samples, thicker Pt layers cannot be used due to increasing reflection and scattering effects at the IDTs which distort the signal and cause additional losses.

3.2 Setup and data evaluation

BAW measurements

The electric impedance at low frequencies is measured using a Solartron 1260 gain-phase analyzer to determine the electrical conductivity. The impedance in the vicinity of the resonance frequency of all BAW resonators is acquired using a HP 5100A network analyzer. For high-temperature measurements in artificial air containing 20 % O_2/Ar, a gas-tight furnace with temperature varying between 20 and 900 °C at the rate of 2 K min^{-1} is used. In case of the conductivity determination, the temperature is held constant before each measurement in order to equilibrate the samples.

The measured impedance data are further fitted using the physical models of the corresponding resonators in order to obtain the components of stiffness and elastic compliance tensors. It is obvious that the number of free parameters for the fit procedure must be much smaller than the total number of parameters, i.e., 16 for the one-dimensional model of thickness shear resonators. Reasonable fits can be, in general, expected if the number of free parameters does not exceed three or four (Fritze, 2011b). As a starting point at room temperature, variables such as the sample dimensions, density and electrode areas are determined precisely. Thermal expansion of the material, which influences the dimensions and density, is taken into account. Further, the thickness of electrodes is measured precisely using the surface profiler. In order to calculate the mass load, their density is assumed to be 100 % of the theoretical Pt density as they are prepared by PLD.

The determination of remaining parameters is performed in several steps:

1. Impedance measurements of X- and Z-cut plates at low frequencies provide information about the dielectric coefficient and the conductivity in the form of complex coefficients $\hat{\varepsilon}_{11}$ and $\hat{\varepsilon}_{33}$.

2. The fit of analytically solved piezoelectric equations uses the dielectric coefficient and conductivity from step (1).

The determination of elastic properties of langasite involves several crystal orientations and modes of vibration. First, the four differently oriented rods shown in Fig. 1 provide the coefficients $s_{11}(T)$, $s_{14}(T)$, $s_{33}(T)$ and $(s_{44}(T) + s_{13}(T))$. Second, the rectangular Y-cut plate vibrating in face shear mode provides the elastic compliance $s_{44}(T)$, and thus, enables the separation of $s_{13}(T)$ compliance from length-extensional measurements. Finally, the round Y-cut resonator excited in the thickness-shear mode of vibration provides the stiffness coefficient $c_{66}(T)$ used to calculate the elastic compliance $s_{12}(T)$.

The data evaluation applied here potentially delivers unreliable absolute values due to, e.g., residual stray capacitance of the high-temperature sample holder. To overcome this problems and to verify the models, fit results are compared with preliminary pulse-echo measurements performed on the same langasite single crystal.

SAW measurements

Two delay lines – one long and one short – are positioned on one chip. They are connected to the same ports and are measured simultaneously. The chip is mounted on a holder and is connected to the RF-flange of the high-temperature furnace via two rigid coaxial cables made of steel tubes with ceramic husked steel wires. The other side of the RF-flange is connected by standard RF coaxial cables to the network analyzer, which measures the transfer function S_{21}. Before starting measurements, the atmosphere inside the furnace is changed to nitrogen in order to reduce oxidation effects. The sample is heated at a slow continuous ramp of 50 K h^{-1} up to the final temperature of 800 °C. The transfer function S_{21} is measured in temperature steps of 5 K.

A specially developed signal processing algorithm utilizing cross-correlations, is used for the determination of the precise time delay and for extracting SAW propagation parameters as described in detail in (Shrena et al., 2008, 2009). As shown in (Shrena et al., 2009), using a cross-correlation algorithm gives better accuracy for the determination of time delays in comparison to other methods.

4 Results and discussion

4.1 Materials parameters

4.1.1 Electric properties at elevated temperatures

The conductivity of the X- and Z-cut langasite plates is summarized in Fig. 5. The anisotropy of the conductivity is clearly visible.

4.1.2 Full set of piezoelectric and elastic properties

As a result of the procedure described in Sect. 3.2, the complete elastic compliance tensor and the piezoelectric tensor as a function of temperature, $s(T)$ and $d(T)$, are available.

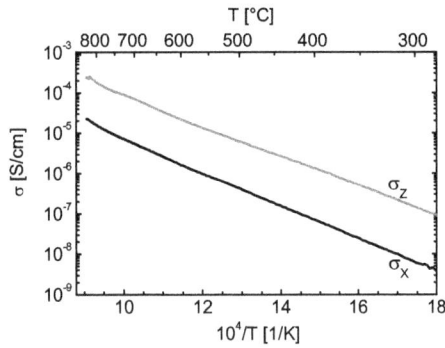

Figure 5. Conductivity of LGS as function of temperature.

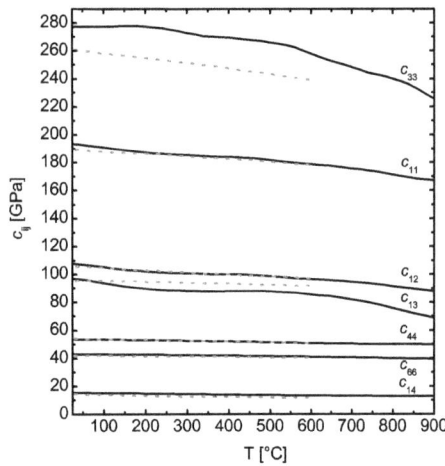

Figure 6. The full set of elastic stiffness coefficients of langasite as function of temperature calculated from BAW measurements (solid line) in comparison with the data obtained by Weihnacht et al. (2012) (dotted line).

They are used to calculate the stiffness tensor $c(T)$ and the piezoelectric tensor $e(T)$ using the respective relations. The elastic stiffness and the piezoelectric coefficient as function of temperature are summarized in Figs. 6 and 7, respectively.

A comparison of elastic data from this work with data obtained by Weihnacht et al. (2012). Weihnacht et al. (2012) show a very good agreement except for the stiffness coefficient c_{33} which differs by about 5 % from the referenced work. In case of the purely acoustic method as used by (Weihnacht et al., 2012) it is possible to measure the phase velocity along the x_3 axis directly. However, in case of resonant measurement, the excitation of resonant vibration in this direction is not possible. As shown in Eq. (7), all components of piezoelectric tensor d or e related to the x_3 direction equal zero. Therefore, the stiffness coefficient c_{33} as well as the elastic compliance s_{33} can be determined using the effective stiffness or compliance resulting from a vibration of a rotated crystal. In the related Eq. (10), the term $\sin^4\varphi$, dominates only for 45° rotated rod and nearly vanishes for all other rotation angles. For this reason the coefficients s_{33} and c_{33} are

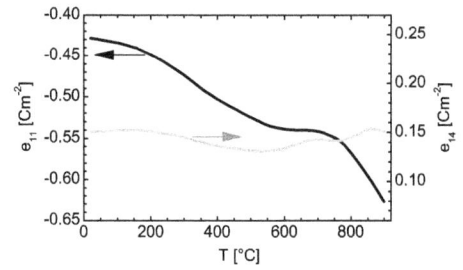

Figure 7. The piezoelectric coefficient of langasite as function of temperature.

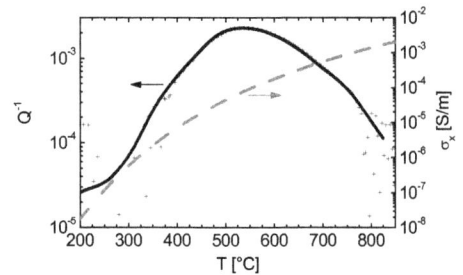

Figure 8. The inverse resonant quality factor Q^{-1} and the electric conductivity σ_X of langasite rod operated at 220 kHz in length extensional mode of vibration.

most error-prone which explains the small discrepancy mentioned above.

4.1.3 Temperature-dependent loss

As already mentioned in Sect. 2.1, the electrical conductivity and mechanical properties such as viscosity contribute to the losses in high-temperature piezoelectric materials. The viscosity of the langasite rod with the angle $\varphi = 0$ (see Fig. 1) expressed in form of the inverse resonant quality factor Q^{-1} (see Eq. 2) is calculated from the physical fit. These data and the electric conductivity σ_X are shown in Fig. 8. Here, a maximum of loss around 500 °C is clearly visible.

4.2 SAW properties

The characterization of SAW devices is carried out up to about 730 °C where the Zr / Pt-metallization failed through de-wetting and decomposition into droplets as visualized in Fig. 9.

The phase velocities v_0 and v_m on free and metallized surfaces for the cut (0, 138.5, 26.6°) are shown in Fig. 10. The obtained phase velocities at room temperature and coefficients of frequency for a second order polynomial fit are given together with literature data in Table 1.

Figure 11 shows the electromechanical coupling coefficient calculated from Eq. (28) for the propagation geometry with Euler angles (0, 138.5, 26.6°) and from the tensor data of BAW resonators. The difference between BAW and

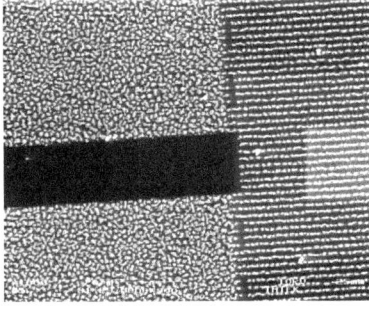

Figure 9. Dewetted metal layer after heating up to 800 °C.

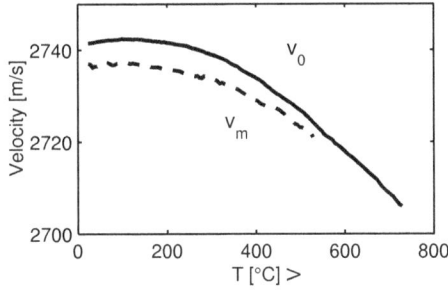

Figure 10. Phase velocity of Rayleigh waves on free surface v_0 (solid line) and on metallized surface v_m (dotted line), cut (0, 138.5, 26.6°).

SAW data for k^2 are caused by the uncertainty of the SAW method. As shown in Eq. (28), k^2 results from subtraction of two large numbers which were obtained by an extrapolation procedure. Furthermore, room temperature data are in satisfactory agreement with the corresponding electromechanical coupling coefficient, $k^2 = 0.33$, calculated from the tensor data given in (Kosinski et al., 2001), and with data obtained by other authors, $k^2 = 0.5$ (Fachberger et al., 2004) and $k^2 = 0.34$ (Naumenko and Solie, 2001).

The propagation losses α on free surfaces for both investigated cuts of langasite are shown in Fig. 12. Similar to the langasite rods operated in length-extensional mode of vibration, SAW devices exhibit a local propagation loss maximum at around 520 °C for all measured frequencies.

The propagation loss on the metallized surface shown in Fig. 13 increases monotonically. However, they exhibit sharp changes of the slope approximately at the same temperature as for the free surface.

4.3 Stability of the electrodes for BAW and SAW resonators

LGS-based SAW and BAW devices exhibit an operation temperature limit caused by the stability of the platinum electrodes. It is found, that their lifetime drops drastically as their thickness decreases. This effect is described by, e.g., Firebaugh et al. (1998). Due to this effect the SAW devices used in this work failed at about 730 °C, whereas the BAW res-

Table 1. Comparison of our measurement results with literature data. In case of (Kosinski et al., 2001), the values are calculated using tensor data.

cut	v_0, m s^{-1}	TCF1, ppm K^{-1}	TCF2, ppb K^{-2}
(0, 138.5, 26.6°)	2741.9	1.5	−41
(0, 30.1, 26.6°)	2464.6	28	−39
	results of other authors		
	2741.8[a]	−7[a]	−51[a]
(0, 138.5, 26.6°)	2734[b]	1[b]	−
	2743[c]	−	−
(0, 30.1, 26.6°)	2462.1[a]	31[a]	−69[a]
	2392.1[c]	−	−

[a] Fachberger et al. (2004), [b] Naumenko and Solie (2001), [c] Kosinski et al. (2001).

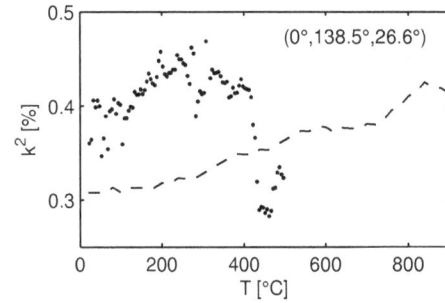

Figure 11. Coupling coefficient k^2 for Rayleigh waves on cut (0, 138.5, 26.6°), calculated with tensor data from BAW measurements (dashed lines) and extracted from SAW measurements (points).

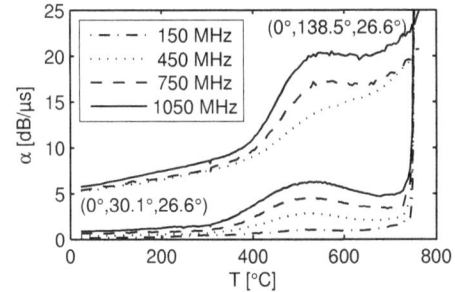

Figure 12. The propagation loss in [dB/μs] of Rayleigh waves on free surface on both cuts (0, 138.5, 26.6°) and (0, 30.1, 26.6°), measured at different frequencies.

onators wearing thicker electrodes could be operated up to 900 °C. If more stable electrodes, obtained by, e.g., screen printing, are used, even higher operation temperatures for BAW devices are demonstrated (Fritze, 2011b).

Above the mentioned temperatures, de-wetting of the metal layer is observed, as shown in Fig. 9 for SAW structures. Several attempts have been made to investigate and possibly reduce this effect. Da Cunha et al. conducted a series of experiments with different layer combinations for the metallization to extend the temperature limits for the SAW devices (da Cunha et al., 2007; Behanan et al., 2015). Most

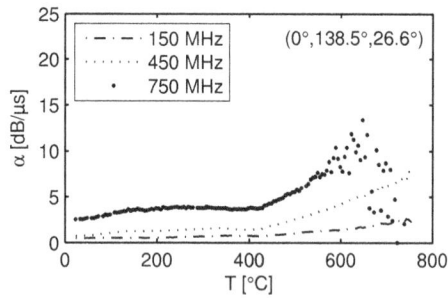

Figure 13. The propagation loss in [dB/μs] of Rayleigh waves on metallized surface on cut $(0, 138.5, 26.6°)$ for different frequencies.

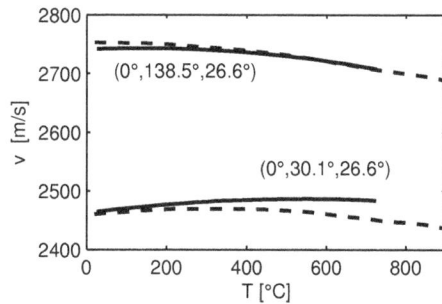

Figure 14. Comparison of Rayleigh wave velocity on free surface v_0 obtained from SAW and BAW measurements (solid and dashed lines, respectively).

recent attempts to minimize the degradation of electrodes and prolong the lifetime of SAW elements include use of metal alloys as well as high-temperature ceramic electrodes. Richter et al. report about successful operation of SAW elements at 800 °C for several hours using Ti / Pt coated langasite (Richter et al., 2011). Further attempts include application of thin protective layers, like Al_2O_3. An example of successful application of multi-layer stack metallization consisting of Al / Al_xO_y / Pt for the high-temperature sensors is given in (Wall et al., 2015).

4.4 Correlation of SAW-measured phase velocities with calculated BAW data

In order to compare the bulk and surface acoustic wave properties, the full set of temperature-dependent materials data is determined using BAW resonators and applied to calculate the phase velocity v_0 of SAW devices with Euler angles of $(0, 138.5, 26.6°)$ and $(0, 138.5, 26.6°)$. As seen in Fig. 14, the results of the calculation and the measured data are in good agreement especially for the $(0, 138.5, 26.6°)$ cut, which confirms the validity of the fitted parameters and the calculated materials data.

5 Conclusions

A method of materials data extraction from measured data of BAW resonators is developed for piezoelectric crystals. All components of the stiffness, elastic compliance and piezoelectric tensors for langasite are determined in the temperature range from 20 to 900 °C. Furthermore, the conductivity of the crystal in X and Y directions and the quality factor Q for resonant deformations in the X direction are obtained as function of temperature. For these experiments, BAW resonators with frequencies in the range 200–5000 kHz are used. The obtained elastic stiffness tensor is in good agreement with data published by other authors, except for the stiffness coefficient c_{33}, which exhibits a deviation of about 5 % from the literature data.

Based on the obtained tensor data, the phase velocity and the coupling coefficient for SAW propagation of the crystal cuts $(0, 138.5, 26.6°)$ and $(0, 30.1, 26.6°)$ is calculated. The same parameters are extracted from measurements of SAW delay lines up to 730 °C, which are fabricated using the crystal cuts mentioned above and operated at frequencies from 150 up to 1050 MHz.

Values for the velocity of SAW propagation and coupling coefficients obtained from SAW and from BAW measurements show a good agreement. SAW devices exhibit a local maximum of propagation loss at around 520 °C. BAW rods exhibit a maximum of losses at the same temperature.

Thin film Pt electrodes for BAW and SAW devices limit the maximum temperature of measurement to 900 and 730 °C, respectively. The operating temperature may be significantly increased by application of protective layers and thick electrodes. The latter is, however, feasible in the case of BAW resonators only.

Acknowledgements. The authors thank the German research foundation (Deutsche Forschungsgemeinschaft, DFG) for financial support and the Energy Research Center Niedersachsen.

References

Bardong, J., Schulz, M., Schmitt, M., Shrena, I., Eisele, D., Mayer, E., Reindl, L., and Fritze, H.: Precise measurements of BAW and SAW properties of Langasite in the temperature range from 25 C to 1000 C, in: 2008 IEEE International Frequency Control Symposium, 326–331, 2008.

Bechmann, R.: Contour Modes of Square Plates excited Piezoelectrically and Determination of Elastic and Piezoelectric Coefficients, P. Phys. Soc. Lond., B64, 323–337, 1951.

Behanan, R., Moulzolf, S., Call, M., Bernhardt, G., Frankel, D., Lad, R., and da Cunha, M.: Thin films and techniques for SAW sensor operation above 1000 °C, in: 2013 Joint UFFC, EFTF and PFM Symposium, 1013–1016, 2015.

Bungo, A., Jian, C., Yamaguchi, K., Sawada, Y., Uda, S., and Pisarevsky, Y.: Analysis of Surface Acoustic Wave Properties of Rotated Y-Cut Langasite Substrate, Jpn. J. Appl. Phys., 38, 3239–3243, 1999.

da Cunha, M. P. and Fagundes, S. D. A.: Investigation on recent quartz-like materials for SAW applications, IEEE Trans. on Ultrasonics, IEEE Int. Ferro., 46, 1583–1590, 1999.

da Cunha, M. P., Moonlight, T., Lad, R., Bernhard, G., and Frankel, D. J.: Enabling very high temperature acoustic wave devices for sensor & frequency control applications, in: 2007 IEEE Ultrasonics Symposium, 2107–2110, 2007.

Engan, H.: Surface acsoustic wave multilevel transducers, IEEE T. Son. Ultrason., 6, 395–401, 1975.

Fachberger, R., Holzheu, T., Riha, E., Born, E., Pongratz, P., and Cerva, H.: Langasite and langatate nonuniform material properties correlatedto the performance of SAW devices, in: Frequency Control Symposium and PDA Exhibition, 2001, Proceedings of the 2001 IEEE International, 235–239, 2001.

Fachberger, R., Bruckner, G., Hauser, R., Ruppel, C., Biniasch, J., and Reindl, L.: Properties of radio frequency Rayleigh waves on langasite at elevated temperatures, in: Proc. 2004 IEEE Ultrasonics Symposium, 1223–1226, 2004.

Firebaugh, S. L., Jensen, K. F., and Schmidt, M. A.: Investigation of High-Temperature Degradation of Platinum Thin Films with an In Situ Resistance Measurement Apparatus, J. Microelectromech. S., 7, 128–135, 1998.

Fritze, H.: High temperature piezoelectric materials: Defect chemistry and electro-mechanical properties, J. Electroceram., 17, 625–630, 2006.

Fritze, H.: High-temperature piezoelectric crystals and devices, J. Electroceram., 26, 122–161, 2011a.

Fritze, H.: High-temperature bulk acoustic wave sensors, Meas. Sci. Technol., 22, 012002, 2011b.

Fritze, H. and Tuller, H. L.: Langasite for High Temperature Bulk Acoustic Wave Applications, Appl. Phys. Lett., 78, 976–977, 2001.

Fritze, H., Schulz, M., Seh, H., and Tuller, H.: Sensor application-related defect chemistry and electromechanical properties of langasite, Solid State Ionics, 177, 2313–2316, 2006.

Fukuda, T., Takeda, P., Shimamura, K., Kawanaka, H., Kumatoriya, M., Murakami, S., Sato, J., and Sato, M.: Growth of new langasite single crystals for piezoelectric applications, in: Applications of Ferroelectrics, 1998. ISAF 98, Proceedings of the Eleventh IEEE International Symposium on, 315–319, 1998.

Ikeda, T.: Fundamentals of piezoelectricity, Oxford University Press, Oxford, UK, 1990.

Ingebrigtsen, K. A.: Analysis of interdigital transducers, IEEE T. Son. Ultrason., 403–407, 1972.

Kosinski, J. A., Pastore Jr., R. A., Bigler, E., Pereira da Cunha, M., Malocha, D. C., and Detaint, J.: A Review of Langasite Material Constants from BAW and SAW Data: Toward and Improved Data Set, in: IEEE International Frequency Control Symposium, 278–286, 2001.

Koskela, J.: Modeling SAW devices including mass-loading effects, Helsinki University of Technology, Helsinki, Finland, 1998.

Malocha, D., da Cunha, M., Adler, E., Smythe, R. C., Frederick, S., Chou, M., Helmbold, R., and Zhou, Y. S.: Recent Measurements of Material Constants versus Temperature for Langatate, Langanite and Langasite, in: Frequency Control Symposium and Exhibition, 2000. Proceedings of the 2000 IEEE/EIA International, 200–205, 2000.

Mason, W. P.: Physical Acoustics, Principles and Methods, vol. 1A, Academic Press, New York, USA, 1964.

Meitzler, A., Tiersten, H., Warner, A., Berlincourt, D., Couqin, G., and Welsh III, F.: IEEE Standard on Piezoelectricity, ANSI/IEEE Std, IEEE, 1988.

Naumenko, N. and Solie, L.: Optimal cuts of langasite, $La_3Ga_5Sio_{14}$ fpr SAW Devices, IEEE T. Ultrason. Ferr., 38, 530–537, 2001.

Peach, R.: A general Green function analysis for SAW devices, in: Proc. IEEE Ultrason. Symp., 221–225, 1995.

Peach, R.: On the existence of surface acoustic waves on piezoelectric substrate, IEEE T. Ultrason. Ferr., 49, 1308–1320, 2001.

Plessky, V., Koskela, J., Lehtonen, S., and Salomaa, M.: Surface transverse waves on langasite, in: 1998 IEEE Ultrasonics Symposium Proceedings, 139–142, 1998.

Richter, D., Sakharov, S., Forsén, E., Mayer, E., Reindl, L., and Fritze, H.: Thin Film Electrodes for High Temperature Surface Acoustic Wave Devices, Procedia Engineering, 25, 168–171, 2011.

Schneider, T., Richter, D., Doerner, S., Fritze, H., and Hauptmann, P.: Novel impedance interface for resonant high-temperature gas sensors, Sensor. Actuat. B-Chem., 111, 187–192, 2005.

Schulz, M. and Fritze, H.: Electromechanical properties of langasite resonators at elevated temperatures, Renew. Energ., 33, 336–341, 2008.

Sherrit, S. and Mukherjee, B. K.: The Use of Complex Material Constants to Model the Dynamic Response of Piezoelectric Materials, in: IEEE Ultrasonics Symposium, 633–640, 1998.

Shrena, I., Eisele, D., Mayer, E., Reindl, L., Bardong, J., and Schmitt, M.: SAW-relevant material properties of langasite in the temperature range from 25 to 750 °C: New experimental results, in: IEEE Ultrasonics Symposium, 209–212, 2008.

Shrena, I., Eisele, D., Bardong, J., and Reindl, L.: High precision signal processing algorithm to evaluate the SAW properties as a function of temperature, in: IEEE Ultrasonics Symposium, 863–866, 2009.

Smits, J. G.: Influence of Moving Domain Walls and Jumping Lattice Defects on Complex Material Coefficients of Piezoelectrics, IEEE T. Son. Ultrason., SU-23, 168–174, 1976.

Thiele, J. A. and Pereira da Cunha, M.: High temperature LGS SAW gas sensor, Sensor. Actuat. B-Chem., 113, 816–822, 2006.

Tuller, H. and Fritze, H.: High-temperature balance, US Patent 6.370.955, 2002.

Wall, B., Gruenwald, R., Klein, M., and Bruckner, G.: A 600 °C Wireless and Passive Temperature Sensor Based on Langasite SAW-Resonators, in: AMA Conferences 2015 – SENSOR 2015 and IRS2 2015, 19–21 May 2015, Nuremberg, Germany, 390–395, doi:10.5162/sensor2015/C3.3, 2015.

Weihnacht, M., Sotnikov, A., Schmidt, H., Wall, B., and Grünwald, R.: Langasite: High Temperature Properties and SAW Simulations, in: 2012 IEEE International Ultrasonics Symposium Proceedings, 1549–1552, doi:10.1109/ULTSYM.2012.0387, 2012.

8

Investigating the influence of Al-doping and background humidity on NO_2 sensing characteristics of magnetron-sputtered SnO_2 sensors

A. A. Haidry[1], **N. Kind**[1,a], **and B. Saruhan**[1]

[1]Institute of Materials Research, German Aerospace Center (DLR) Linder Hoehe, 51147 Cologne, Germany
[a]currently at: Laboratoire de Tribologie et Dynamique des Systèmes (LTDS) – UMR 5513 Bâtiment D4 Ecole Centrale de Lyon 36 Avenue Guy DE COLLONGUE 69134 Ecull, France

Correspondence to: A. A. Haidry (azhar.haidry@dlr.de) and B. Saruhan (bilge.saruhan@dlr.de)

Abstract. Elevated temperatures and humidity contents affect response, lifetime and stability of metal-oxide gas sensors. Remarkable efforts are being made to improve the sensing characteristics of metal-oxide-based sensors operating under such conditions. Having versatile semiconducting properties, SnO_2 is prominently used for gas sensing applications. The aim of the present work is to demonstrate the capability of the Al-doped SnO_2 layer as NO_2 selective gas sensor working at high temperatures under the presence of humidity. Undoped SnO_2 and Al-doped SnO_2 (3 at. % Al) layers were prepared by the radio frequency (r.f.) reactive magnetron sputtering technique, having an average thickness of 2.5 μm. The sensor response of Al-doped SnO_2 samples was reduced in the presence of background synthetic air. Moreover, under dry argon conditions, Al doping contributes to obtain a stable signal and to lower cross-sensitivity to CO in the gas mixtures of $CO + NO_2$ at temperatures of 500 and 600 °C. The Al-doped SnO_2 sensors exhibit excellent chemical stability and sensitivity towards NO_2 gas at the temperature range of 400–600 °C under a humid environment. The sensors also showed satisfactory response ($\tau_{res} = 1.73$ min) and recovery ($\tau_{rec} = 2.7$ min) towards 50 ppm NO_2 in the presence of 10 % RH at 600 °C.

1 Introduction

Semiconducting metal oxides (MOX) are the dominant material in the gas sensor industry with SnO_2 being a widely employed material for various emission gases (Göpel and Schmierbaum, 1995; Yamazoe and Shimanoe, 2007; Tricoli et al., 2010; Bochenkov et al., 2010). However, its use is limited to low operating temperatures in the range of 150–350 °C, because of low stability and poor selectivity obtained at higher temperatures (Yamazoe and Shimanoe, 2007; Tricoli et al., 2010). Among other MOX sensors, SnO_2 is widely studied as it is sensitive to many gases. Many metal additives, such as Pt, Pd, Ag and Au, have been used to improve the SnO_2 sensing properties (Göpel and Schmierbaum, 1995; Bochenkov et al., 2010). In the case of Al doping, Al-doped SnO_2 thin sensor films were produced by sputtering Sn on the substrate and using rheotaxial growth and a thermal oxi-

dation process (Faglia et al., 1996). It has been reported that Al-doped SnO_2 sensor layers show better sensitivity to NO_2 when the voltage is swept from direct current (DC) to 4 MHz at impedance measurements (Faglia et al. 1994).

It is well known that operating conditions of a sensor play a significant role in defining the sensor properties such as sensitivity, response/recovery time, stability, selectivity, etc. (Pavelko et al., 2010). While the effect of background water and oxygen species on gas sensing properties of SnO_2 materials has recently been reported (Korotcenkov et al., 2007; Großmann et al., 2013), this issue is still a subject of debate due to the complexity of this matter. A significant decrease in sensor response has been reported when the SnO_2 layer was exposed to a reducing gas (CO) in the presence of humidity, possibly due to pre-adsorbed oxygen being the only available ions for both CO and humidity (Großmann et al., 2013). Moreover, an increase in the CO sensing was reported in the presence of dry air, relying on the availability of more surface oxygen adsorbed from the air. Molecular water and oxygen species are adsorbed on the sensor surface below 150 °C under atmospheric pressure, while at higher temperatures (e.g., above 250 °C for water and 150 °C for oxygen) the hydroxyl groups and ionic oxygen are reported to be present on the sensor surface (Batzill et al., 2006). These reports indicate that the role of background oxygen and water in sensing can be significant at a wider temperature range, especially for application areas where the presence of these species cannot be ruled out.

In this report, we demonstrate the effect of aluminium doping on NO_2 sensing of SnO_2 sensors at relatively higher temperatures (>400 °C) and analyze the effect of humidity on NO_2 sensing. The sensor coatings were prepared by means of a radio frequency (r.f.) reactive magnetron sputtering technique on alumina-based sensor platforms decorated with platinum inter-digital electrodes. The sensors were subsequently annealed at 800 °C to achieve a crystalline structure. The sensitivity of the undoped and Al-doped SnO_2 coatings for NO_2 in dry as well as in humid argon at working temperatures $T_w = 400$–600 °C are presented. The sputtering technique allows for flexibility in adjusting composition and layer thickness yielding fine columnar structured morphology. Doping of SnO_2 with inexpensive aluminum and its deposition by sputtering on simple, designed sensor platforms will yield sensors competing with sophisticated and expensive nanowire/nanotube sensor layers. This paper aims at the validation of this concept.

2 Experimental

2.1 Preparation of sensing layers

The sensing layers were deposited by means of r.f. magnetron reactive sputtering on sensor platforms of 4.2 mm × 24.5 mm dimension. These sensor platforms were of Al_2O_3 substrates which contained inter-digital platinum electrodes (IDEs) previously patterned with screen printing methods, (see Fig. 1a). Each IDEs circuit contains five parallel 300 μm wide fingers with a 300 μm gap between two successive electrodes. The thickness of screen-printed Pt is about 2 μm. The sputter equipment (from Co. von Ardenne Anlagentechnik GmbH, Germany) can contain two metallic or ceramic targets. In the present case, Sn and Al metallic targets of same diameter (⌀ = 90 mm) were used and placed opposite to each other. Both targets were adjusted with different applied powers (P_{Sn}, P_{Al}). The sputtering process was carried out without any substrate heating and under high purity Argon + O_2 gaseous mixture (purity of Argon and O_2 were 99.9990 % and 99.9995 %, respectively). The partial pressures of oxygen (p_{O_2}) and argon (p_{Ar}) were controlled by mass flow controllers from MKS Instruments GmbH. During the deposition of undoped SnO_2 coatings, no rotation of substrate holders was applied. During the coating

process of Al-doped SnO_2, hereafter denoted as SnO_2 : Al, the substrates were rotated at a rate of 13 rpm. This rotation is necessary to achieve homogenous distribution of aluminium in the SnO_2 matrix. The sputtering conditions for both types of sensing layers are listed in Table 1. Under the given sputtering conditions, sensing layers of thickness of approximately 2.5 μm were obtained. Both sensor types (i.e., undoped SnO_2 and SnO_2 : Al) were manufactured during three different coating runs having 4–6 sensor platforms. At least three of these sensors were tested for their sensing properties.

Single crystal Al_2O_3 sapphire circular disks (⌀ = 13 mm) and silicon (Si) substrates (20 × 20 mm) were also placed side by side to the sensor platforms during each deposition run to be used for XRD (X-ray diffraction) measurements and EDX (energy dispersive X-ray) analysis, respectively, in order to avoid Pt interference from the inter-digital circuitry and Al from Al_2O_3 substrates. After deposition, the layers are annealed in static air at 800 °C for 3 h with a heating rate of 100 °C 15 min^{-1} in a M110 muffle furnace from Heraeus instruments. This ex situ annealing is necessary to obtain crystalline structures and also to stabilize the morphology and phase conditions prior to high temperature sensor tests.

2.2 Structural, morphological and compositional characterization

The structural investigation of the sensors was performed in Bragg–Brantano geometry by the XRD method, using s Siemens D5000 X-ray diffractometer with a CuKα radiation ($\lambda_{CuK\alpha} = 0.15418$ nm) and the graphite-curved monochromator. For phase analysis, $\theta/2\theta$- spectra is measured with an acceleration voltage of 40 kV at a step of 0.020° and a step rate of 3 s step^{-1}. The obtained data were compared with the JCPS database (Joint Committee on Powder Diffraction Standards) via EVA software from Bruker AXS in the range of 10–80° . The analysis of microstructure, surface morphology and chemical composition was carried out by using a field emission scanning electron microscope (FE-SEM, Carl Zeiss NTS Ultra 55) equipped with an EDX spectrometer.

2.3 Gas sensing characterization

Gas sensing characterization was carried out in a computer-controlled gas sensing experimental setup. This sensor and catalyst characterization unit (SESAM) consists of an eight-channel flow controller from MKS Instruments GmbH (MFC-647b) followed by a gas mixing chamber consisting of a CARBOLITE tube furnace with a quartz-glass recipient and DC-measurement unit from a Keithley 2635A Sourcemeter. The sample was mounted on a sample holder and then placed in a quartz-glass recipient with no heater at the back of the sensor sample. This setup allows us to accurately measure resistance R up to 10 GΩ at various

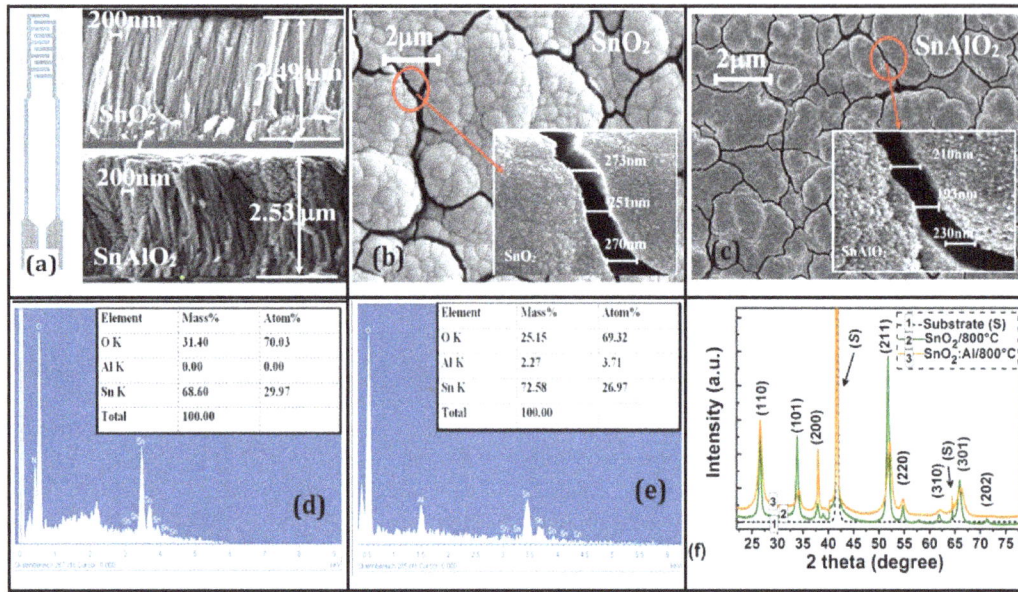

Figure 1. Structural and morphological characteristics of the sensing layers. (**a**) Schematic of Al_2O_3 sensor platforms with inter-digital electrodes (IDEs) and FE-SEM micrographs of both layers at the cross section confirming the columnar structure. (**b**) and (**c**) FE-SEM micrographs of SnO_2 and SnO_2 : Al layers, respectively, showing grain size differences and the insets showing the cracks generated during annealing. (**d**) and (**e**) EDX spectra of SnO_2 and SnO_2 : Al layers, respectively. (**f**) The X-ray diffractograms of c-cut single crystal sapphire Al_2O_3 substrate (1), SnO_2 (2) and SnO_2 : Al (3). The sensing layers were ex situ annealed at a temperature of 800 °C in static air, the peak from (0006) the plane of c-cut sapphire Al_2O_3 is denoted as "S".

Table 1. Deposition conditions for the undoped and Al-doped SnO_2 coatings.

Layer	$p_{chamber}$ (bar)	p_{O_2} (sccm)	p_{Ar} (sccm)	P_{Sn} (W)	P_{Al}(W)	Sputter time (h)	Layer thickness (μm)
SnO_2	6.10×10^{-3}	45	70	100	0	3	2.49
SnO_2 : Al	5.3×10^{-3}	35	70	70	150	12.50	2.53

gas concentrations and various operating temperatures up to 1200 °C. Sensor tests in SESAM are carried out in a quartz-glass recipient heated by a tube furnace, the temperature of which is controlled over three cascades and adjusted by a thermocouple positioned in the quartz recipient at a distance of 2 cm from the sensor surface. The quartz-glass recipient has a 3 m long spiral through which the test gas is sent and heated by the same furnace in order to avoid cooler gas contacting the sensor surface. The whole system including the gas inlet and sensor chamber is kept heated during the sensor test. Relying on the longer soaking times before and during the gas exposure (e.g., 30–60 min), the expected temperature difference between the thermocouple and the sensor surface will be negligibly small. Prior to gas-sensing measurements, a warm-up heating at the required testing temperature (i.e., 400–600 °C) was employed to all samples for 1 h under argon flow. This heating is necessary (Kim et al., 2013) to achieve electrical and chemical equilibrium at the sample surface. During the warming-up period, adsorption and desorption of

gas molecules results in chemical stabilization of the sensor surface and a steady baseline resistance is obtained.

All measurements were done in argon flow at a rate of $400 \, mL \, min^{-1}$ with a constant current of $1 \times 10^{-6} \, A$. Considering the ideal cases, the gas mixing process in the chamber is governed by the differential equation as given in Haidry et al. (2011). The gas mixing time τ_{mix} can be estimated as the chamber volume divided by the flow rate. The volume of the cylindrical chamber for pre-mixing the gases is $110 \, cm^3$ and the flow rate is $400 \, mL \, min^{-1}$, which gives a mixing time constant of 16.2 s. Further delay of a few seconds must be added to residence time as the gas mixture flows through approximately 4.5 m long tubes of 4.0 mm diameter before entering the sensing chamber. For calculation of the reaction/recovery times, this delay has been taken into account. The NO_2 concentrations were varied in the range of 50–200 ppm in dry and humid (1–10 % RH) argon (purity ≥ 99.998 %) for a continuous cycle of 30 min.

3 Results and discussion

3.1 FE-SEM and EDX analysis

As the cross-sectional micrographs of undoped and Al-doped SnO_2 in Fig. 1a show, the morphologies of both layers are very similar and columnar, despite the fact that the SnO_2 : Al layer was rotated during deposition. As mentioned earlier, the purpose of rotation was to obtain a uniform Al distribution in Sn bulk. The columnar structure and layer thicknesses of about 2.5 μm were checked using cross sections by SEM and presented in Fig. 1a. The top-view SEM micrograph of the SnO_2 layer deposited on sensor platforms (Fig. 1b) indicates fine columns leading to dense layer morphology with relatively wide cracks. The surface micrograph of the heat-treated layer does not exhibit any granular structure and XRD results confirm this (see Sect. 3.2 below). For SnO_2 : Al layers, the top-view SEM micrograph confirms the presence of more finer columns (i.e., grains) with a denser morphology and narrow cracks (Fig. 1c). The chemical composition of the SnO_2 is found to be 1 : 2, while the average Al^{3+} contents lie at about 3 at. % as confirmed by the EDX analysis (Fig. 1d, e).

3.2 XRD analysis

The XRD diffractograms of sapphire substrate without any layer, SnO_2 and SnO_2 : Al are shown in Fig. 1f. The XRD patterns of the both sensing layers, i.e., SnO_2 and SnO_2 : Al, are found to be normalized with a peak from (0006) the plane of c-cut sapphire Al_2O_3 at $2\theta = 41.7°$. After annealing, the polycrystalline sensing layers exhibit an orientation along (110), (101), (200), (211), (220), (310), and (301) at $2\theta = 26.6, 33.9, 37.9, 51.8, 54.7, 61.9$ and $65.9°$, respectively. These values are given in Table 2. These peaks are those which have three extra-low-intensity peaks at $2\theta = 38.9, 57.8$, and $71.8°$ at the undoped SnO_2 sensing layers. The strong-intensity peak for SnO_2 (211) corresponds to $2\theta = 37.9°$ and for SnO_2 : Al (200) to $2\theta = 37.9°$. Although SnO_2 shows good lattice match with the Al_2O_3 substrate, a slight shift of $2\theta = \pm0.02°$ can be observed comparing the above listed peaks with the standard XRD powder diffraction file (PDF2 #00-041-1445) of tetragonal SnO_2. According to Bazargan et al. (2013), this shift is most probably caused due to the difference in surface structure of rhombus Al_2O_3 and tetragonal SnO_2/SnO_2 : Al layers. No extra peak(s) for Al or its oxides are observed. Based on similar results (Gürakar et al., 2014), the authors conclude that no extra Al peaks in the XRD diffractogram can be observed, indicating a homogenous distribution of Al atoms within the SnO_2 matrix. Moreover, they report that intensities decrease significantly and the peaks are broadened. This may possibly be caused by the replacement of Sn^{4+} ions by Al^{3+} ions, as already mentioned in Mohagheghi and Saremi (2004). The average grain size of SnO_2 : Al layers (12 nm) is smaller than that of SnO_2

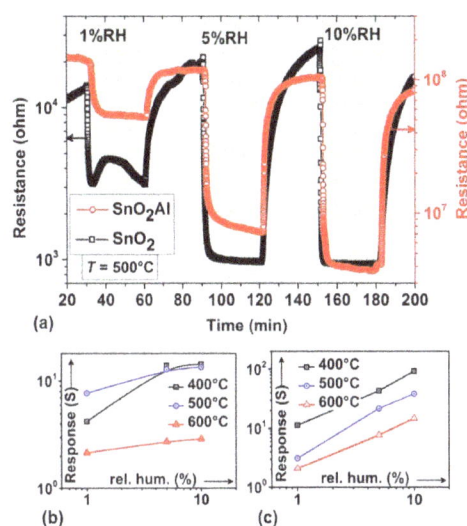

Figure 2. (a) The dynamic responses of SnO_2 (□) and SnO_2 : Al (○) layers towards 1–10 % RH at $T_w = 500$, (b) response versus relative humidity correlation of SnO_2 and (c) SnO_2 : Al. The response was recorded for a cyclic exposure of 30 min to dry and humid argon.

(45 nm); the grain size was estimated by means of the Scherrer formula.

3.3 Humidity sensing properties

Dynamic testing of both types of sensors under relative humidity (RH) is performed in the temperature range of 400–600 °C in order to understand the chemistry involved in NO_2 sensing under a humid environment. Both sensing layers show reproducible dynamic responses (Fig. 2), where the resistance of the layers decrease on exposure to various RH. This decrease is linear in the relative humidity range of 1–10 % RH as shown in Fig. 3b. The response (τ_{res}) and recovery (τ_{rec}) times are estimated for the 90 % change of the final saturated sensor resistance when switching the gas ON and OFF. The response ($S_{\%RH} = R_{argon}/R_{\%RH}$) and response times for the undoped SnO_2 layers are better at $T_w = 500$ °C ($\tau_{res} = 1.2, 1.5$ and 1.9 min towards 1, 5 and 10 % RH, respectively). However, as given in Table 3, longer recovery times are recorded reaching to several tens of minutes ($\tau_{rec} = 24$–30 min at $T_w = 400$ °C and $\tau_{rec} = 6.6, 13.9$ and 24.8 min towards 1, 5 and 10 % RH, respectively, at $T_w = 500$ °C). On the other hand, the increase of temperature to $T_w = 600$ °C yields relatively shorter recovery times ($\tau_{rec} = 2.9, 3.1$ and 5.5 min towards 1, 5 and 10 % RH).

Interestingly, the Al doping of SnO_2 results in more stable and reproducible sensing characteristics towards various RH (Fig. 2a). The response times are somewhat in the same range as those of undoped SnO_2 at $T_w = 400$ °C and become shorter with further increase of the working temperature, i.e., at $T_w = 500$ and 600 °C. Meanwhile, a significant shortening

Table 2. XRD-obtained peaks and orientations, where hkl are the Miller indices. The cells with × indicate NO peak observed at given 2θ.

Layer	$2\theta°$/hkl	$2\theta°$/hkl	$2\theta°$/hkl	$2\theta°$/hkl	$2\theta°$/hkl	$2\theta°$/hkl	$2\theta°$/hkl	$2\theta°$/hkl	$2\theta°$/hkl	$2\theta°$/hkl
SnO_2	26.6/(110)	33.9/(101)	37.9/(200)	38.9/(111)	51.8/(211)	54.7/(220)	57.8/(002)	61.9/(310)	65.9/(301)	71.8/(320)
SnO_2 : Al	26.6/(110)	33.9/(101)	37.9/(200)	×	51.8/(211)	54.7/(220)	×	61.9/(310)	65.9/(301)	×

Table 3. Comparison of recovery times for SnO_2 and SnO_2 : Al sensing layer towards 1, 5 and 10 % RH (without NO_2 exposure).

T_w (°C)		400			500			600	
Humidity (%)	1 %	5 %	10 %	1 %	5 %	10 %	1 %	5 %	10 %
τ_{rec} (min) for SnO_2	24.30	26.5	30	6.65	13.9	24.8	2.95	3.16	5.55
τ_{rec} (min) for SnO_2 : Al	4.68	8.9	12	5.83	5.11	3.85	4.03	2.11	2.23

of recovery time is recorded at each working temperature; see Table 3. The response to increasing humidity concentrations (% RH) shows the similar decreasing trend (Fig. 2b, c) with increasing temperature (T_w) as that with undoped SnO_2 layers (see Table 3). The reaction and recovery time constants in most cases can be correlated to operating temperature, as sensors produce different sensing properties depending on the operating conditions (Chen and Lu, 2005). It is observed in Fig. 2 that sensitivity and sensor response towards humidity is improved by Al doping. The response of the undoped SnO_2 layer towards lower humidity concentrations (e.g., 1 % RH) is very unsteady and recovery is very slow.

3.4 NO_2 sensing properties

The dynamic responses of SnO_2 and SnO_2 : Al sensors towards various NO_2 gas concentrations (50, 100 and 200 ppm) in dry argon background were recorded at $T_w = 400$, 500 and 600 °C; see Fig. 3. We define the sensor response/signal to NO_2 by the following equation: $S_{NO_2} = R_{NO_2}/R_{argon}$. A decrease in the sensor response of SnO_2 layers with increase in T_w was observed with incomplete baseline recovery at $T_w = 400$–500 °C, which causes a drift in sensor signal (Fig. 3a). Similar observations were reported in a previous study (Barsan et al., 2007), where decrease in sensitivity and short-term stability of SnO_2 was recorded when temperature increased above $T_w = 350$ °C. In contrast, as seen in Fig. 3b, SnO_2 : Al layers show a stable and reproducible response at $T_w = 600$ °C with reasonable recovery, although a weak or no response is obvious at $T_w = 400$ and 500 °C. These results indicate that Al doping leads to better short-term stability and sensitivity at 600 °C. The short-term stability means the ability of a sensor to produce the same dynamic response and to reach the same baseline resistance after switching OFF the gas; this also means a reproducible sensor response. The long-term stability defines, on the other hand, the sensor response on repetitive measurements of several months.

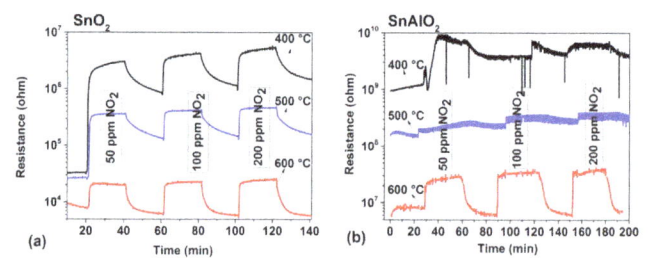

Figure 3. Dynamic responses of **(a)** SnO_2 and **(b)** SnO_2 : Al sensors when exposed to 50, 100 and 200 ppm of NO_2 gas concentrations in dry argon background, measured at $T_w = 400$, 500 and 600 °C.

3.5 Effect of background humidity and oxygen on NO_2 sensing

Several dynamic responses were recorded in order to check the effect of background humidity. A comparative plot of the two layers towards 100 ppm of NO_2 gas in 10 % RH at 500 °C is given in Fig. 4a. This plot shows that the SnO_2 : Al response to NO_2 in humidity background becomes better and stable with full baseline resistance recovery. On the other hand, in the case of undoped SnO_2 layers, the baseline is not recovered. Figure 4b shows one of the typical dynamic responses of SnO_2 : Al towards 50 ppm NO_2 in 10 % RH background for three temperatures: 400, 500 and 600 °C. The sensitivity of SnO_2 in a humidity background is highest only at 400 °C and decreases with increasing temperature and even then reaction time constants (τ_{res} and τ_{rec}) are mostly longer at 400 °C with no baseline recovery (see Table 4). On the other hand, the sensitivity of SnO_2 : Al sensors increases with increasing humidity, exhibiting a better response and recovery times at 600 °C. The response and recovery times are in the range of 1.5–5 and 6.6–20 min, respectively, in 1–10 % RH for the applied NO_2 concentrations at 400 and 500 °C. Meanwhile, the response and recovery times of SnO_2 : Al at 600 °C towards the lowest applied NO_2 concentration of 50 ppm are $\tau_{res} = 1, 1.56$ and 1.73 min and $\tau_{rec} = 7.9, 4.8$ and 2.7 min in the presence of 1, 5, and 10 % RH, respectively

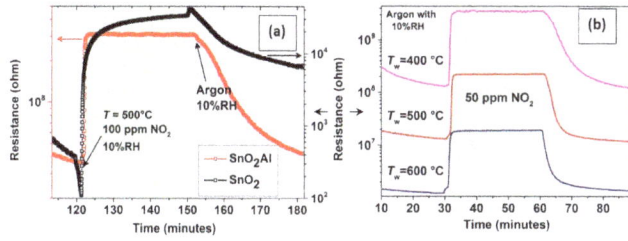

Figure 4. (a) Comparison of dynamic responses of SnO_2 ($\square$) and SnO_2 : Al ($\bigcirc$) layers towards 100 ppm NO_2 gas in 10 % RH at 500 °C. (b) Typical dynamic response of the SnO_2 : Al sensor towards 50 ppm NO_2 with a cycle of 30 min switching the gas ON/OFF at $T_w = 400, 500$ and 600 °C in 10 % RH.

(Table 4). The procedure used for definition of the response and recovery times is presented in Fig. 5b and c. As parallel experiments confirm, in the case of oxygen background gas, the response decreases further. One of the typical responses is shown in Fig. 5a.

3.6 NO2 cross-sensitivity

In order to check the cross-sensitivity, SnO_2 : Al sensors were tested for NO_2 and CO in combination as a single dynamic measurement towards 50 and 100 ppm gas concentrations separately and simultaneously as shown in Fig. 6. As it is demonstrated with the response curves in Fig. 6, SnO_2 : Al sensors exhibit responses to both NO_2 and CO when the gases are separately released into the test chamber. However, it is also clear that the Al-doped SnO_2 sensing layer yields a low CO cross-sensitivity when both gases are simultaneously present in the test chamber. This fact holds because the simultaneous action of NO_2 and CO yields a resistance value ($R = 5.18 \times 10^7 \Omega$; in 50 ppm NO_2+CO) near to that obtained only with NO_2 ($R = 5.5 \times 10^7 \Omega$; in 50 ppm NO_2), while the value of resistance in the presence of only 50 ppm CO is corresponding to $R = 6.49 \times 10^6 \Omega$. Meanwhile, it is observed that the baseline resistance also decreases to some extent and sensor response increases somewhat when both gases were simultaneously introduced to the SnO_2 : Al sensor ($S = 2.9$; in 50 ppm NO_2 and $S = 3.9$ in 50 ppm $NO_2 + CO$). In Sect. 4, a discussion on the possible explanation of the sensing mechanism is presented based on the results given in Sects. 3.3–3.6.

4 Sensing mechanism and discussion

4.1 Humidity sensing mechanism

It is generally accepted that during heating processes, oxygen is adsorbed on the surface of metal oxides in its molecular O_2^- form ($T_w < 150$ °C) and atomic O^-/O^{2-} forms ($T_w > 150$ °C) (Zakrzewska, 2003; Haidry et al., 2012a). However, as metal-oxide sensors operate at temperatures

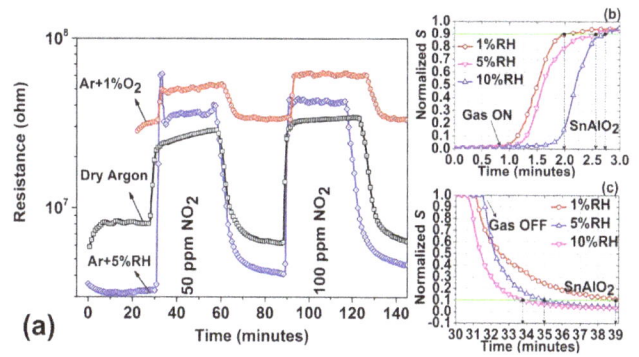

Figure 5. (a) The dynamic responses of SnO_2 : Al sensors when exposed to 50 and 100 ppm of NO_2 gas concentrations in dry argon, argon with 1 % oxygen and argon with 5 % RH in the background, measured at $T_w = 600$ °C. (b) and (c) provide a detailed view of the time-dependent normalized sensor response given on the right-hand side (Normalized $S = [R_{(t)} - R_{(Ar)} / R_{(NO_2)} - R_{(Ar)}]$) after the gas was switched ON and OFF. Here, $R_{(t)}$, $R_{(Ar)}$ and $R_{(NO_2)}$ denote real-time resistance, resistance in argon and NO_2 gas, respectively.

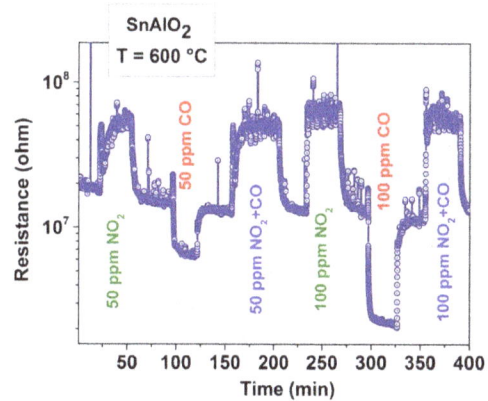

Figure 6. Dynamic response of the Al-doped SnO_2-based sensor to individual NO_2 or CO and in combination of 50 and 100 ppm of $NO_2 + CO$ gases.

above 150 °C, the atomic forms of oxygen are proven to influence the sensor's resistance and signal (Yamazoe and Shimanoe, 2008) according to the equation Eq. (1):

$$O_2 \,(gas) + 2e^- \rightarrow 2O^- (ads). \tag{1}$$

Through adsorbed oxygen species which trap electrons and form a depletion layer, the electrical conductivity of the n-type metal-oxide layers decreases causing an upwards band bending. In case of Al doping, it is likely that the Fermi level moves towards the middle of the energy gap and ultimately creates the acceptor level (Mohagheghi and Saremi, 2004; Scanlon and Watson, 2012) below E_F similar to the case of TiO_2 (Fig. 7a). In the presence of water, H_2O can get adsorbed on the SnO_2 surface as a result of (i) a physisorption process in its molecular form at relatively lower temperatures and (ii) a chemisorption process in its ionized form at

Table 4. Comparison of response and recovery times of SnO_2 and $SnO_2 : Al$ sensing layers towards NO_2 in the presence of 1, 5 and 10 % RH. The cells with "×" indicate incomplete baseline resistance recovery.

Humidity (%)	1 %			5 %			10 %		
NO_2 concentration (ppm)	50	100	200	50	100	200	50	100	200
SnO_2 at $T_w = 400\,°C$									
τ_{res} (min)	11	8.5	16	6.85	6.9	7.5	6.25	4.86	7.1
τ_{rec} (min)	×	×	×	9.1	×	×	21	×	×
$SnO_2 : Al$ at $T_w = 600\,°C$									
τ_{res} (min)	1.0	1.75	2.3	1.56	2.2	2.9	1.73	1.6	3.03
τ_{rec} (min)	7.9	10.2	12.7	4.8	8.6	9.1	2.7	8.4	8.8

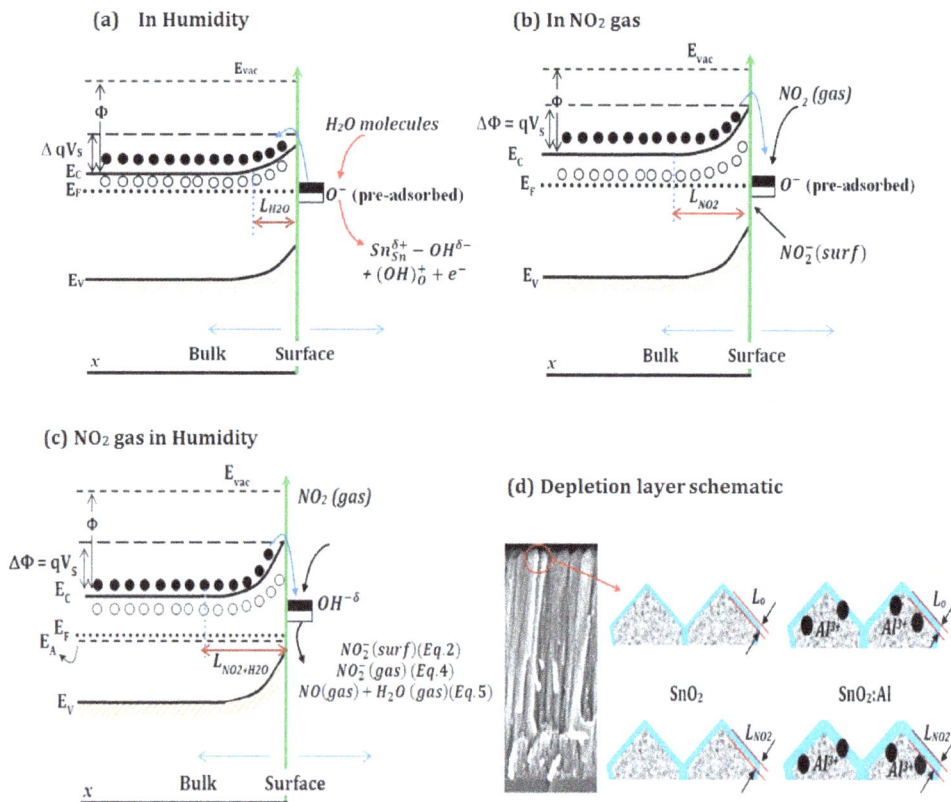

Figure 7. The illustration of the band bending due to the adsorption of oxygen species for an n-type semiconductor such as a SnO_2 layer. The schematic of chemisorption (**a**) NO_2 and (**b**) water molecules at the surface of an n-type SnO_2. (**c**) The surface reaction of $SnO_2 : Al$ columnar structures with NO_2 under a humid environment is presented, where NO_2 reacts with pre-adsorbed $OH^{\delta-}$ groups for improved sensitivity. (**d**) The schematic of the depletion layer due to adsorption of oxygen from the environment and reaction with oxidizing gases such as NO_2; L_o and L_{NO_2} represent the depletion region on columnar morphology in argon and NO_2 environments, respectively; here $L_o < L_{NO_2}$ and $L_o(SnO_2) \ll L_o(SnO_2 : Al)$. Electrons and holes are represented by (●) and (○), respectively.

higher temperatures. Generally, water vapors react with both surface Sn sites and surface pre-adsorbed oxygen ions. At higher temperatures e.g., above 400 °C, water molecules are adsorbed on the SnO_2 surface in the form of hydroxyl groups ($OH^{\delta-}$ groups) being active surface species and rooted OH groups (Barsan and Weimer, 2003; Cao et al., 1999). In this

case, $OH^{\delta-}$ behaves like a donor on the surface (see Fig. 7a).

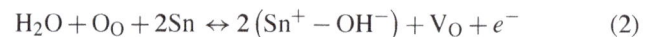

$$H_2O + O_O + 2Sn \leftrightarrow 2\left(Sn^+ - OH^-\right) + V_O + e^- \qquad (2)$$

The surface reaction, as described by Eq. (2), causes some release of the captured electrons back to the metal-oxide surface resulting in a decrease of the upward band bending and ultimately a decrease in the resistance (Batzill, 2006; Haidry et al., 2012b). This is valid also for reducing gases (see e.g., Fig. 6 for CO).

4.2 NO$_2$ sensing mechanism

The surface reaction of oxidizing and reducing gases with undoped SnO$_2$-based gas sensors is well documented (Yamazoe et al., 2003, 2007; Korotcenkov, 2005); still, it is a complex mechanism, especially at higher temperatures. In this case, working temperature determines the appropriate reactions involved in sensor response as thermodynamic equilibrium of the reactive gas may deviate from the primary one. For instance, NO$_2$ gas transformation into a mixture of NO and NO$_2$ at just above 600 °C has been reported in Saruhan et al. (2013), with a fair amount of oxygen. The authors observed only NO as the temperature increased to 800 °C, while NO$_2$ gas was supplied in argon carrier gas. In the present case, it is believed that NO$_2$ is adsorbed on the SnO$_2$ surface as NO$_{2\,surf}^-$ anions, in accordance of the following reaction in Eq. (2), as described in Cho et al. (2011):

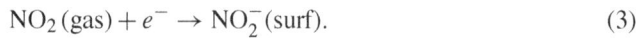

$$NO_2\,(gas) + e^- \rightarrow NO_2^-\,(surf). \qquad (3)$$

Eventually, the strong adsorption of NO$_2$ gas dominates over O$^-$/O^{2-} and creates an extended electron depletion layer ($L_o < L_{NO_2}$) leading to an increase of SnO$_2$ resistance (Cho et al., 2011); a schematic illustration is presented in Fig. 7b. Here, the reaction of NO$_2$ (gas) with pre-adsorbed oxygen O$^-$/O^{2-} is unlikely to occur. A superior adsorption capability of NO$_2$ over oxygen can also be observed in Fig. 5a, where the sensor showed a response towards NO$_2$ with an abundant amount of oxygen present in the background. Moreover, the response in an oxygen-rich background is lower than that in an argon background. This is also an indication that the sensing mechanism of NO$_2$ gas is governed by the surface adsorption of NO$_2^-$ anions rather than the reaction through pre-adsorbed oxygen at the surface.

4.3 Effect of Al doping in the NO$_2$ sensing mechanism

The partial substitution of Al^{3+} in the SnO$_2$ lattice may result in (i) generation of more oxygen vacancies, (ii) ultimately more pre-adsorbed oxygen species, and (iii) higher resistance with $L_o(SnO_2) \ll L_o(SnO_2\!:\!Al)$ as suggested by Xu et al. (1991a) and schematically shown in Fig. 7d. This causes the different adsorption/desorption kinetics of gases on the sensor surface. In fact, the substitution of Al adds more acceptor levels resulting in lowering the Fermi level down towards center of band gap. The reaction between NO$_2$ and SnO$_2$:Al surface reduces the surface electron concentration by trapping more electrons from the surface and hence an increase of resistance is noted (Sayago et al., 1995). Our results exhibit a resistance value of SnO$_2$:Al, which is about 5 orders of magnitude higher than that of pure SnO$_2$, i.e., $R_{SnO_2} = 2.5 \times 10^4\,\Omega \ll R_{SnO_2:Al} = 1.2 \times 10^9\,\Omega$ at 400 °C in pure argon (see Fig. 2). It has been previously reported that SnO$_2$:Al

doped with 1–5 at. % Al had a 4 order-of-magnitude higher resistance than that undoped, although further increase in the dopant level does not cause much change in the sensor resistance (in the range of a few $10^9\,\Omega$) (Xu et al., 1991). In addition to that, the improved sensitivity of our sensors towards NO$_2$ (i.e., in general towards oxidizing gases) in the presence of humidity can be explained in the light of attached surface hydroxyl groups OH$^{\delta-}$ coming from humidity. The NO$_2$ gas hitting the sensor surface reacts with already attached active hydroxyl groups OH$^{\delta-}$. These donors provide an increased number of adsorption sites for NO$_2$, thus leading to the improvement of sensor response. This explanation is valid for both sensors, undoped SnO$_2$ and Al-doped SnO$_2$. But in the case of Al-doped SnO$_2$ and in the presence of humidity, three possible reactions can occur simultaneously: the first is already given by Eq. (2) and the other two are expressed by Eqs. (3) and (5).

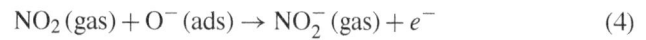

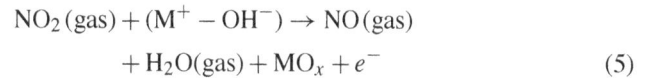

$$NO_2\,(gas) + O^-\,(ads) \rightarrow NO_2^-\,(gas) + e^- \qquad (4)$$

$$NO_2\,(gas) + (M^+ - OH^-) \rightarrow NO\,(gas)$$
$$+ H_2O\,(gas) + MO_x + e^- \qquad (5)$$

Both Eqs. (3) and (5), rely strongly on the availability of pre-adsorbed oxygen on the surface; here, M is the metal part of the reaction, in our case, that could be Sn or Al. The occurrence of the surface reactions given in Eqs. (3) and (5) is generally not possible in the case of undoped SnO$_2$. As a result of the reactions given in Eqs. (3) and (5), more adsorption sites on the sensor surface for NO$_2$ gas become available (see Fig. 7c, d). These kinds of simultaneous reactions clearly indicate the increase of sensitivity and shortening of reaction/recovery time constants for Al-doped SnO$_2$ sensors. It can also be noticed from our results presented in Fig. 4 that the sensor response of Al-doped SnO$_2$ sensors improves and becomes faster.

Low cross-sensitivity to CO in the presence of NO$_2$ and fast response to NO$_2$ achieved with an Al-doped SnO$_2$ sensing layer, as demonstrated in Fig. 6, can be partially associated with the Al incorporation into SnO$_2$. This association may be due to the doping-related electronic structure alteration (Hübner, 2011) as well as the microstructure and morphology controlled condition of the sensing layer. Al doping of SnO$_2$ by the sputtering process yields finer SnO$_2$ grains. Sintering of these fine grains, on further heat treatment of the sensing layer, results in formation of a finer crack network with higher density leading to increased grain boundary and surface area (see Figs. 1b, c and 7d). Similar observations have previously been reported in the literature (Choi et al., 2013; Liewhiran and Phanichphant, 2007; Xu at al., 1991b). As mentioned in Sect. 3.4, slower recovery and drift under dry conditions can be a result of the so-called surface poisoning effect. In the literature it is reported that the interaction of higher NO$_2$ concentrations with the sensor surface can cause poisoning due to the generation of doubly charged N$_2$O$_4^{2-}$ ions that stick to the surface of the sensor firmly and are hard to remove. This leads to longer recovery times of the sensors at relatively low or intermediate temperatures (Ruhland et al., 1998). We assume that such poisoning may occur at lower temperatures ($<500\,°C$). However, this effect is eliminated as the test temperatures are increased above $500\,°C$ and thus the sensor signal improves at $600\,°C$ as shown in Figs. 3 and 4.

5 Conclusions

The effect of background humidity and Al doping in SnO_2 layers was investigated for the NO_2 gas response at temperatures above $400\,°C$. A response of similar size but in opposite direction is recorded for individual gas exposures to NO_2 and CO at $600\,°C$ while the sensor becomes more selective towards NO_2 in a gas mixture of $CO + NO_2$. Enhanced sensitivity of SnO_2 : Al was observed towards various concentrations of NO_2 in a humid background environment, while the sensitivity is reduced in an oxygen-rich environment. The response ($\tau_{res} = 1.73$ min) and recovery ($\tau_{rec} = 2.7$ min) times towards 50 ppm NO_2 in 10 % RH are much shorter than in pure argon. The chemical stability and microstructure of SnO_2-based sensors were significantly improved by Al doping. Moreover, SnO_2 : Al layers exhibit finer grain size with a denser morphology as a result of annealing. This largely interconnected crack network leads to faster response and recovery times. In addition, the improved sensing characteristics of Al-doped SnO_2 in a humid environment were supported by the presence of hydroxyl groups, $OH^{\delta-}$, on the surface. We propose that these surface hydroxyl groups, $OH^{\delta-}$, provide more surface adsorption sites for oxidizing gas such as NO_2. Hence, these sensors would be promising candidates for monitoring NO_2 at higher temperature under humid environments with fast response rates. In addition, considering the uncomplicated processing of the material, the compatibility of thin-layer technology with modern electronics will make this sensing material more suitable for mass production.

Acknowledgements. This work has been partially supported by DAAD-DLR fellowship funding under the fellowship number 165. The authors thank Uwe Schulz, the head of the department High Temperature and Functional Coatings, for his valuable support and G. C. Mondragon Rodriguez for his assistance with FE-SEM.

References

Barsan, N., Koziej, D., and Weimer, U.: Metal oxide-based gas sensor research: How to, Sensor. Actuat. B-Chem., 121, 18–35, 2007.

Barsan, N. and Weimar, U.: Understanding the fundamental principles of metal oxide based gas sensors; The example of CO sensing with SnO_2 sensors in the presence of humidity, J. Phys. Con. Matter, 15, R813–R83, 2003.

Batzill, M.: Surface science studies of gas sensing materials: SnO_2, Sensors, 6, 1345–1366, 2006.

Batzill, M., Bergermayer, W., Tanaka, I., and Diebold, U.: Tuning the chemical functionality of a gas sensitive material: Water adsorption on SnO_2 (101), Surface Science Letters, 600, 29–32, 2006.

Bazargan, S., Thomas, J. P., and Leung, K. T.: Magnetic interaction and conical self-reorganization of aligned tin oxide nanowire array under field emission conditions, J. Appl. Phys. 113, 234305, 2013.

Bochenkov, V. E. and Sergeev, G. B.: Sensitivity, Selectivity, and Stability of Gas-Sensitive Metal-Oxide Nanostructures, in: Metal Oxide nanostructures and their applications, edited by: Ahmad, U. and Hanh, Y. B., American Scientific Publishers, 3, 31–52, 2010.

Cao, L., Spiess, F. J., Huang, A., and Suib, S. L.: Heterogeneous photocatalytic oxidation of 1-Butene on SnO_2 and TiO_2 films, J. Phys. Chem. B, 103, 2912–2917, 1999.

Chen, Z. and Lu, C.: Humidity sensors: a review of materials and mechanisms, Sens. Lett., 3, 274–295, 2005.

Cho, N. G., Yang, D. J., Jin, M. J., Kim, G. G., Tuller, H. L., and Kim, I. D.: Highly sensitive SnO_2 hollow nanofiber-based NO_2 gas sensors, Sensor. Actuat. B-Chem., 160, 1468–1472, 2011.

Choi, S.-W., Katoch, A., Sun, G.-J., Wu, P., and Kim, S. S.: NO_2-sensing performance of SnO_2 microrods by functionalization of Ag nanoparticles, J. Mater. Chem. C, 1, 2834–2841, 2013.

Faglia G., Nelli, P., and Sberveglieri, G.: Frequency effect on highly sensitive NO_2 sensors based on RGTO SnO_2(Al) thin films, Sensor. Actuat. B-Chem., 19, 497–499, 1994.

Faglia, G., Benussi, G., Depero, L., Dinelli, G., and Sberveglieri, G.: NO_2 sensing by means of SnO_2(Al) thin films grown by the rheotaxial growth and thermal oxidation technique, Sensor. Mater., 8, 239–249, 1996.

Göpel, W. and Schierbaum, K. D.: SnO_2 sensors: current status and future prospects, Sensor. Actuat. B-Chem., 26, 1–12, 1995.

Großmann, K., Wicker, S., Weimaer, U., and Barsan, N.: Impact of Pt additives on the surface reactions between SnO_2, water vapour, CO and H_2; An operando investigation, Phys. Chem. Chem. Phys., 15, 19151–19158, 2013.

Gürakar, S., Serin, T., and Serin, N.: Electrical and microstructural properties of (Cu, Al, In)-doped SnO_2 films deposited by spray pyrolysis, Adv. Mat. Lett., 5, . 309–314, 2014.

Haidry, A. A., Schlosser, P., Ďurina, P., Mikula, M., Tomášek, M., Plecenik, T. Roch, T., Pidík, A., Štefečka, M., Noskovič, J., Zahoran, M., Kúš, P., and Plecenik, A.: Hydrogen gas sensors based on nanocrystalline TiO_2 thin films, Cent. Eur. J. Phys., 9, 1351–1356, 2011.

Haidry, A. A., Puškelová, J., Plecenik, T., Ďurina, P., Greguš, J., Truchlý, M., Roch, T., Zahoran, M., Vargová, M., Kúš, P., Plecenik, A., and Plesch G.: Characterization and hydrogen gas sensing properties of TiO_2 thin films prepared by sol – gel method, Appl. Surf. Sci., 259, 270–275, 2012a.

Haidry, A. A., Ďurina, P., Tomášek, Greguš, J., Schlosser, P., Mikula, M., Truchly, M., Roch, T., Plecenik, T. Pidík, A., Zahoran, M., Kúš, P., and Plecenik, A.: Effect of post-deposition annealing treatment on the structural, optical and gas sensing properties of TiO_2 thin films, Key Eng. Mat., 510–511, 467–474, 2012b.

Hübner, M.: New Approaches for the Basic Understanding of Semiconducting Metal Oxide Based Gas Sensors: Sensing, Transduction and Appropriate Modeling , Dissertation, Uni. Tübingen, 102–108, 2011.

Kim, B., Lu, Y., Hannon, A., Mayyappan, M., and Li, J.: Low temperature Pd/SnO_2 sensors for CO detection, Sensor. Actuat. B-Chem., 177, 770–775, 2013.

Korotcenkov, G.: Gas response control through structural and chemical modification of metal oxides: State of the art and approaches, Sensor. Actuat. B-Chem., 107, 209–232, 2005.

Korotcenkov, G., Blinov, I., Brinzari, V., and Stetter, J. R.: Effect of air humidity on gas response of SnO_2 thin films ozone sensors, Sensor. Actuat. B-Chem., 122, 519–526, 2007.

Liewhiran, C. and Phanichphant, S.: Improvement of Flame-Made ZnO nanoparticle thick film morphology for Ethanol sensing, Sensors, 7, 650–675, 2007.

Mohagheghi, M. M. B. and Saremi, M. S.: The influence of Al doping on the electrical, optical and structural properties of SnO_2 transparent conducting films deposited by the spray pyrolysis technique, J. Phys. D: Appl. Phys., 37, 1248–1253, 2004.

Pavelko, R. G., Daly, H., Hardacre, C., Vasiliev, A. A., and Llobet, E.: Interaction of water, hydrogen and their mixtures with SnO_2 based materials: the role of surface hydroxyl groups in detection mechanism, Phys. Chem. Chem. Phys., 12, 2639–2647, 2010.

Ruhland, B., Becker, T., and Müller, G.: Gas-kinetic interactions of nitrous oxides with SnO_2 surfaces, Sensor. Actuat. B-Chem., 50, 85–94, 1998.

Saruhan, B., Yüce, A., Gönüllü, Y., and Kelm, K.: Effect of Al doping on NO_2 gas sensing of TiO_2 at elevated temperatures, Sensor. Actuat. B-Chem., 187, 586–597, 2013.

Sayago, I., Gutierrez, J., Ares, L., Robla, J. I., Horrillo, M. C., Getino, J., Rino, J., and Agapito, J. A.: The effect of additives in tin oxide on the sensitivity and selectivity to NO_x and CO, Sensor. Actuat. B-Chem., 26–27, 19–23, 1995.

Scanlon, D. O. and Watson, G. W.: On the possibility of p type SnO_2, J. Mater. Chem., 22, 25236–25245, 2012.

Tricoli, A., Righettoni, M., and Teleki, A.: Semiconductor gas sensors: Dry synthesis and application, Angew. Chem. Int. Edit., 49, 7632–7659, 2010.

Xu, C., Tamaki, J., Miura, N., and Yamazoe, N.: Promotion of tin oxide gas sensor by aluminum doping, Talanta, 38, 1169–1175, 1991a.

Xu, C., Tamaki, J., Miura N., and Yamazoe, N.: Grain size effects on gas sensitivity of porous SnO_2-based elements, Sensor. Actuat. B-Chem., 3, 147–155, 1991b.

Yamazoe, N. and Shimanoe, K.: Theory of power law for metal oxide, Sensor. Actuat. B-Chem., 128, 566–573, 2008.

Yamazoe, N., Sakai, G., and Shimanoe, K.: Oxide semiconductor gas sensors, Catal. Surv. Asia, 1, 63–75, 2003.

Yamazoe, N., Shimanoe, K., and Sawada, C.: Contribution of electron tunneling transport in semiconductor gas sensors, Thin Solid Films, 515, 8302–8309, 2007.

Yamazoe, N. and Shimanoe, K.: Overview of gas sensor technology, in: Science and Technology of Chemiresistive Gas Sensors, edited by: Aswal, D. K. and Gupta, S. K., Nova Science Publisher, New York, USA, 1–32, 2007.

Zakrzewska K.: Titanium Dioxide Thin Films for Gas Sensors and Photonic Applications, AGH Ucelniane Wydawnictwa Naukowo-Dydadaktyczne, Kraków, 10–35, 2003.

9

A micro-capacitive pressure sensor design and modelling

Ali E. Kubba[1], Ahmed Hasson[1], Ammar I. Kubba[2], and Gregory Hall[1]

[1]Fusion Innovations Ltd., Research and Innovation Services, Birmingham Research Park, Vincent Drive, Edgbaston, Birmingham, B15 2SQ, UK
[2]School of Engineering, Mechanical Engineering, University of Birmingham, Edgbaston, Birmingham, B15 2TT, UK

Correspondence to: Ali E. Kubba (a.e.s.kubba@bham.ac.uk)

Abstract. Measuring air pressure using a capacitive pressure sensor is a robust and precise technique. In addition, a system that employs such transducers lies within the low power consumption applications such as wireless sensor nodes. In this article a high sensitivity with an elliptical diaphragm capacitive pressure sensor is proposed. This design was compared with a circular diaphragm in terms of thermal stresses and pressure and temperature sensitivity. The proposed sensor is targeted for tyre pressure monitoring system application. Altering the overlapping area between the capacitor plates by decreasing the effective capacitance area to improve the overall sensitivity of the sensor ($\Delta C/C$), temperature sensitivity, and built-up stresses is also examined in this article. Theoretical analysis and finite element analysis (FEA) were employed to study pressure and temperature effects on the behaviour of the proposed capacitive pressure sensor. A MEMS (micro electro-mechanical systems) manufacturing processing plan for the proposed capacitive sensor is presented. An extra-low power short-range wireless read-out circuit suited for energy harvesting purposes is presented in this article. The developed read-out circuitry was tested in terms of sensitivity and transmission range.

1 Introduction

This article presents a proposed design of a MEMS capacitive pressure sensor for tyre inflation pressure measurement purposes. A diaphragm-based pressure sensing mechanism is employed and analysed in both analytical and numerical techniques, including Matlab and finite element analysis (FEA), which are employed to examine the proposed micro-pressure sensor response under a certain range of pressure and temperature conditions. A low power read-out and transmission circuit for the capacitive pressure sensor is also illustrated.

1.1 Motivation

This research is part of a TSB funded project which is led by Fusion Innovations Ltd. ECORR (Variable Rolling Resistance Tyre System) is an Innovate UK collaborative project as part of the Low Carbon Vehicles Disruptive technologies

field. This project is currently partnered by Randle Engineering, Potenza Technology, and the University of Birmingham[1].

According to the Pressure Sensor Market report (MarketsandMarkets, 2015) published by MarketsandMarkets, the global pressure sensor market was valued at USD 6.7 billion in 2014 and is expected to grow to USD 9.48 billion by 2020. In 2014, around 36.7 % of commercial pressure sensors were piezoresistive, followed by capacitive pressure sensors as the second dominant type, accounting for a market share of 27.6 %. The automotive industry is the largest application market for pressure sensors and accounted for a 25 % share of the total revenue of 2014 (MarketsandMarkets, 2015). Sizable growth of the pressure sensor market in the Asia–Pacific region, particularly in the automotive industry, is predicted due to the introduction of new regulations and legislations re-

---

[1]Fusion Innovations Ltd, available at: http://www.fusion-innovations.com

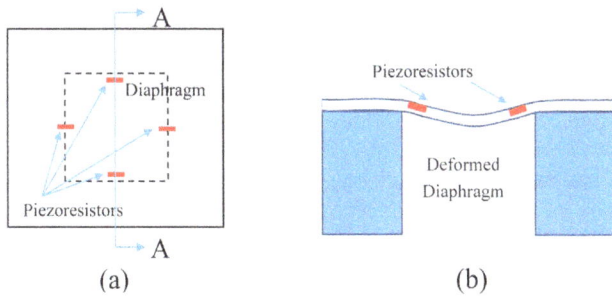

Figure 1. Typical piezoresistive pressure sensor structure. (**a**) Top view of the piezoresistive pressure sensor. Four piezoresistors are placed on each edge forming a Wheatstone bridge circuit. (**b**) Cross section A–A showing deflected diaphragm with piezoresistors at maximum stress locations (Barlian et al., 2009).

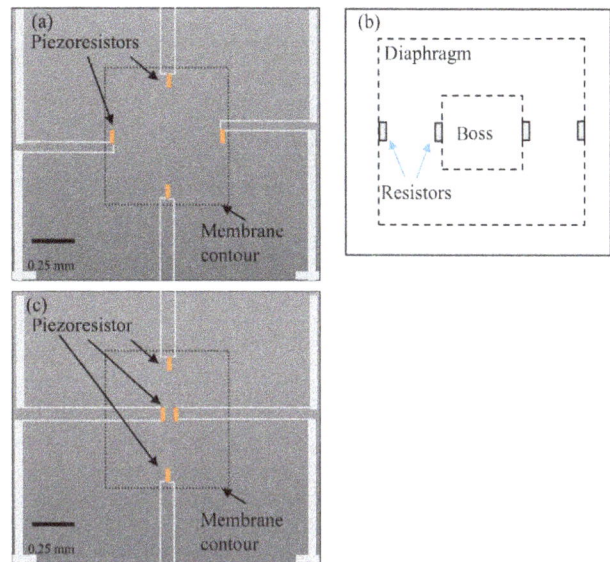

Figure 2. Resistor placement on diaphragm-based piezoresistive pressure sensors: (**a**) four resistors placed on the diaphragm edges and (**b**) two resistors on the diaphragm edges and the other two at the centre of the diaphragm. In both cases, the resistors are connected in Wheatstone bridge (**c**) resistors mounted on the inner and outer perimeters on a bossed diaphragm (Beeby et al., 2004; Chien-Hung et al., 2006).

garding the use of such sensors in vehicles. More TPMS (tyre pressure monitoring system) market analysis can be found in Technavio (2015).

A listing for pressure sensors, manufacturers, and a technical comparison between both piezoresistive and capacitive pressure sensors can be found in Fragiacomo (2012).

1.2 Piezoresistive pressure sensors

Piezoresistive pressure sensors are based on the piezoresistive effect, which is a change in resistance with the applied stress. By applying piezoresistor(s), with known values and where a maximum variation of stresses occurs, to a pressure sensitive diaphragm, and monitoring the change in that resistor as the applied pressure varies, in most cases the change in the applied pressure can be detected by using a Wheatstone bridge circuit. The fact that single-crystal silicon has a piezoresistive nature, and that silicon is mainly used in micro-machining, enabled production of semi-conductor-based sensors (Eaton and Smith, 1997). Examples of industrial piezoresistive pressure sensors are shown in Figs. 1 and 2.

Piezoresistive pressure sensors have the advantages of low cost, linearity, a simple read-out circuit, and the possibility of being batch produced. The piezoresistive coefficients of each mounted resistor depend upon the orientation of the wafer and diaphragm, the type and amount of doping, and the temperature (Beeby et al., 2004). The latter is a main disadvantage in applying piezoresistive pressure sensors to tyre monitoring systems, as tyre temperature changes over a wide range. Power consumption is highly related to the sizes of the mounted resistors' values, which vary according to the size of the piezoresistive pressure sensor overall. However, most existing commercially available TPMSs include a piezoresistive pressure sensor for their relatively low price. A typical commercial piezoresistive pressure sensor developed by Intersema Sensoric, a Measurement Specialties company (2009), is shown in Fig. 3.

Scaling down the mounted resistors creates an increase in power consumption – providing the same piezoresistive coefficient, mounting accuracy, and noise problems (Cullinan et al., 2012). In addition, the limitation of the resistors' size might reduce the sensitivity of the piezoresistive pressure sensor by averaging the developed stress on the sensor diaphragm. Another problem with piezoresistive pressure sensors is that piezoresistance is susceptible to junction leakage and surface contamination (Bao, 2000). This is true when using silicon as the main substrate to fabricate a MEMS pressure sensor, and can be overcome by using SOI (silicon on insulator) instead (Li Sainan et al., 2015), though it would double the sensor's substrate cost. A good review of the working principles and sensitivity of piezoresistive pressure sensors is explained in Gad-el-Hak (2002). With capacitive and piezoresistive measurement principles, there are competing concepts on the market. Whereas the piezoresistive pressure sensor is found in the vast majority of commercial TPMS (Fragiacomo, 2012; Nwagboso, 2012), it does not meet the requirement of battery-less systems anymore due to the fact that the power consumption is much higher, and since sleep modes for measurement acquisition are mandatory (Thurau and Ruohio, 2004).

Figure 3. Miniature SMD (surface-mounted device) pressure sensor mounted on a read-out PCB (printed circuit board) developed by Intersema Sensoric (now Measurement Specialties, Inc., a TE Connectivity company; photo taken with permission from Measurement Specialties).

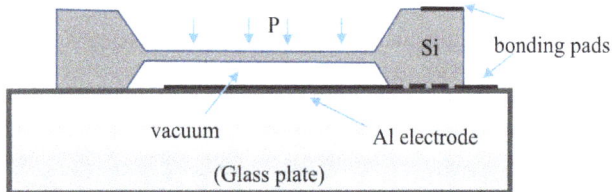

Figure 4. A schematic view of a basic capacitive pressure transducer (Bao, 2000).

1.3 Capacitive pressure sensors

This type of micro-machined pressure sensor was first developed in the late 1970s and early 1980s (Gad-el-Hak, 2002). It can be classified as the simplest in principle among all pressure sensing mechanisms. Capacitive pressure sensors primarily consist of two parallel conductive plates, so-called electrodes, separated by a dielectric material. Usually one of the electrodes is pressure sensitive, whereas the other electrode is located on a rigid substrate beneath it. However, a capacitive pressure sensor with two sensitive diaphragms has been developed (Fonseca et al., 2002). When pressure is applied onto the sensitive diaphragm, the cavity enclosed between the two parallel plates reduces in volume as the sensitive diaphragm deflects and approaches the stationary one, resulting in a detectable change in the capacitance between the electrodes. A schematic diagram for a typical capacitive pressure sensor is shown in Fig. 4.

The principle advantages of capacitive pressure sensors over piezoresistive pressure sensors are lower power consumption, increased pressure sensitivity, and lower temperature cross-sensitivity (Lee and Wise, 1982; Eaton and Smith, 1997; Eddy and Sparks, 1998; Bever et al., 2003; Beeby et al., 2004). In addition, because there is no need to mount any resistors on the sensor diaphragm, scaling down the device dimensions is easier because concerns about stress averaging

and resistor tolerance are eliminated. Variation of temperature has a major contribution to the sensitivity of piezoresistive pressure sensors, mainly due to the resistance dependence on temperature. This factor is eliminated in capacitive sensors (Gad-el-Hak, 2002). There is virtually no power consumption in the sensing element in capacitive pressure sensors due to the dc current component being zero (Gad-el-Hak, 2002) and the low power capability of the capacitive measuring principle (Kolle et al., 2004).

Capacitive pressure sensors have been applied to TPMS devices due to their low power consumption, high accuracy and low temperature sensitivity (Rudolf and Hoogerwerf, 1997; APOLLO, 2003; Thurau and Ruohio, 2004; Kolle et al., 2004). Bracke et al. (2007) presented the power consumption of various ultra-low power capacitive sensors' interfaces and showed experimentally that it is possible to achieve an average power consumption of $7.3\,\mu$W for a 10 Hz sampling frequency and 8 bit accuracy in the 100 to 130 kPa pressure range. This amount of power consumption is lower than the average piezoresistive pressure sensor power consumption if a similar sampling rate and digitalization level is used. A new patented capacitive sensor measurement chip (Pico-Cap)[2] could be utilized for TPMS devices as it offers signal digitalization at a high sampling rate (up to 500 k sample s^{-1}) under an extra low power requirement.

The main limitations of capacitive pressure sensors in general are nonlinearity between the applied pressure, the change in capacitance, and the large impedance of the sensor output signal (Tian et al., 2009), which also affects the interface circuit design, and the parasitic capacitance between the interface circuit and the device output can have a significant negative impact on the read-out, which means that the circuit must be placed in close proximity to the device in a hybrid or monolithic implementation. To address the nonlinearity problem in capacitive pressure sensors, different approaches have been used to mitigate this drawback, e.g. bossed diaphragms, centrally clamped diaphragms and partial electrode patterned diaphragms; however, these approaches reduce the overall sensor sensitivity and limit the measured pressure range. Touch mode capacitive pressure sensors, as shown in Fig. 5, are commonly used to employ linearity as regards the nonlinearity problems in ordinary capacitive pressure sensors. One potential drawback of touch mode devices is hysteresis arising from friction between the surfaces as they move together and apart, as well as the risk of stiction (Beeby et al., 2004).

Another important issue in the making of capacitive pressure sensors is sealing, which is a fabrication process complexity (Tian et al., 2009). Extra care should be taken when fabricating and packaging absolute pressure sensors and especially in sealing the vacuum cavity beneath the sensitive diaphragm, as it is the reference pressure for the sensor (Gad-

---

[2]acam messelektronik gmbh, available at: http://www.acam.de/de/produkte/picocap/

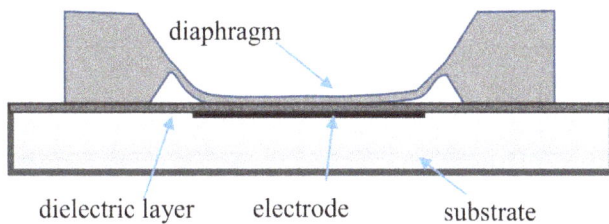

Figure 5. A touch mode capacitive pressure sensor (Gad-el-Hak, 2002).

el-Hak, 2002). Proper sealing in absolute capacitive pressure sensors is crucial to achieve long term stability. Due to cost issues in both wafers and fabrication processes, and the match between Pyrex glass and silicon thermal expansion coefficients, Pyrex glass substrate is a common material for the sensor constraint base when making capacitive pressure sensors to support the sensor silicon dies. Anodic bonding is usually used to attach Pyrex glass to silicon wafers, which is essential to produce a reliable and hermetically sealed MEMS capacitive pressure sensor under medium bonding temperature (Frank, 2000; Hsu, 2008). Glass is therefore a well-known material, but also has some specific characteristics like outgasing and absorption of residual gas on the glass surface which would lead to a change in the internal pressure. In order to address this problem, an interesting approach was used in fabricating a capacitive pressure sensor – which is especially designed for TPMS applications, as shown in Thurau and Ruohio (2004). In this design, a wet etched vacuum reservoir, also called a gas pocket, was added and connected to the reference pressure volume underneath the sensor diaphragm enabling good long term stability. A good review of a MEMS capacitive pressure sensor, using a technology developed by a group of engineers in Finland who originally developed pressure sensors targeted for weather measurement application, and which were especially designed to withstand harsh environment conditions, is explained in detail in Thurau and Ruohio (2004). A comparison of various high-temperature pressure sensors and passive wireless pressure sensors is well presented by Fonseca (2007).

1.4 Resonant pressure sensors

Resonance micro-machined pressure sensors were first developed in the early 1980s by Greenwood (1984). The resonant sensing mechanism is based on containing a resonant structure in which its resonance frequency is a function of a mechanical quantity, such as pressure, strain, temperature, etc. This type of sensor can be considerably more robust than the measurement of a resistor or capacitor, and therefore it is mainly affected by the mechanical qualities rather than electrical qualities of the device (Gad-el-Hak, 2002).

The main advantages of this type of sensing technique over capacitive and piezoresistive techniques are its high accuracy and quality factors. In addition, resonant pressure sensors'

output is frequency; therefore, interfacing to a digital system can be easier. However, the fabrication of such sensors has some technical challenges. More details about the advantages and disadvantages are explained in Gad-el-Hak (2002) and Beeby et al. (2004). Other main issues with resonant pressure sensors are power consumption, which is considerably higher than conventional capacitive and pressure sensors, and temperature sensitivity, which affects the resonator structure material properties and geometry, and therefore changes its natural frequency.

Having vehicle tyre inflation pressure measurement, powered by energy harvesting, as the main thrust in this work, the dynamic environment of the measurement system forces onto the system a wide range of random vibration excitation, mainly from tyre rolling (Tsujiuchi et al., 2005; Brusarosco et al., 2008, 2010), and also from various vehicle components or vibrations transferred through the road originating from other vehicles or nearby machinery and high-impact and centrifugal acceleration components (APOLLO, 2005), which means that the pressure measurement system must not be sensitive to or affected by such conditions. This is particularly a major problem when using a resonant sensing system, as it is not possible to isolate the tyre pressure sensing system from all the vibrations occurring within the tyre structure. For this reason, and due to the high power consumption of resonant pressure sensors, this type of sensor is out of the scope of this research. However, Grossmann (1999) claims that developing a quartz-based resonant passive tyre pressure sensor, in which its natural frequency lies in the megahertz range, is extremely stable. On the other hand, the system is battery free and powered by an RF signal generated by the read-out system, which has to be mounted within a short distance from the resonator and is powered by the vehicle battery and consumes 10 mW of power. It is not clear whether the resonator was tested in a moving vehicle. Hannan et al. (2008) state that this type of tyre pressure sensor has a main disadvantage of low robustness in a harsh environment during vehicle operation.

Preliminary resonant devices are rather complex in terms of design, fabrication, and packaging. In addition, resonant devices contain piezoresistors for strain measurements and piezoelectric materials for vibration excitation, either of which can be used directly in the sensing structure for measuring most mechanical quantities. As such, using the resonant approach in the sensing mechanism is only justified in high-performance sensing applications (Beeby et al., 2004). An example of a commercialized resonant pressure sensor, developed by Druck Limited, is shown in Fig. 6 below (Beeby et al., 2004; Druck, 2012).

In the previous paragraphs, the three most commercially available micro-machined pressure sensors were explained and compared. In the following paragraphs within this section, other pressure sensing techniques are briefly covered (Gad-el-Hak, 2002; Beeby et al., 2004).

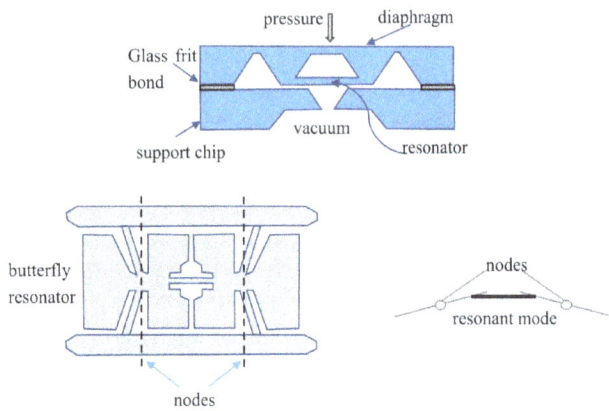

Figure 6. Druck resonant pressure sensor (Beeby et al., 2004).

1.5 Surface acoustic wave

Surface acoustic wave (SAW) pressure sensors are well suited for automotive application because they are designed to be passive, rugged, and extremely small. SAW sensors are different in their sensing mechanisms compared to other resonant sensors in which a mechanical resonating structure is employed. In SAW sensors, a piezoelectric crystal is used as the resonating element. This element resonates acoustic wave resonance by applying an oscillating voltage through the electrodes due to the inverse piezoelectric effect (Cavalloni et al., 2002).

SAW technology was originally used in electronic analogue signal processing in the 1970s as filters in the field of radio transmission in wireless systems (Yurish and Gomes, 2005). However, SAW sensor structure is sensitive to some physical quantities like strain, temperature, humidity, acceleration, etc., and therefore can be applied to measure these quantities (Reindl et al., 1996; Kalinin, 2004; Chin et al., 2010; Hew et al., 2011). Figure 7 shows different strain measurement mechanisms that can be employed when using SAW sensors.

A SAW measurement system is capable of wirelessly monitoring applications and acts as a passive sensor in such a way that the oscillation frequency of the transmission signal passing through the SAW sensor changes in speed or phase with the measured value. Consequently, applying SAW sensors in a tyre pressure and temperature monitoring system has been explored for their advantages in the sense that they have very small size and weight, can pass signals wirelessly, receive the required power wirelessly by the energy of the RF field (and therefore no local battery is needed), and can resist harsh environment conditions (Pohl et al., 1997, 1998; Xiangwen et al., 2004; Zhang et al., 2004; Li et al., 2010). The working principle of SAW sensors is explained in detail in chapter five in Yurish and Gomes (2005). One of the problems associated with a SAW sensing system is its short transmission range; therefore, the transmitting and receiving antennas of the read-out system have to be mounted close to

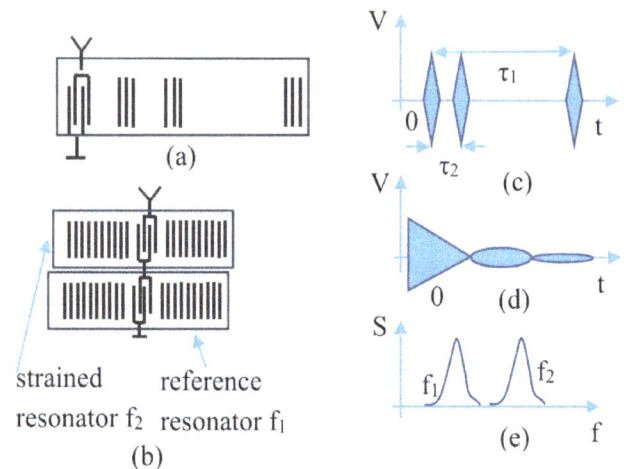

Figure 7. Reflective delay line sensor (**a**), resonant sensor (**b**), impulse responses of the delay line (**c**), the resonant (**d**) sensors, and the spectrum of the resonator response (**e**) (Kalinin 2004).

the SAW sensors, with a maximum distance for detection of 40 cm (Oh et al., 2008). In the TPMS application, for example, cables from the central read-out circuit need to be installed with antennas in their ends nearby each tyre, which increases the overall weight and the installation cost of the system. Although SAW sensors do not need batteries, the system, particularly the read-out part, is not self powered and therefore consumes power from the vehicle battery.

Transense, a company based in the UK, has developed a SAW-based tyre pressure monitoring system (Stack TPMS) which is supposed to be available on the market for motorsport applications[3]. SAW sensors in this system are attached to the tyre's valve and communicate with the read-out circuit antenna at 433 MHz via a 5 mm diameter antenna with an optional length to be specified by the customer. Transense claims that their SAW tyre pressure sensor has 1 m of reading range. The system performance and characteristics are clearly explained by Dixon et al. (2006).

SAW transponders can also be designed in such a way that an external sensor, e.g. a capacitive sensor, can be connected to a reflector within the SAW chip and act as an impedance load on that reflector. Schimetta et al. (2000) present a SAW hybrid tyre pressure sensor in which a capacitive pressure sensor is used for pressure measurement. In this system, the reflectivity of the impedance loaded reflector will be a function of the variable sensor impedance, which is a function of the measured value, e.g. pressure. Such a system has the advantage of a high signal to noise ratio and a larger modulation factor compared to ordinary SAW sensors. However, this type of sensor is more complex to model and detect. The

---

[3]Transense, available at: http://www.transense.co.uk/technologies/temppressure/motorsport-tpms and http://www.stackltd.com/tpms.html

Figure 8. Schematic of a general capacitive sensor.

principle and design of a TPMS based on a hybrid SAW sensor are presented in Schimetta et al. (2000).

In a previous publication (Kubba and Jiang, 2014), the author made a comparison between the above-listed pressure sensing mechanisms and a justification was made for choosing the capacitance sensing technique. Therefore, the following sections focus on capacitive pressure sensors.

2 Transduction mechanism of capacitive pressure sensor

In this section the capacitive transduction is explained in detail. Traditionally, a basic capacitor consists of two parallel plates, or electrodes, of equal area A and at a distance of d_o apart (Fig. 8), and a dielectric material in-between. The mathematical expression of the electrical capacitance of the two plates can be obtained as follows (Beeby et al., 2004):

$$C = \varepsilon_o \varepsilon_r \frac{A}{d_o}, \tag{1}$$

where ε_o is the free space permittivity and equals 8.854×10^{-12} in Farad per metre ($\mathrm{F\,m^{-1}}$), ε_r is the relative permittivity of the dielectric material in between the capacitor electrodes, A is the area of overlap between the electrodes in metres squared ($\mathrm{m^2}$), and d_o is the separation between the electrodes in metres (m).

In order to achieve capacitive transduction, one of the independent variables in Eq. (1) has to vary with a measured quantity. This can be done either by moving the dielectric material or by moving the electrodes linearly with respect to each other. The latter is quite common in capacitive sensors due to its simplicity.

A diaphragm-based capacitive sensor has one of the electrodes acting as a diaphragm that is sensitive to pressure, resulting in a change in the height distribution between the electrodes when pressure is applied, and therefore change in the capacitance of the sensor can be detected. The usual way that capacitive absolute pressure sensors are designed is illustrated in Fig. 9. The lower plate of the capacitive sensor is rigid and stationary, while the upper one is flexible and pressure sensitive. When pressure is applied, the upper plate deforms and the average distance to the stationary plate decreases, causing an increase in capacitance. The space between the two plates is in a vacuum. A schematic diagram of the proposed capacitive pressure sensor is shown in Fig. 10. In this design, the lower and upper plates are stationary. The vacuum cavity is underneath the deformable diaphragm and

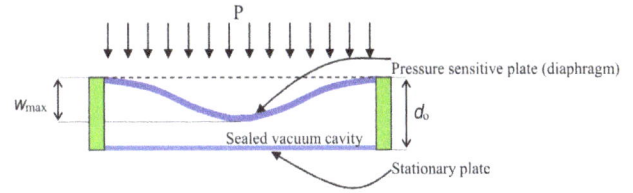

Figure 9. Schematic diagram of a general capacitive pressure sensor.

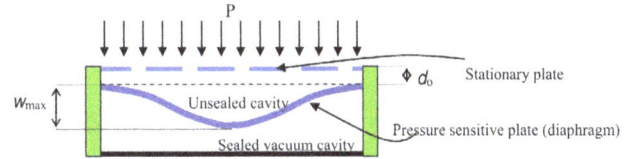

Figure 10. Schematic diagram of the proposed capacitive pressure sensor.

the upper plate is not sealed, allowing air pressure to transfer to the diaphragm.

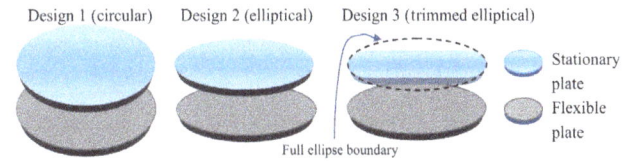

Figure 11. Capacitive sensor's functional plates of the three designs.

Unlike the conventional capacitive pressure sensors, the proposed design operates in such a way that the average distance between the capacitor plates increases when pressure is applied to the deformable diaphragm. The main advantage of this design is the avoidance of having one of the sensor electrodes within the sealed vacuum cavity and to avert the need to transfer the output signal from the inside of the sealed cavity, which can increase the long term leaking rate of the vacuum cavity. However, in the proposed design the sensor sensitivity decreases slightly with the increase in applied pressure, while conventional capacitive pressure sensors operate in the opposite way. For the application of this pressure sensor, higher sensitivity at low pressure is desirable, as in most cases detecting low tyre inflation pressure is targeted.

The following sections analyse the proposed micro-pressure sensor in both mechanical and electrical aspects.

3 Sensor geometry

This section examines three possible designs of the diaphragm-based micro-pressure sensor shown in Fig. 10. It is assumed that sensors in these designs have the same pressure sensitive diaphragm area in either circular or elliptical geometries as shown in Fig. 11. The first two designs have their upper capacitor electrode fully covering the lower pressure sensitive electrode, while the third has the upper plate partially covering the lower capacitor plate. The upper plate in design 3 has two segments of chords parallel to the major axis removed, that is, the area between chords and dashed lines in design 3 in Fig. 11, making up 70 % of the full ellipse upper plate area in design 2. The aspect ratio (major to minor axis ratio) of the elliptical diaphragm in the second and third designs is 2. These designs are then compared in terms of pressure sensitivity, temperature dependence, and stress levels.

In the literature, comparisons between different basic diaphragm geometries with equal areas are presented in Wang and Ko (1999) and Muhammad (2012). The comparisons show that a circular diaphragm has the lowest maximum stress and the highest sensitivity among the basic shapes, that is, square, rectangular and circular diaphragms with constant area. Elliptical diaphragms that have the same area as a circular diaphragm have higher stiffness, and therefore cover a higher pressure detection range. In addition, having higher stiffness means higher dynamic stability when included within a dynamic domain (Swett, 2012). In this section, a comparison between a circular and elliptical diaphragm is made in order to determine which is more suited for the TPMS application.

4 Sensor analysis

4.1 Analytical mechanical analysis

This section focuses on the analytical calculations of the deformable diaphragm within the micro-capacitive pressure sensor for both circular and elliptical diaphragms. The obtained analytical diaphragm deformation expressions, for both the circular and elliptical geometries, can be used to determine the enclosed capacitance between the deformable diaphragm and a stationary plate located on top of it, separated by distance d_o as illustrated in Figs. 10 and 11.

Despite the lack of accuracy associated with most analytical calculations usually resulting from assumptions and approximations, analytical calculations can still offer a rather nimble and useful way of gaining a valuable answer, or of comprehending the physical process behind a particular problem. In the case of designing a pressure sensitive diaphragm, analytical techniques can be used to estimate the deflection response when subjected to uniform pressure. The analytical formulae applied in this section are quoted from "Theory of plates and shells" by Timoshenko (1940). These formulae are based on the following assumptions (Beeby et al., 2004).

- The diaphragm is not curved when unloaded and has a uniform thickness.

- The diaphragm substance is homogenous and isotropic.

- The pressure on the diaphragm is uniformly distributed and perpendicular to the diaphragm plan.

- The applied pressure is within the elastic loading limits and does not cause any plastic deformation.

- The diaphragm thickness to diameter ratio is less than 20 %.

- All deformation occurring in the diaphragm is due to bending only, and the neutral axis of the diaphragm stays at a zero stress level.

The mathematical expression of the deflection of a circular and elliptical plates under uniform pressure with clamped edges can be obtained using the expression given in Timoshenko (1940).

As mentioned earlier, and in order to compare between circular and elliptical diaphragms, thickness and surface area in both geometries are identical.

Hence

$$A_o = A_0, \tag{2}$$

where A_o is the circular diaphragm area in m^2 and A_0 the elliptical diaphragm area in m^2. Then

$$\pi a_o^2 = \pi ab. \tag{3}$$

As such, the radii of both geometries can be expressed as

$$a_o = \sqrt{ab}, \tag{4}$$

where a_o is the diaphragm outer radius in m, and a and b are major and minor axes respectively, both in m.

If

$$a = 2b, \tag{5}$$

Eq. (4) can be re-written as

$$a = \sqrt{2a_o}. \tag{6}$$

The proposed design uses Eq. (6) to relate between the circular and elliptical diaphragm diameters.

In terms of the stresses acting on the diaphragm edge as a result of the diaphragm deflection, the elliptical geometry offers a lower stress level near the major axis edge compared to the circular diaphragm geometry (Timoshenko, 1940). This fact can be exploited when bonding the capacitor terminals; that is, the bonding and isolating substance (SU-8) can be developed over that low stress area. Feng and Farris (2003) show that the process of fabricating SU-8 can change its mechanical properties. In this research, it is assumed that the SU-8 mechanical properties and shear bonding strength are as presented by Pang et al. (2008) and Guerin (2008) respectively. This technique improves the overall sensor robustness and reduces the temperature sensitivity of the transducer. As

Table 1. Circular and elliptical diaphragms' design parameters.

Parameter	Value
a_o	1.25 mm
a	1.767767 mm
b	0.883883 mm
h	100 µm
$E_{\text{Si[plate bending]}}{}^{*}$	170 GPa
$E_{\text{SU-8}}$	4.4 GPa
$\nu_{\text{Si[plate bending]}}{}^{*}$	0.064
$\nu_{\text{SU-8}}$	0.22
α_{Si}	2.6 (ppm °C^{-1})
$\alpha_{\text{SU-8}}$	52 (ppm °C^{-1})
$(\varepsilon_{\text{r}})_{\text{SU-8}}$	3
* for analytical calculations only.	

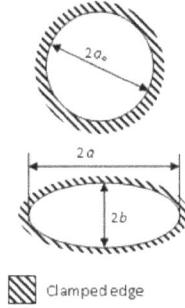

these stresses are rather difficult to calculate manually, particularly the calculations of the thermal stresses' effect, numerical simulations with COMSOL software were used. The numerical results are presented in Sect. 4.3.

The dimensions and bi-dimensional silicon elastic properties for plate bending (Hopcroft et al., 2010) of the circular and elliptical diaphragms are shown in Table 1. The radii of the elliptical diaphragm are calculated using Eqs. (5) and (6), where α_{Si} and $\alpha_{\text{SU}-8}$ are the silicon and SU-8 coefficients of thermal expansion (CTEs) respectively. Provided that silicon is an anisotropic material, and that a silicon wafer orientation of (100) is the most common case of MEMS fabrication, silicon mechanical properties presented in Hopcroft et al. (2010), which are also proven experimentally by Boyd and Uttamchandani (2012), were applied in COMSOL simulations. However, in the analytical calculations, values were used assuming plate bending for <110>in (100) silicon to simplify initial design calculations.

4.2 Analytical capacitance modelling and calculations

By considering the sensor schematic illustrated in Fig. 10, the capacitance enclosed between the stationary plate and a clamped circular deformable diaphragm subjected to a uniformly distributed pressure can be obtained:

$$C(p) = 8\pi\varepsilon_o\varepsilon_{\text{r}}\sqrt{\frac{D}{d_o p}}\tan^{-1}\left(-\frac{a_o^2}{8}\sqrt{\frac{p}{d_o D}}\right), \quad (7)$$

where ε_o is the free space permittivity and equals 8.854×10^{-12} in Farad per metre (F m^{-1}), ε_{r} is the relative permittivity of the dielectric material in between the capacitor electrodes, D is the flexural rigidity in N m^2, and p is the pressure in Pa.

In the case of the elliptical diaphragm, the capacitance expression can be obtained by defining the diaphragm radius

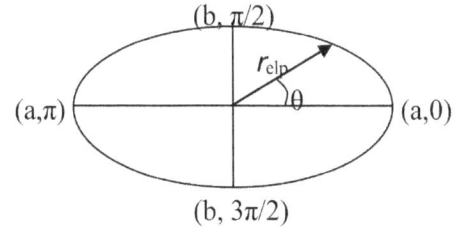

Figure 12. Ellipse border representation using the polar coordinate system.

using the polar coordinate system with the origin at the centre of the ellipse (Fig. 12), and then by integrating over the whole oval area using double integration. This results in a rather complicated mathematical expression:

$$C(p) = \varepsilon_o\varepsilon_{\text{r}}\int_0^{2\pi}\int_0^{r_{\text{elp}}}\frac{r\,\mathrm{d}r\,\mathrm{d}\theta}{d_o + w_0\left(1 - \frac{(r\cos\theta)^2}{a^2} - \frac{(r\sin\theta)^2}{b^2}\right)^2}, \quad (8)$$

where r and θ are the polar coordinates at which diaphragm deflection is calculated in m and rad respectively, r_{elp} is the ellipse radius at a certain angular position θ in m, and w_0 is the central deflection of the diaphragm.

Wolfram Mathematics software[4] was used to solve the first integral of Eq. (8), and it can be seen that a numerical integration could be a better choice to achieve a useful result (see Appendix A). Attempts to solve Eq. (8) with MATLAB tools were not successful for the whole pressure range, and therefore FEA tools (COMSOL) were used to evaluate the integral over a range of pressures. In order to find out COMSOL's accuracy, the circular diaphragm results from both analytical and FEA analyses are compared.

Due to the symmetrical nature of the three designs studied in this research, COMSOL simulation was undertaken for a quarter of the sensors' geometries to accelerate simulation time and to improve simulation accuracy. Three-dimensional views of a quarter of the three diaphragm designs' capacitive pressure sensor configurations are shown in Figs. 13, 14, and 15 respectively.

In order to evaluate the total capacitance of the sensor shown in Fig. 13, the parasitic capacitance has to be calculated, that is, the capacitance enclosed between the upper plate and the sealing plate excluding the deformable diaphragm. Details of the analytical calculations of the parasitic capacitance, 19.59 pF, are presented in Appendix A. The elliptical diaphragm design has the same area as the circular design but with an aspect ratio of 2 : 1 (Fig. 14); therefore, both designs share the same parasitic capacitance value. The capacitance–pressure curve of the circular diaphragm pressure sensor presented in Fig. 13 is shown in Fig. 16. These data were calculated both analytically using MATLAB and

---

[4]http://integrals.wolfram.com

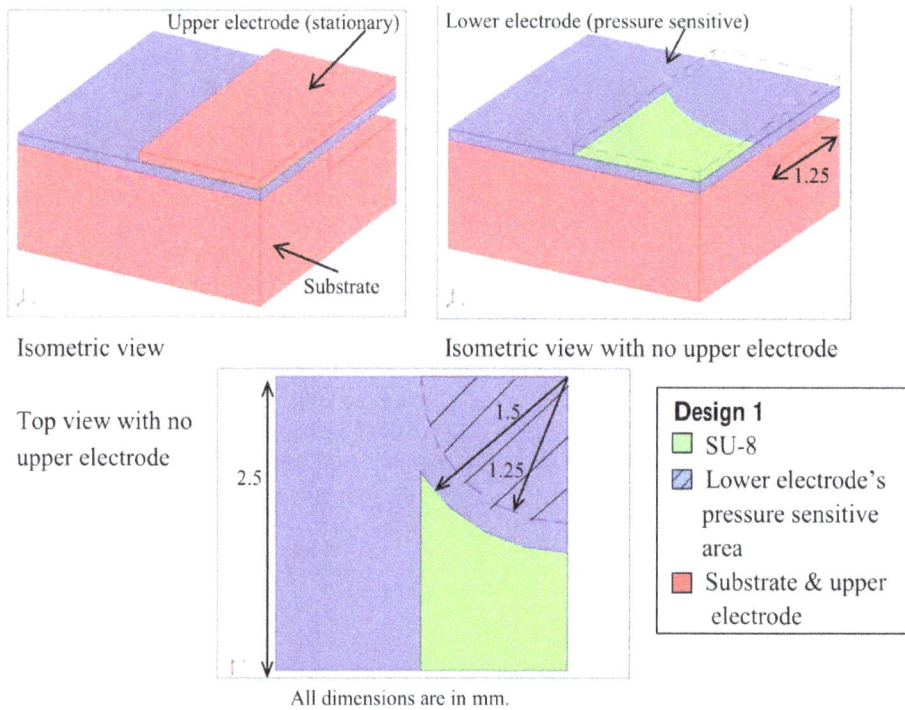

Figure 13. A three-dimensional view of a quarter of the circular diaphragm pressure sensor.

numerically using COMSOL – to be illustrated in the following section. Having the same area in both the elliptical and the circular diaphragm pressure sensor designs, the initial – 0 Pa – capacitance values are identical and equal to 41.32 pF. In the elliptical diaphragm design the MATLAB code only converges at 1.5 MPa applied pressure and results in a capacitance value of 36.63 pF. This shows that the elliptical diaphragm design has less change in the capacitance value within the same pressure range, that is, less sensitivity and high sensor bias to the output signal range ratio. To increase the sensor's output signal range to bias ratio, the area of the sensor upper electrode can be reduced. That is the case in the third design, shown in Fig. 15, where only 70 % of the elliptical diaphragm is the functional capacitor area. Thermal response and stress levels in all the designs have to be investigated. This can be a rather complicated task for manual calculation, and therefore numerical analysis is employed as the second best alternative. The following section presents numerically simulated results, using COMSOL software, of the mechanical and electrical variables for all three designs' parameters.

4.3 Numerical analysis

In this section, initially, the first design (Fig. 13) simulation results are compared with the analytical results to check the reliability of the FEA results. Then, the second and third designs are simulated using FEA, and all three designs are compared in order to select the most suitable one for the TPMS

application. Having the three designs symmetrically around $x - z$ and $y - z$ planes, there are two main quantities between the first design simulation results and the analytical results: the parasitic capacitance and the diaphragm deflection.

The parasitic capacitance was found analytically to be equal to 19.59 and 17.46 pF using FEA. This is due to the approximation made by the software by replacing arcs with sets of straight lines giving slightly smaller areas for both the diaphragm and SU-8 layer. This difference can be treated as bias, or offset, error, and therefore it will not affect the sensor sensitivity calculations.

Regarding the diaphragm deflection, it is worth mentioning that the analytical diaphragm deformation analysis for the first design is based on a clamped-edge mounting condition, which might not be the exact case for how the diaphragm is fixed onto the sensor substrate. Having the diaphragm bonded from one side, its bending stiffness under uniform pressure is less than the clamped-edge mounting stiffness. For this reason, FEA results might give a higher diaphragm deflection under a certain pressure than that obtained using the analytical analysis described in Sect. 4.1 (see Fig. 16). It can be seen that a slightly higher central deflection (w_o) is obtained compared to the calculated one using clamped-edge mounting, in which higher flexural rigidity exists.

The temperature sensitivity of the sensor-resulting capacitance for the three designs is shown in Fig. 19.

The reasonable agreement between the analytical results and the numerical results obtained for design 1 supports the use of FEA simulations for designs 2 and 3. Diaphragm de-

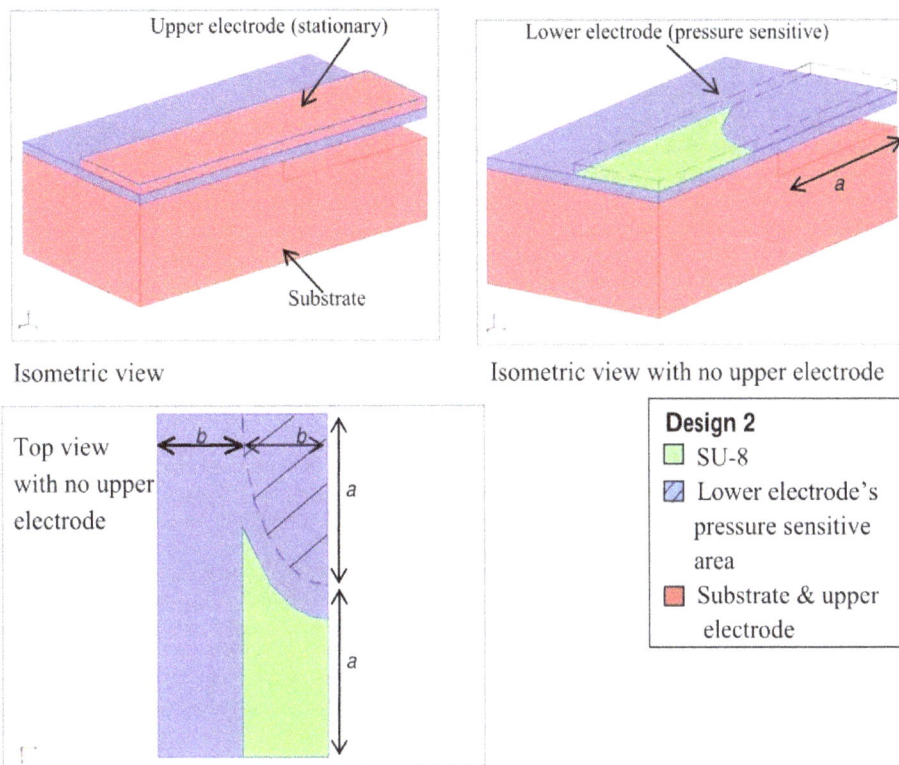

Figure 14. A three-dimensional view of a quarter of the elliptical diaphragm pressure sensor.

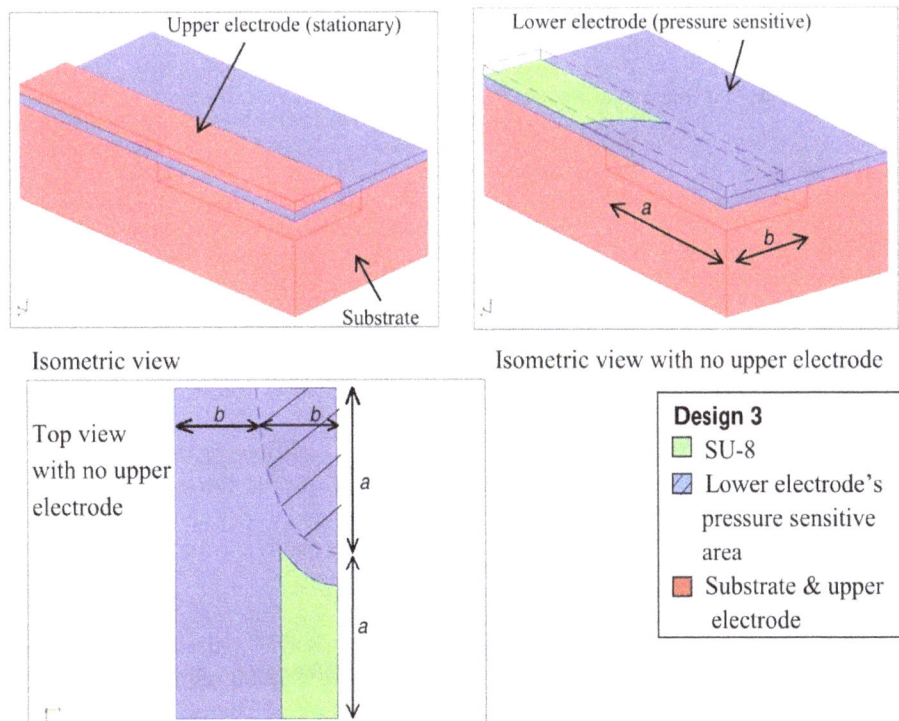

Figure 15. A three-dimensional view of a quarter of the narrowed elliptical diaphragm pressure sensor.

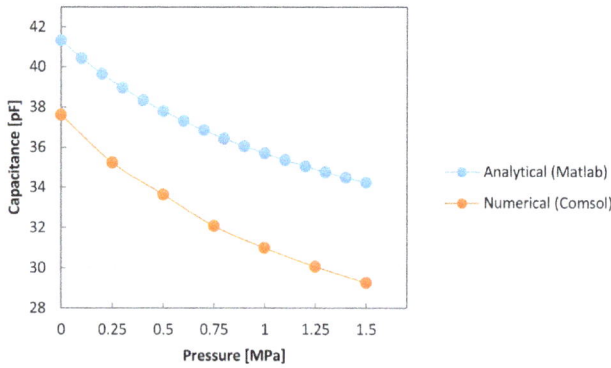

Figure 16. Capacitance–pressure curve of the circular diaphragm in design 1 using FEA.

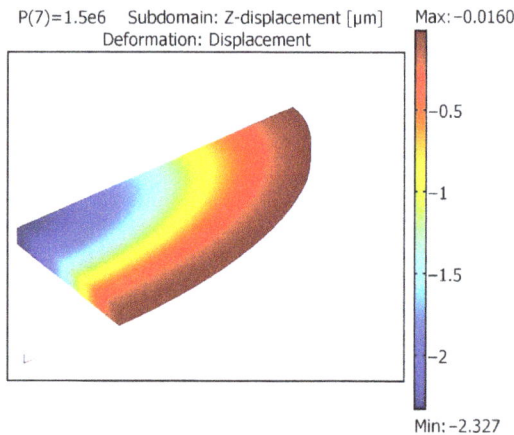

Figure 17. A quarter of the elliptical diaphragm displacement in design 2 using FEA.

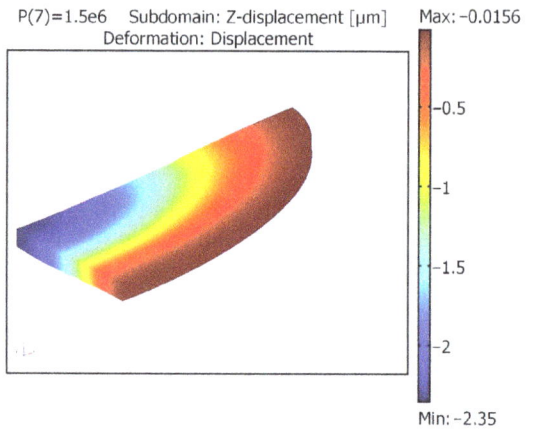

Figure 18. A quarter of the elliptical diaphragm displacement in design 3 using FEA.

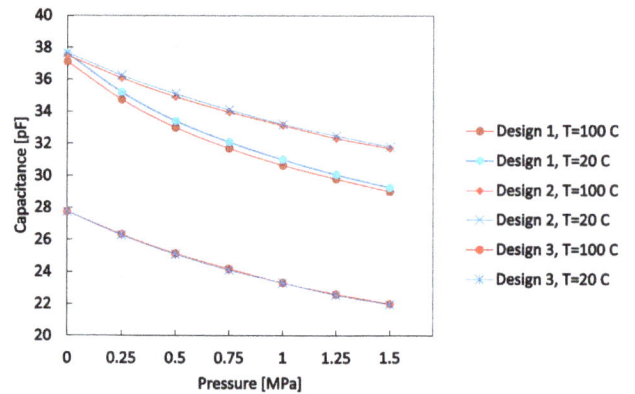

Figure 19. Temperature effect on the output signal in design 3 using FEA.

formation and the sensor output signal at 20 and 100 °C for designs 1 and 2 are shown in Figs. 17–26. The misalignment between the analytical and numerical results for design 1 is attributed to the assumption of isotropic behaviour in the diaphragm deflection calculations, whereas COMSOL applies the actual anisotropic mechanical properties of a (100) plane oriented silicon wafer as illustrated in Hopcroft et al. (2010).

By comparing the output signal of design 2 with design 3, it can be seen that design 3 offers lower bias and lower temperature sensitivity on the output capacitance. This is attributed to the smaller parasitic capacitance caused by the SU-8 layer. In addition, the simulation results listed in Table 2 show that design 3 has the lowest stress levels among all three designs, particularly the bonding shear stress between Si and SU-8, which should not exceed 17.15 MPa (Pang ct al., 2008). Bonding surface shear stress in the presented designs results from the difference in thermal expansion between SU-8 and Si. It is determined within the surface using COMSOL after increasing the temperature by 80 °C.

5 MEMS capacitive pressure sensor micro-fabrication

This section presents suggested micro-fabrication processes for making the designed MEMS pressure sensor, that is, design 3. However, due to the limited budget of this research project and the lack of essential micro-fabrication equipment needed for this design, there were no micro-fabrication processes involved in the study. Therefore, the following process flow gives only the general, and not detailed, micro-fabrication steps involved in making the designed micro-pressure sensor, as it is not the objective of this research.

Silicon is a dominant material in the semi-inductor industry due to its availability, low costs, suitability to the requirements of a wide range of mechanical and electrical properties, readiness of mature batch fabrication techniques, and the potential for integration with electronic circuitry (Petersen, 1982; Beeby et al., 2004; Hsu, 2008). For these reasons, silicon is employed in the proposed micro-pressure sensor in the diaphragm, the conductive plates (electrodes) of the capacitor, and the vacuum chamber of the sensor.

Table 2. A summarized comparison of the three capacitive pressure sensor designs.

Item	Property	Design 1 $T = 20\,°C$	Design 1 $T = 100\,°C$	Design 2 $T = 20\,°C$	Design 2 $T = 100\,°C$	Design 3 $T = 20\,°C$	Design 3 $T = 100\,°C$
SU-8	σ_{max} (Von Mises) (MPa)	1.48	58	2.4	44.44	0.12	40.17
	Bonding surface τ_{max} (MPa)	0.05	9.85	0.11	2.35	0.002	0.827
Si	σ_{max} (Von Mises) (MPa)	171.17	191	111	120.23	113.58	119.86
	Average sensitivity (pF bar^{-1})[a]	0.558		0.39		0.386	
	$\Delta C / C$	−0.22		−0.155		−0.208	

[a] Sensitivity was averaged by considering the slope of the best fitting straight line, i.e. by linear regression.

The photolithography processes required in the proposed design are mainly etching the vacuum cavity in the lower part of the sensor structure and etching the upper stationary plate in order to allow air pressure to pass to the pressure sensitive diaphragm and to determine the overlapping surface area between the two capacitor plates.

In order to construct an absolute capacitive pressure sensor, bonding between the sensor layers is essential. The vacuum chamber has to be sealed hermitically and be able to withstand temperature variation. This can be done by first sealing the vacuum chamber and then isolating the capacitive electrodes. From the literature, one reliable bonding technique is anodic bonding, which is the most suited bonding method for sealing the vacuum chamber in the proposed micro-pressure sensor. Silicon-to-silicon anodic bonding can be obtained by applying sputtered borosilicate glass as a thin film layer on either of the silicon surfaces to be bonded, as demonstrated in Chapter 9 in Halbo and Ohlckers (1995). In the proposed design, a cross section is shown in Fig. 20, the vacuum chamber is sealed hermetically using anodic bonding, and the electrodes are bonded, and isolated, using SU-8 photoresist. SU-8 is low cost, easy to process, and a good isolation material for capacitive pressure sensor application, and it can work as an adhesive bonder too (Chang and Allen, 2004; Pang et al., 2008).

The main micro-fabrication steps include wet etching for Si, Si to Si anodic bonding, and Si to Si adhesive bonding using SU-8 (see Fig. 20).

6   Sensor read-out circuitry

This section presents a read-out and low power short-range transmission circuit suited to capacitive sensors. This circuit is based on a negative resistance lambda diode RF oscillator which contains a tuned LC circuit for both measurement and transmission purposes.

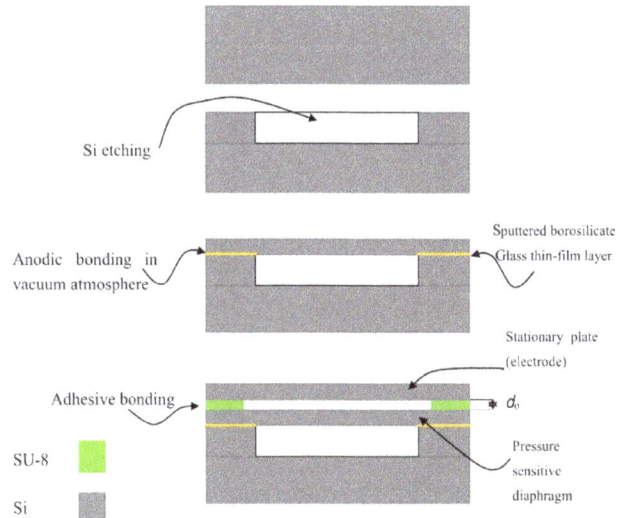

Figure 20. A suggested micro-machining process flow of the designed MEMS pressure sensor.

In the literature, a tunnel diode oscillator was employed by Suster et al. (2002) for a high-temperature wireless pressure sensor node. The simplicity of this system offers very low transmission power (< 1 mW), which is very attractive for energy harvesting systems. Suster et al. (2003) demonstrated an optically powered wireless pressure sensor node based on a tunnel diode oscillator using a photo-diode as the energy harvesting unit. As tunnel diodes were not commercially available, equivalent combinations of JFET transistors acting as a negative resistance, so-called lambda diodes, were employed. When a pair of JFETs (N-channel and P-channel) is connected in a certain way, they can serve as a negative resistance within a negative resistance oscillator (Chua et al., 1985). A schematic diagram of the lambda diode oscillator is shown in Fig. 21a. In this circuit, the LC circuit oscillates at

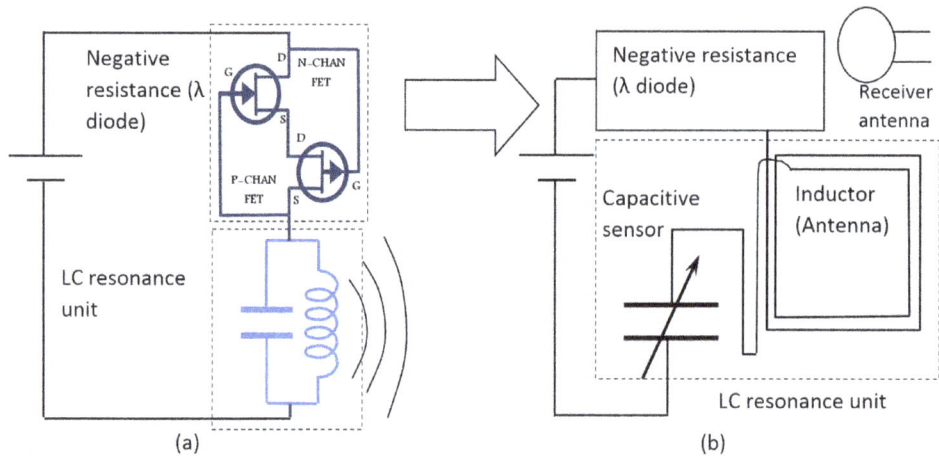

Figure 21. (**a**) Negative resistance oscillator schematic; (**b**) capacitive sensor read-out and transmission circuit. Electronic circuit design published by Butler (1997).

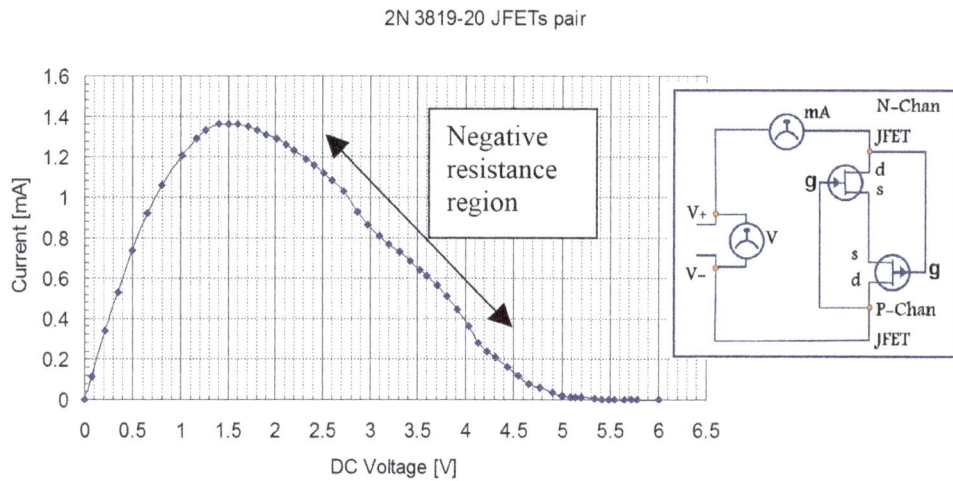

Figure 22. The 2N3819 and 2N3820 JFET pair lambda diode characteristic. Electronic circuit design published by Butler (1997).

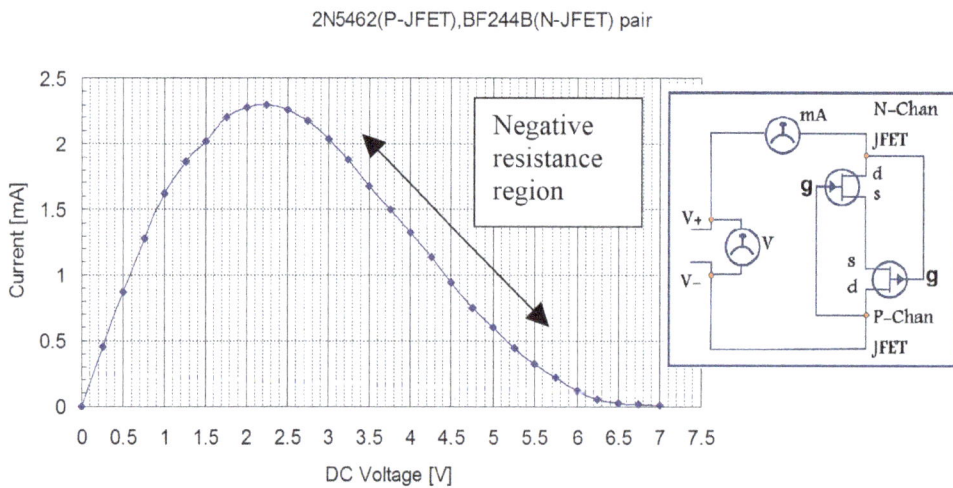

Figure 23. The 2N5462 and BF244B JFET pair lambda diode characteristic. Electronic circuit design published by Butler (1997).

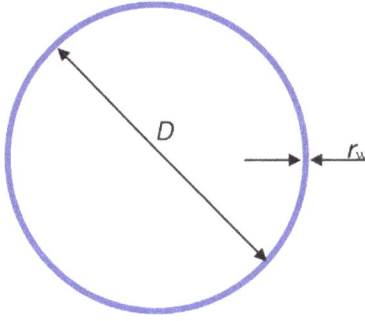

Figure 24. Loop antenna schematic.

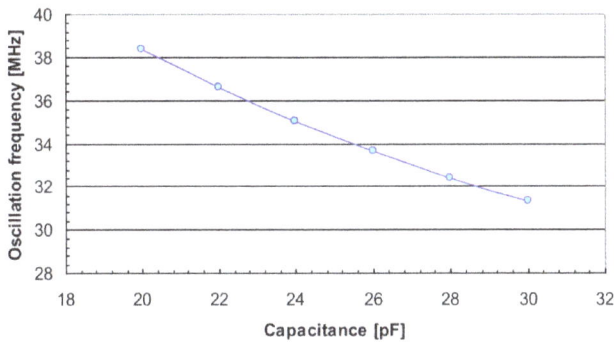

Figure 25. Predicted first harmonic oscillation frequency of the lambda diode oscillator when its tuning circuit contains the design 3 capacitive pressure sensor and the loop antenna of Fig. 24.

Figure 26. Lambda diode oscillator transmitted and received signals conducted using a digital storage oscilloscope: (**a**) transmitted signal; (**b**) received signal.

resonance frequency and its harmonics (multiples), as shown in the equation below:

$$f = n\frac{1}{2\pi\sqrt{LC}}; n = 1, 2, 3, \ldots \infty. \tag{9}$$

When the applied voltage is within the negative resistance regain of the contained lambda diode, which is in parallel with the LC circuit, oscillation frequency might not however follow Eq. (9) exactly. Two JFETs pairs were tested experimentally and their voltage-current characteristics are shown in Figs. 22 and 23.

A variable capacitor may act as the capacitor in the LC circuit as shown in Fig. 21b; the tuning frequency will vary according to the capacitance values and the capacitance value can be found by measuring the oscillator resonance frequency. The variable capacitor can be a capacitive sensor, and therefore the oscillation frequency can reflect the measured pressure value. The second part of the LC tuning circuit is the inductor, which in this design acts as the antenna of the oscillator, or the transmitter, and an external receiver with an identical antenna can detect the oscillation frequency.

The inductance of a circular loop antenna contains N identical loops, shown in Fig. 24, and can be approximated using the following formula (Paul, 2010):

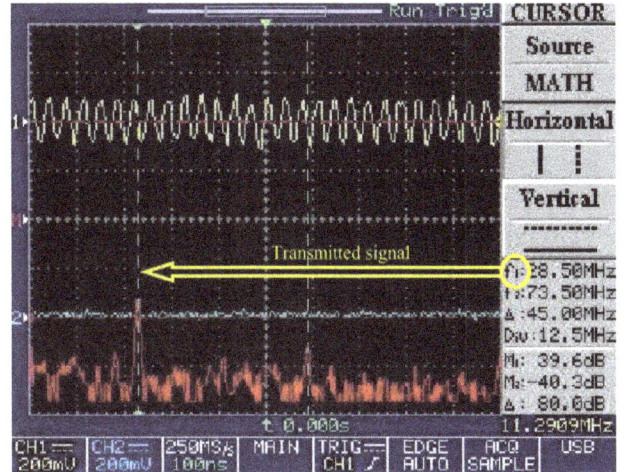

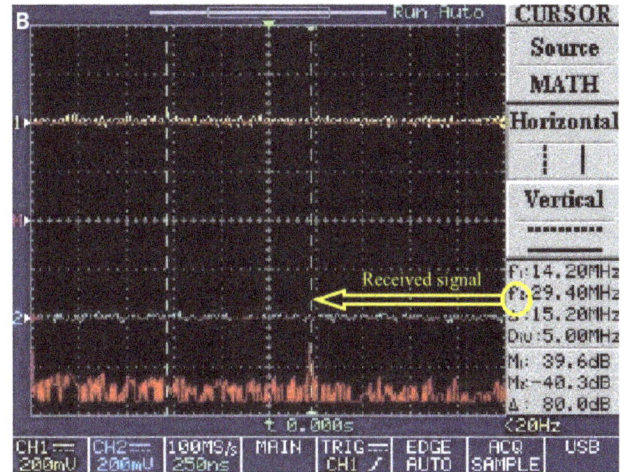

$$L_{N\,\text{loops}} \cong N^2 \frac{\mu_r \mu_0 D}{2}\left(\ln\frac{4D}{r_w} - 2\right), \tag{10}$$

where μ_r is the relative permeability, in this case essentially for air, and equals 1, μ_0 is the permeability of the free space and equals $4\pi \times 10^{-7}$ (H m^{-1}), D is the loop diameter in m, and r_w is the wire diameter in m.

Given that N equals 2, r_w equals 0.6 mm and D equals 80 mm, using Eq. (10), the inductance of the loop antenna will be approximately 860 (nH).

If the simulated capacitive pressure sensor (design 3) is substituted as the variable capacitor in the tuning circuit of the lambda diode oscillator shown in Fig. 21a, and using the estimated inductance of the loop antenna, that is, 860 (nH), the first harmonic oscillation frequency from Eq. (9) of the oscillator will be obtained as shown in Fig. 25. This graph shows an average sensitivity of 1.4 MHz pF^{-1}, meaning a sensitivity of 0.543 MHz bar^{-1} for the simulated (design 3) MEMS capacitive pressure sensor. It is worth mentioning that the parasitic capacitance of the loop antenna was neglected.

Figure 26 shows the lambda diode oscillator output signal as measured using a digital storage oscilloscope (DSO), manufactured by TENMA model no. 72-7240. The tuning circuit contains a trimmer capacitor (approximate range 10–50 pF) to simulate the variable capacitance of the MEMS pressure sensor and the loop antenna shown in Fig. 24. The conditions at which these results were obtained were that both the transmitting and receiving antennas were identical (0.6 mm wire diameter, 80 mm coil diameter, two turns) and mounted coaxially and separated by 110 mm. The receiver antenna was connected directly to the DSO probes with no amplification or filtering circuits.

Despite the approximate value of the trimmer capacitor capacitance range, it was observed that the oscillation frequency changes noticeably, approximately by 10 MHz over the trimmer capacitor capacitance range, and can be employed for measuring the variable capacitor value and ultimately the pressure value when a capacitive pressure transducer is employed. It was also noticed that there is a frequency difference of approximately 900 kHz between the transmitted and received signals, which can be treated as an offset in the very basic receiver unit.

7 Conclusions

In this article, three different micro-capacitive pressure sensors are compared and a read-out and transmission circuit for a capacitive transducer are presented.

The comparison between the three simulated diaphragm-based pressure transducers shows that although a circular shaped diaphragm has higher pressure sensitivity than an elliptical one, an elliptical shaped diaphragm can offer higher thermal stability and less bonding stresses compared to a circular diaphragm, which can be a useful property for a pressure sensor, particularly when employed in a harsh environment, e.g. tyre pressure monitoring.

The presented read-out circuit is selected for its simplicity and its extremely low power consumption is suitable as part of a wireless sensor node powered by energy harvesting.

In a separate article (Kubba and Jiang, 2013), the author developed a vibration-based energy harvester which was eventually used to power the presented lambda diode read-out and transmission unit.

Appendix A: Analytical integration formulae

The following integrations were solved with the aid of an open-access online integral solver[5].

$$\int \frac{e\,x}{d+w\left(1-\frac{(x\cos(t))^2}{a^2}-\frac{(x\sin(t))^2}{b^2}\right)^2}\,dx = \frac{\left(e\,a^2\,b^2\,\tan^{-1}\left(\frac{\left(2\,a^2\,b^2\,\sqrt{d}\left((b^2-a^2)\cos 2t+a^2+b^2\right)\right)}{\sqrt{w\left((b^2-a^2)\cos 2t+a^2+b^2\right)^2\left(x^2(a^2-b^2)\cos 2t+a^2(2b^2-x^2)-b^2\,x^2\right)}}\right)\right)}{\left(\sqrt{d}\,\sqrt{w\left((b^2-a^2)\cos 2t+a^2+b^2\right)^2}\right)}$$

where e is the permittivity of free space, $x=r$, d is the capacitance initial air gap, $w=w$ max, $t=$ theta, and a and b are the major and minor radii respectively.

Appendix B: Calculations for the parasitic capacitance and its surface areas

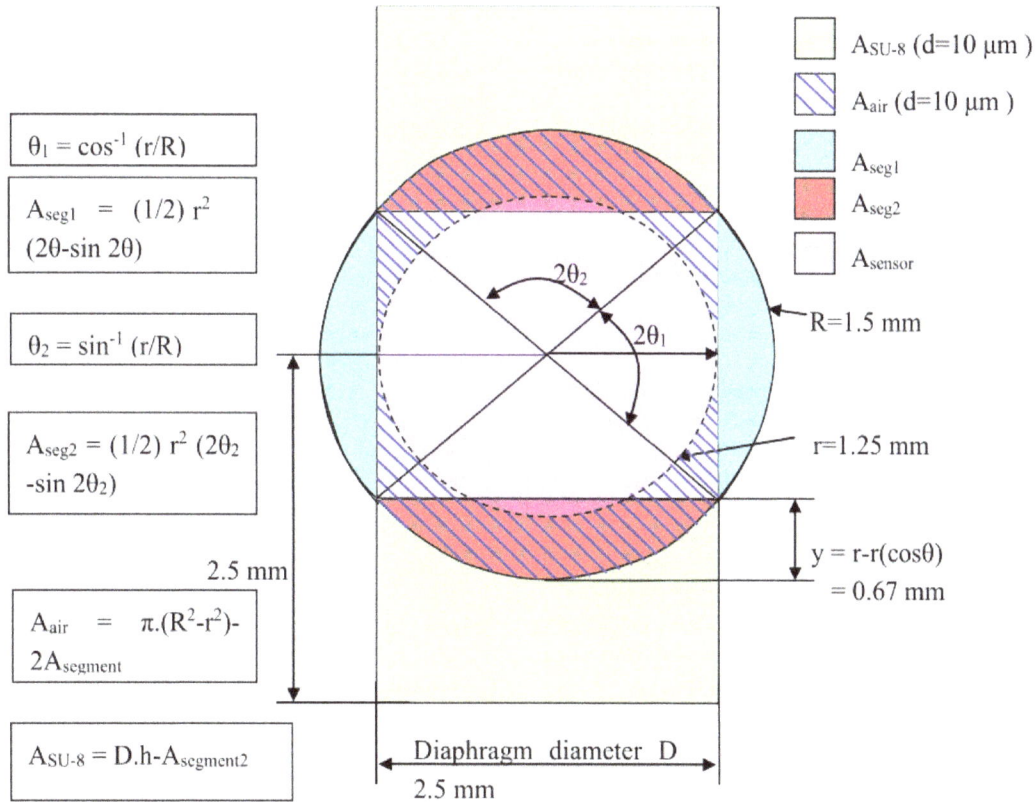

Figure B1.

$$C_p = C_{\text{SU-8}}+C_{\text{air}}+ = \varepsilon_{\text{freespace}}\varepsilon_{\text{SU-8}}\,(A_{\text{SU-8}}/10\,\mu m)+\varepsilon_{\text{air}}\,(A_{\text{air}}/10\,\mu m) = 19.59\,\text{pF}$$

$$C_{\text{sensor}} = \varepsilon_{\text{air}}\,(A_{\text{air}}/2\,\mu m) = 21.7\,\text{pF}$$

$$C_o = C_{\text{sensor}} + C_p = 41.3\,\text{pF}$$

---

[5]Wolfram, available at: http://integrals.wolfram.com/index.jsp,

Acknowledgements. The author would like to show his appreciation for all research and staff members at Mechanical Engineering for their valuable assistance and guidance towards accomplishing this project.

References

A Measurement Specialties company Sensoric, Miniature SMD pressure sensor, Series Datasheet, available at: http://www.nickbelsondesign.co.uk/MS5401-AM.pdf (last access: 18 April 2012), 2009.

APOLLO: Intelligent Tyre Systems – State of the Art and Potential Technologies, IST-2001-34372, Deliverable D7, 2003.

APOLLO: Final Report, Intelligent Tyre for Accident–Free Traffic, Technical Research Centre of Finland (VTT), IST-2001–3437, Deliverable 22/23, 2005.

Bao, M. H.: Micro mechanical transducers: pressure sensors, accelerometers, and gyroscopes, Elsevier, Amsterdam, the Netherlands, 2000.

Barlian, A. A., Park, W. T., Mallon, J. R., Rastegar, A. J., and Pruitt, B. L.: Review: Semiconductor Piezoresistance for Microsystems, Proceedings of the IEEE, 97, 513–552, 2009.

Beeby, S., Ensell, G., Kraft, M., and White, N.: MEMS Mechanical Sensors, Artech House Publishers, Norwood, MA, USA, 2004.

Bever, T., Kandler, M., and Valldorf, J.: Solutions for tire pressure monitoring systems, 7th Int. Conf. Adv. Microsyst. Automot. Appl., 22–23 May 2003, Teltow/Berlin, Springer, Berlin, Germany, 2003.

Boyd, E. J. and Uttamchandani, D.: Measurement of the Anisotropy of Young's Modulus in Single-Crystal Silicon, J. Microelectromech. S., 21, 243–249, 2012.

Bracke, W., Puers, R., and Hoof, C. V.: Ultra Low Power Capacitive Sensor Interfaces, Springer, Dordrecht, the Netherlands, 2007.

Brusarosco, M., Cigada, A., and Manzoni, S.: Experimental investigation of tyre dynamics by means of MEMS accelerometers fixed on the liner, Vehicle Syst. Dynam., 46, 1013–1028, 2008.

Brusarosco, M., Cigada, A., and Manzoni, S.: Measurement and analysis of tyre and tread block dynamics due to contact phenomena, Vehicle Syst. Dynam., 49, 855–869, 2010.

Butler, L.(VK5BR) – A Dip Meter Using the Lambda Negative Resistance Circuit – Amateur Radio, January 1997, available at: http://users.tpg.com.au/users/ldbutler/NegResDipMeter.htm (last access: 30 December 2015), 1997.

Cavalloni, C., Berg, J., Krueger, S., and Gessner, W.: Overview: Principles and Technologies for Pressure Sensors for Automotive Applications, in: Advanced Microsystems for Automotive Applications Yearbook 2002, edited by: Krueger, S. and Gessener, W., Springer, Berlin-Heidelberg, Germany, 232–276, 2002.

Chang, S.-P. and Allen, M. G.: Demonstration for integrating capacitive pressure sensors with read-out circuitry on stainless steel substrate, Sensor. Actuat. A-Phys., 116, 195–204, 2004.

Chien-Hung, W., Zorman, C. A., and Mehregany, M.: Fabrication and testing of bulk micromachined silicon carbide piezoresistive pressure sensors for high temperature applications, IEEE Sens. J., 6, 316–324, 2006.

Chin, T.-L., Irving, P. Z., Oppenheim, J., and Greve, D. W.: Surface acoustic wave devices for wireless strain measurement, Sensors and Smart Structures Technologies for Civil, Mechanical, and Aerospace Systems 2010, Masayoshi Tomizuka, SPIE, 2010.

Chua, L., Juebang, Y., and Youying, Y.: Bipolar – JFET – MOSFET negative resistance devices, IEEE T. Circuits Syst., 32, 46–61, 1985.

Cullinan, M. A., Panas, R. M., DiBiasio, C. M., and Culpepper, M. L.: Scaling electromechanical sensors down to the nanoscale, Sensor. Actuat. A-Phys., 187, 162–173, 2012.

Dixon, B., Kalinin, V., Beckley, J., and Lohr, R.: A Second Generation In-Car Tire Pressure Monitoring System Based on Wireless Passive SAW Sensors, International Frequency Control Symposium and Exposition, June 2006, Miami, FL, USA, 374–380, 2006.

Druck: Resonant Pressure Transducer, RPT Series Datasheet, available at: http://www.druck.com/, last access: 19 January 2012.

Eaton, W. P. and Smith, J. H.: Micromachined pressure sensors: review and recent developments, Smart Mater. Struct., 6, 530–539, 1997.

Eddy, D. S. and Sparks, D. R.: Application of MEMS technology in automotive sensors and actuators, Proceedings of the IEEE, 86, 1747–1755, 1998.

Feng, R. and Farris, R. J.: Influence of processing conditions on the thermal and mechanical properties of SU8 negative photoresist coatings, J. Micromech. Microeng., 13, 80–88, 2003.

Fonseca, M. A.: Polymer/ceramic wireless mems pressure sensors for harsh environments: high temperature and biomedical applications, PhD, Georgia Institute of Technology, Atlanta, GA, USA, 2007.

Fonseca, M. A., English, J. M., von Arx, M., and Allen, M. G.: Wireless micromachined ceramic pressure sensor for high-temperature applications, J. Microelectromech. S., 11, 337–343, 2002.

Fragiacomo, G.: Micromachined capacitive pressure sensor with signal conditioning electronics, PhD, Technical University of Denmark, Copenhagen, Denmark, 2012.

Frank, R.: Understanding smart sensors, Artech House, Norwood, MA, USA, 2000.

Gad-el-Hak, M.: The MEMS handbook, CRC Press LLC, FL, USA, 2002.

Greenwood, J. C.: Etched silicon vibrating senso, J. Phys. E Sci. Instrum., 17, 650–652, 1984.

Grossmann, R.: Wireless measurement of tire pressure with passive quartz sensors, P. Soc. Photo-Opt. Ins., 3670, 214–222, 1999.

Guerin, L. J.: The SU8 homepage, available at: http://www.oocities.org/guerinlj/, last access: 10 July 2008.

Halbo, L. and Ohlckers, P.: Electronic Components, Packaging and Production, University of Oslo, Oslo, Norway, 1995.

Hannan, M. A., Hussain, A., Mohamed, A., and Samad, S. A.: TPMS Data Analysis for Enhancing Intelligent Vehicle Performance, J. Appl. Sci., 8, 1926–1938,

Hew, Y., Deshmukh, S., and Huang, H.: A wireless strain sensor consumes less than 10 mW, Smart Mater. Struct., 20, 10, 2011.

Hopcroft, M. A., Nix, W. D., and Kenny, T. W.: What is the Young's Modulus of Silicon?, J. Microelectromech. S, 19, 229–238, 2010.

Hsu, T.-R.: MEMS & Microsystems: Design, Manufacture, and Nanoscale Engineering, John Wiley & Sons Inc., Hoboken, NJ, USA, 2008.

Kalinin, V.: Passive wireless strain and temperature sensors based on SAW devices, Radio and Wireless Conference, IEEE, 2004.

Kolle, C., Scherr, W., Hammerschmidt, D., Pichler, G., Motz, M., Schaffer, B., Forster, B., and Ausserlechner, U.: Ultra low-power monolithically integrated, capacitive pressure sensor for tire pressure monitoring, Sensors, 2004, Proceedings of IEEE, 2004.

Kubba, A. and Jiang, K.: Efficiency Enhancement of a Cantilever-Based Vibration Energy Harvester, Sensors, 14, 188–211, 2013.

Kubba, A. and Jiang, K.: A Comprehensive Study on Technologies of Tyre Monitoring Systems and Possible Energy Solutions, Sensors, 14, 10306–10345, 2014.

Lee, Y. S. and Wise, K. D.: A batch-fabricated silicon capacitive pressure transducer with low temperature sensitivity, IEEE T. Electron. Dev., 29, 42–48, 1982.

Li, T., Hu, H., Xu, G., Zhu, K., and Fang, L.: Pressure and Temperature Microsensor Based on Surface Acoustic Wave in TPMS, Acoustic Waves, D. Dissanayake, InTech, 341–357, 2010.

Li Sainan, L. T., Wei, W., Yingping, H., Tingli, Z., and Jijun, X.: A novel SOI pressure sensor for high temperature application, Journal of Semiconductors, 36, 14014–14018, 2015.

MarketsandMarkets: Pressure Sensor Market – Global Forecast to 2020, Vancouver, WA, USA, 2015.

Muhammad, H. B.: Development of a bio-inspired MEMS based tactile sensor array for an artificial finger, PhD, University of Birmingham, Birmingham, UK, 2012.

Nwagboso, C.: Automotive Sensory Systems, Springer, Dordrecht, the Netherlands, 2012.

Oh, J.-G., Choi, B., and Lee, S.-Y.: SAW based passive sensor with passive signal conditioning using MEMS A/D converter, Sens. Actuat. A-Phys., 141, 631–639, 2008.

Pang, C., Zhao, Z., Du, L., and Fang, Z.: Adhesive bonding with SU-8 in a vacuum for capacitive pressure sensors, Sens. Actuat. A-Phys., 147, 672–676, 2008.

Paul, C. R.: Inductance: Loop and Partial, John Wiley & Sons Inc., Hoboken, NJ, USA, 2010.

Petersen, K. E.: Silicon as a mechanical material, Proceedings of the IEEE, 70, 420–457, 1982.

Pohl, A., Ostermayer, G., Reindl, L., and Seifert, F.: Monitoring the tire pressure at cars using passive SAW sensors, Ultrasonics Symposium, 5–8 October 1997, Toronto, ON, Canada, 1, 471–474, 1997.

Pohl, A., Springer, A., Reindl, L., Seifert, F., and Weigel, R.: New applications of wirelessly interrogable passive SAW sensors, IEEE MTT-S., 7–12 June 1998, Baltimore, MD, USA, 2, 503–506, 1998.

Reindl, L., Scholl, G., Ostertag, T., Ruppel, C. C. W., Bulst, W. E., and Seifert, F.: SAW devices as wireless passive sensors, Ultrasonics Symposium, 3–6 November 1996, San Antonio, TX, USA, 1, 363–367, 1996.

Rudolf, S. S. F. and Hoogerwerf, A.: Components for battery-powered wireless tire pressure and temperature monitoring systems, in Proc. SensorExpo, Detroit, USA, 15–20, 1997.

Schimetta, G., Dollinger, F., and Weigel, R.: A wireless pressure-measurement system using a SAW hybrid sensor, IEEE T. Microw. Theory., 48, 2730–2735, 2000.

Suster, M., Young, D. J., and Ko, W. H.: Micro-power wireless transmitter for high-temperature MEMS sensing and communication applications, The Fifteenth IEEE International Conference on Micro Electro Mechanical Systems, 24 January 2002, Las Vegas, NV, USA, 641–644, 2002.

Suster, M., Ko, W. H., and Young, D. J.: Optically-powered wireless transmitter for high-temperature MEMS sensing and communication, TRANSDUCERS, 12th International Conference on Solid-State Sensors, Actuators and Microsystems, 8–12 June 2003, Boston, MA, USA, 2, 1703–1706, 2003.

Swett, D. W.: Apparatus and method for generating broad bandwidth acoustic energy, US Patent 0069708A1, 2012.

Technavio: Global Tire Pressure Monitoring System (TPMS) Market 2015–2019, London, UK, 68 pp., 2015.

Thurau, J. and Ruohio, J.: Silicon Capacitive Absolute Pressure Sensor Elements for Battery-less and Low Power Tire Pressure Monitoring, Advanced Microsystems Automotive Applications, Springer, New York, USA, 2004.

Tian, B., Zhao, Y., Jiang, Z., Zhang, L., Liao, N., Liu, Y., and Meng, C.: Fabrication and Structural Design of Micro Pressure Sensors for Tire Pressure Measurement Systems (TPMS), Sensors, 9, 1382–1393, 2009.

Timoshenko, S.: Theory of plates and shells, McGraw-Hill, New York, USA, 1940.

Tsujiuchi, N., Koizumi, T., Oshibuchi, A., and Shima, I.: Rolling Tire Vibration Caused by Road Roughness, SAE Technical papers, Grand Traverse, MI, USA, 2005.

Wang, Q. and Ko, W. H.: Modeling of touch mode capacitive sensors and diaphragms, Sens. Actuat. A-Phys., 75, 230–241, 1999.

Xiangwen, Z., Zhixue, W., Leifu, G., Yunfeng, A., and Feiyue, W.: Design considerations on intelligent tires utilizing wireless passive surface acoustic wave sensors, Fifth World Congress on Intelligent Control and Automation, 15–19 June 2004, Hangzhou, China, WCICA, 4, 3696–3700, 2004.

Yurish, S. Y. and Gomes, M. T. S. R.: Smart sensors and MEMS, Kluwer Academic in cooperation with NATO Scientific Affairs Division, Dordrecht, the Netherlands, 2005.

Zhang, X., Wang, F., Wang, Z., Wei, L., and He, D.: Intelligent tires based on wireless passive surface acoustic wave sensors, The 7th International IEEE Conference on Intelligent Transportation Systems, 3–6 October 2004, Washington DC, USA, 960–964. 2004.

10

Novel microthermal sensor principle for determining the mixture ratio of binary fluid mixtures using Föppl vortices

B. Schmitt[1], **C. Kiefer**[1,*], **and A. Schütze**[1]

[1]Laboratory for Measurement Technology, Saarbrücken, Germany
[*]now at: Chair of Micromechanics, Microfluidics/Microactuators, Saarbrücken, Germany

Correspondence to: B. Schmitt (b.schmitt@lmt.uni-saarland.de)

Abstract. A novel sensor principle for determining binary fluid mixtures of known components is presented, making use of different thermal and rheological properties of the mixture's components. Using a microheater, a heat pulse is introduced in the mixture. The resulting temperature increase depends on the thermal properties of the mixture, allowing determination of the mixture ratio. Placing a bluff body in the fluid channel causes the formation of a stationary pair of vortices behind the body. The length of the vortex pair depends on the mixture's viscosity and thus its composition. By placing the microheater in the vortex area and making use of forced convection which changes with the size of the vortex, the sensitivity for determination of the mixture ratio can be increased by a factor of 2.5 compared to the direct thermal measurement. The flow velocity is measured independently of the mixture ratio using time-of-flight thermal anemometry.

1 Introduction

In many technical systems, monitoring and control of fluid mixtures is important for proper operation, while the cost of the sensor has to be very low. One potential application is the control of the mixture ratio of methanol in water for Direct Methanol Fuel Cells (DMFC), a technology that is promising to replace common batteries used in mobile applications (Dyer, 2002). Ren et al. (2000) and Dyer (2002) found out that the optimum concentration of methanol to be supplied to the fuel cell is approx. 1 M, corresponding to a volume fraction of approx. 4.2 %, to achieve optimum efficiency and long lifetime of the cell. This fraction is reduced as methanol is consumed in the cell. The optimal fraction then needs to be restored by adding pure methanol from a storage tank to achieve maximum energy density. Thus, Ren et al. (2000) stated that a sensor is required to measure the remaining methanol fraction and to determine the amount of methanol to be added.

Another potential application can be found in the exhaust aftertreatment of diesel cars using selective catalytic reduction (SCR) to reduce NO_x emission. Here, a solution of 32.5 % (W/W) of urea in water is injected into the exhaust stream in front of the catalyst to react with the NO_x to harmless N_2 and water (Trautwein, 2003). According to Trautwein (2003), this is the most promising way of reducing NO_x emission without decreasing the engine's efficiency. A suitable sensor could allow the identification of the correct mixture to alert in case of wrong fuelling or to prevent conscious deception.

In both applications, low cost and long lifetime are main requirements for suitable sensor systems. To ensure this, a simple physical sensor principle for determining the mixture ratio is investigated because of the inherently greater stability and higher robustness compared to chemical sensor principles. First, sensor designs developed and experiments performed by Schmitt et al. (2013a) which were based on direct measurement of thermal fluid properties, have shown promising results, but improved sensitivity and accuracy are required to meet the application requirements. The presented measurement principle is subject to patent application (Kiefer et al., 2015).

2 Measurement principle

When a bluff body or baffle is placed in a fluid flow, different types of flow past the obstacle can occur, depending on the present Reynolds number Re (Föppl, 1913; Lienhard, 1966; Hucho, 2011). This is calculated with the (unimpaired) flow velocity u_0, the characteristic length d of the bluff body and the kinematic viscosity ν, see Eq. (1).

$$Re = u_0 \frac{d}{\nu(\varphi, T)} \tag{1}$$

Here, the dependence of the viscosity on temperature T and volume fraction ϕ of the two components in the mixture is considered. At very low Reynolds numbers close to zero, so-called creeping or unseparated flow past the obstacle is observed, which is characterized by a smooth flow without the presence of turbulence or vortices. With increasing Reynolds numbers, a stationary and symmetric pair of vortices (also called Föppl vortices) develops, forming a wake behind the obstacle, as illustrated in Fig. 1. The resulting total flow velocity along the symmetry axis is also shown schematically. Since the fluid flows in the opposite direction in the wake, a point x_T exists at which the flow is zero. This point denotes the length of the vortex pair and it is used in this work for determining the fluid mixture ratio. Based on the work of Van Dyke (1988), Hucho (2011) showed that in the case of circular cylinders, the length of the vortex pair depends linearly on Re for Re lower than approx. 50. Similar behaviour has been reported for equilateral triangular cylinders by Shademani et al. (2013) based on finite volume simulations, while Zielinska and Wesfreid (1995) determined that for these obstacles, the critical Reynolds number is approx. 38. For higher Reynolds numbers the well-known Kármán vortex street is formed, with its typical repeating pattern of swirling vortices that are caused by the unsteady separation of fluid flow. This effect is used by vortex flow sensors to measure the flow velocity, since the frequency f of the vortex shedding is given by

$$f = St \frac{u_0}{d}, \tag{2}$$

with the Strouhal number St (Webster, 1999). At even higher Reynolds numbers the fluid flow becomes turbulent.

The novel measurement principle investigated here makes use of the changing length of the stationary vortex pair, which depends on the Reynolds number and hence also on flow velocity u_0 and kinematic viscosity $\nu(\phi, T)$. If flow velocity and temperature are known, the viscosity and thus the mixture ratio of two known components in a mixture can be determined based on the length of the vortex pair. Considering methanol / water mixtures, the vortex principle promises a higher sensitivity vs. the methanol fraction than other measurement principles, as the viscosity shows a much higher influence of the mixture ratio compared to the relative permittivity or thermal properties such as the thermal conductivity (cf. Table 1). The only physical property with higher

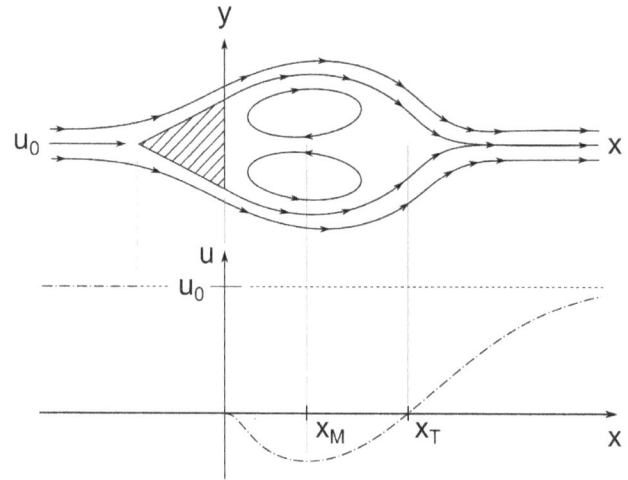

Figure 1. Schematic representation of fluid flow past a triangular bluff body at medium Reynolds numbers, causing Föppl vortices and the resulting total flow velocity along the symmetry axis.

sensitivity is the electrical conductivity σ, which, however, is also highly sensitive to surface contamination and therefore considered less robust.

The determination of the vortex pair length is based on the effect of forced convection which depends on the local flow velocity above a heated area. Here, a resistive microheater is placed at x_T and excited, e.g. using a constant current, resulting in an increase in temperature. Since the local flow velocity at the end of the vortex pair is close to zero, heat is mainly dissipated by conduction, which is influenced by the thermal properties like thermal conductivity and heat capacity, as shown by Schmitt and Schütze (2014). If the length of the vortex pair changes due to a change in viscosity and/or flow velocity, the magnitude of the local flow velocity over the heater increases. Thus, additional heat is dissipated due to forced convection and the resulting heater temperature decreases. By using multiple microheaters located as an array along the symmetry axis behind the bluff body, as shown in Fig. 2, the heater showing the highest temperature increase indicates the length of the vortices. If the flow velocity is known, knowledge of the vortex length allows the determination of the mixture ratio. To address applications with varying flow velocity this can be determined simultaneously by a time-of-flight method investigated by Schmitt et al. (2013a) in an earlier work. Here, the temperature sensor (depicted in green on the right side at the end of the heater array in Fig. 2) is designated to be used to detect the temperature pulse induced by one of the heaters of the array. The time required for the pulse to reach the temperature sensor is inversely proportional to the flow velocity and unaffected by the mixture ratio which influences the thermal properties of the fluid, as shown by Schmitt et al. (2013a). The second temperature sensor (depicted in green on the left side in front of the bluff body; cf. Fig. 2) is used to determine the reference temperature of the

Table 1. Relative sensitivity vs. changing methanol fraction of some physical properties for methanol / water mixtures (at 25 °C).

Property	Change in % per % (W/W)	Ref.
Electrical conductivity σ	$+5.1...10.9^a$	Thomas and Mandeville (1937)
Kinematic viscosity ν^b	$+3.1$	Mikhail and Kimel (1961)
Thermal conductivity λ	-0.87	Bates et al. (1938)
Relative permittivity ε	-0.55	Albright and Gosting (1946)
Density ρ	-0.17	Thomas and Mandeville (1937)

[a] The cited reference does not specify the exact concentration notation. [b] Calculated from dynamic viscosity and density.

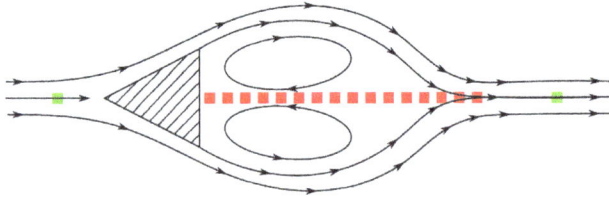

Figure 2. Schematic presentation of the array of microheaters (depicted in red) and two additional temperature sensors (green).

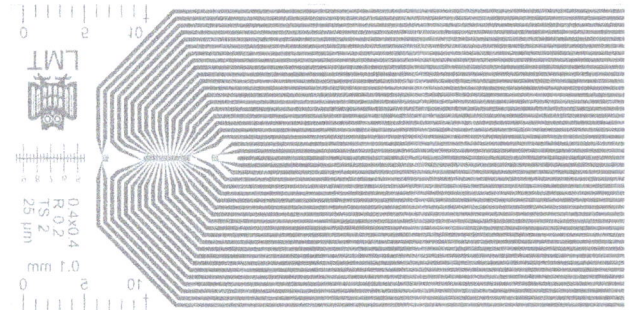

Figure 3. Developed sensor layout with heater array and two distinct temperature sensors. The chips are contacted using FPC connectors (Schmitt et al., 2013b).

fluid. This is especially required since the viscosity of water shows a very high sensitivity to temperature of approx. $-2\% \mathrm{K}^{-1}$, according to Mikhail and Kimel (1961).

3 Experimental setup

3.1 Sensor layout

The sensor layout used and an enlarged view of the active area are shown in Figs. 3 and 4, respectively. The meander-shaped resistive sensing elements are made of sputtered aluminium on a polyimide foil with a thickness of $125\,\mu\mathrm{m}$. The array in the middle can be heated in individual areas with a size of $0.4 \times 0.4\,\mathrm{mm}^2$ each which are overlapping the neighbouring heaters by 50%, resulting in an array of 16 heaters with a total length of $3.4\,\mathrm{mm}$. Each individual heater is excited with a constant current while its temperature increase is measured. As depicted in Fig. 4, the red wires are used to excite the heater while the rest of the meanders and the green contact wires are used to determine its exact resistance with four-terminal sensing and with its temperature. The temperature sensor placed in front of the bluff body is used to measure the reference temperature of the fluid. To realize the aforementioned time-of-flight method to determine the flow velocity, the last heater of the array, where the fluid flow is again undisturbed, and the temperature sensor at the end of the chip are used. Note that the two additional temperature sensors also have four contact wires each, allowing four-terminal sensing for exact temperature measurements. The accuracy achieved for the temperature measurement of the heaters and temperature sensors is approx. $0.1\,\mathrm{K}$.

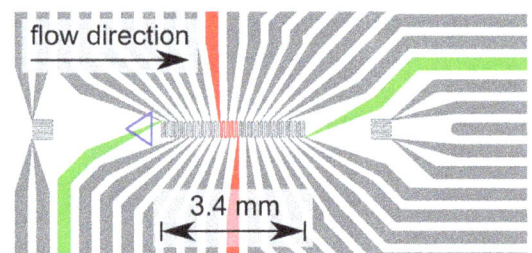

Figure 4. Enlarged view of the developed sensor layout with triangular bluff body (Schmitt et al., 2013b).

The thickness of the aluminium structures is approx. $0.5\,\mu\mathrm{m}$. In addition, a $40\,\mathrm{nm}$ thick titanium layer was used to improve the adhesion of the polyimide foil. The structures are passivated by spin-coating with a polyamide–imide resin (Durimide 32A, Fujifilm). A photograph of a realized sensor chip is shown in Fig. 5. The resistance of a single heater area is approx. $12\,\Omega$ at room temperature; the temperature coefficient was determined to be $3.25 \times 10^{-3}\,\mathrm{K}^{-1}$ for a reference temperature of $0\,°\mathrm{C}$.

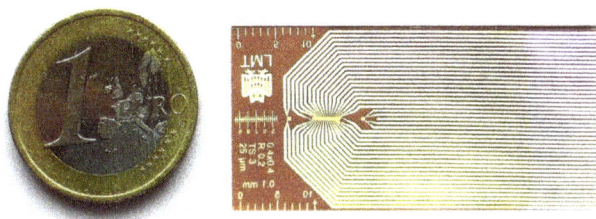

Figure 5. Photograph of sensor chip realized on polyimide foil (Schmitt et al., 2013b).

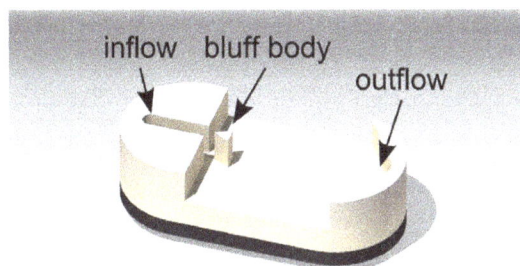

Figure 6. Insert manufactured using 3-D printing with integrated triangular bluff body and sealing on the bottom side (Schmitt et al., 2013b).

3.2　Measurement setup

To generate the stationary vortices, the insert with the integrated triangular bluff body, shown in Fig. 6, was fabricated using 3-D printing technology. This part is inserted into the cavity of the measurement chamber, shown in Fig. 7. The insert has a length of 17.5 mm and a width of 8.5 mm. The inflow is realized by a channel with a cross section of 1×1 mm^2 to increase the flow velocity in front of the bluff body, and hence increase the Reynolds number, in order to have a longer vortex pair. The triangular bluff body used has a base length and height of 1 mm each and is placed with a gap of 0.5 mm between its tip and the end of the inflow channel. The bottom side of the insert is covered with a softer material to act as a sealing. The bore holes which allow in- and outflow at the beginning of the inflow channel and at the end of the insert through its base are hidden in this view. Syringe pumps (C3000, TriContinent) were used to feed the system with the methanol / water mixtures to achieve very low pulsation and exact control of the flow rate. The control of the system and data acquisition using a USB-6009 data acquisition system (National Instruments) was implemented in Lab-VIEW (National Instruments).

3.3　Measuring process

The investigated mixtures of methanol in deionized water with volume fractions of 0, 5 and 10 % were prepared 1 day

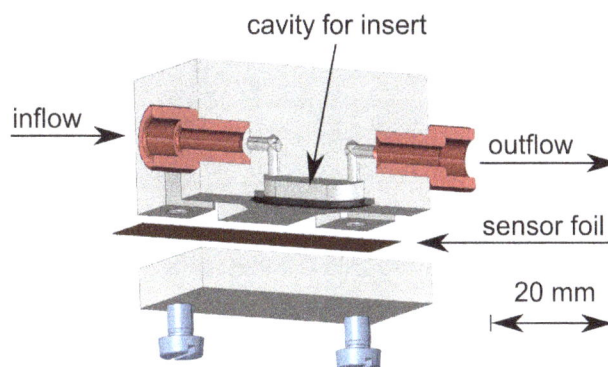

Figure 7. 3-D cross-sectional view of the measurement chamber used (according to Schmitt et al., 2013b).

before the measurements due to the mixing enthalpy, which causes a temperature increase when mixing methanol and water. The sensor substrate was positioned underneath the insert at the bottom of the measurement chamber, resulting in a small gap of approx. 0.1 mm between the first heater of the array and the back side of the bluff body. During the measurements at an ambient (reference) temperature of 28 °C, the applied flow rate was kept constant at values between 0.01 and 0.1 mL s^{-1}. Due to the cross section of the insert inflow channel of 1×1 mm^2, these flow rates correspond to a mean flow velocity in this channel between 0.01 and 0.1 m s^{-1}. The temperatures of the two separate temperature sensors were measured by applying a constant current of 5 mA corresponding to approx. 0.4 mW. Using this low power, no temperature increase was observed after applying the current, even without fluid flow. For the heaters in the array, a heating current of approx. 65 mA was applied for 3 s to each heater of the array successively, beginning with the heater with the largest distance from the bluff body. This corresponds to an initial heating power of approx. 50 mW, which increased to approx. 58 mW during heating due to the increase of the heater resistance with temperature. To take the slight differences of the individual heater resistances R_n of the n elements into account, the current I_n of each element was adapted so that the same heating power $P = R_n \times I_n^2$ is applied at room temperature. Note that the heating power increases during the heat pulse due to the increase of the heater resistance, but the resulting heating power curve is the same again for all heater elements. After changing the flow rate, a break of 5 s was applied to allow the vortex pair to fully develop before starting the next measurement. Also, the first heater was preheated for 5 s to compensate for the fact that the following ones are affected by the previous heaters due to their overlapping. Finally, the heater temperature was averaged over the last 2 s of heating for every individual heater. Since every individual heater is heated for 3 s, the measurement time for a full sweep of all 16 individual heaters re-

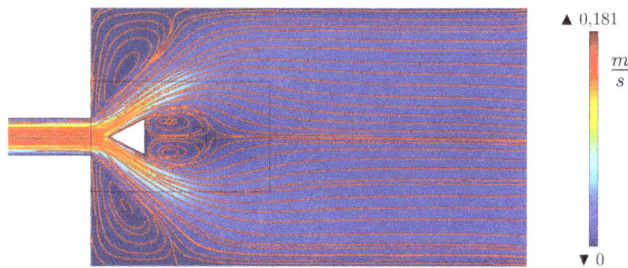

Figure 8. Simulated fluid flow past the triangular bluff body for water with a flow rate of $0.07 \, \mathrm{mL \, s^{-1}}$, with colours indicating velocity magnitude and streamlines depicted in red.

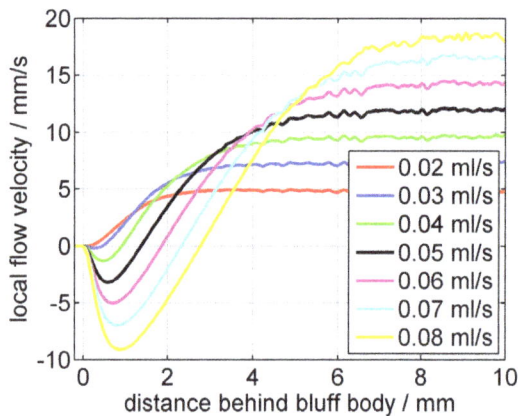

Figure 9. Simulation of local flow velocity behind the triangular bluff body along the symmetry axis for different flow rates of water.

quires 50 s, which is too long for many applications. However, this can be shortened, as discussed in the conclusion.

4 Simulations

Finite element simulations of the measurement principle were performed in Comsol Multiphysics (Comsol AB) using the conjugate heat transfer interface. Figure 8 shows the simulated velocity magnitude and streamlines for water at a flow rate of $0.07 \, \mathrm{mL \, s^{-1}}$ at an ambient temperature of $25\,^{\circ}\mathrm{C}$. As expected, a pair of stationary vortices forms behind the bluff body. Figures 9 and 10 show the simulated local flow velocity along the symmetry axis behind the triangular bluff body for different flow rates of water and for different methanol fractions at a constant flow rate of $0.08 \, \mathrm{mL \, s^{-1}}$, respectively. From Fig. 9 we can deduce that a flow rate of at least $0.04 \, \mathrm{mL \, s^{-1}}$ is required to cause the formation of a vortex pair as evidenced by the negative flow velocity behind the bluff body. As expected from the analytical model, in the simulated flow rate and fraction ranges the simulations show a nearly linear dependence of the length of the vortex pair on the flow rate and an inversely proportional dependence on the methanol fraction (which causes a nearly linear change in viscosity), respectively, as shown in Fig. 11. In Fig. 12 the

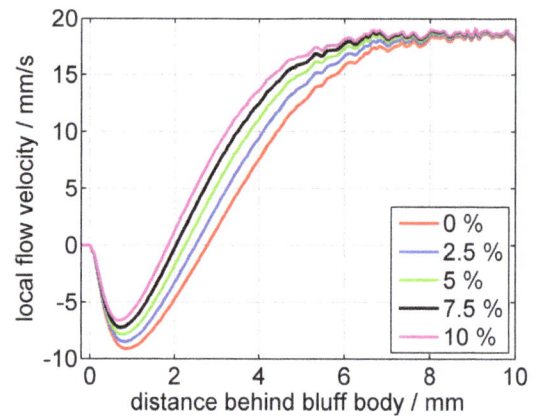

Figure 10. Simulation of the local flow velocity behind the triangular bluff body along the symmetry axis for different fractions of methanol in water at a flow rate of $0.08 \, \mathrm{mL \, s^{-1}}$.

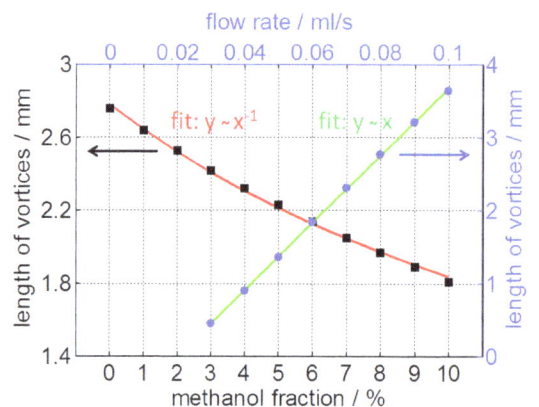

Figure 11. Simulated dependence of the stationary vortex pair length on flow rate of water and methanol fraction at a flow rate of $0.08 \, \mathrm{mL \, s^{-1}}$.

simulated temperature increase of individual heaters with diameters of 0.5 mm, heated with a constant power of 30 mW, is shown for different positions behind the bluff body for a flow rate of $0.08 \, \mathrm{mL \, s^{-1}}$ for two volume fractions of 5 and 10 % methanol in water. The different shape of the heaters to the realized layout was used due to former simulations without using stationary vortices. As expected, the resulting local heater temperatures show a maximum which is located at the point where the local flow velocity is zero, cf. Fig. 10. It has to be noted here that the heaters were not simulated as an array of successively heated areas, but as single heaters. Thus, the influence of a former heated area on the temperature increase of other heaters is not considered here. Furthermore, the heat dissipation into the substrate underneath the heaters is neglected; thereby the simulated temperatures of the heaters are only qualitatively comparable with the measurements.

The simulations show that placing a heater inside the wake decreases the length of the vortices by less than about 3 %.

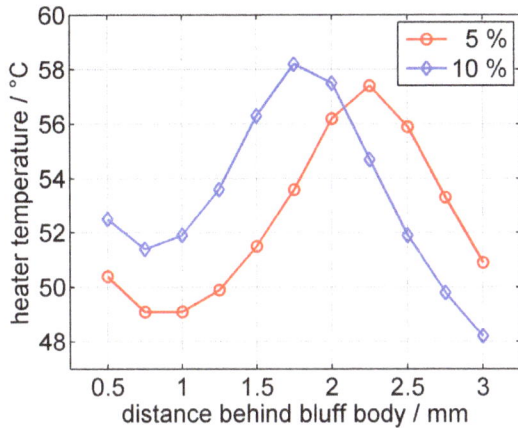

Figure 12. Simulated local temperature increase of individual heaters located at different positions behind the bluff body for 5 and 10 % methanol in water, respectively, for a constant flow rate of $0.08\,\mathrm{mL\,s^{-1}}$.

On the other hand, the local flow velocity inside the wake is strongly affected and shows an increase of up to 20 %, depending on the apparent flow rate. The penetration depth of the temperature field into the fluid is less than 0.5 mm.

5   Measurement results and discussion

In Fig. 13 the measured temperature increase of the individual heaters of the array as a function of their distance from the bluff body is shown for different flow rates of water. At a flow rate of $0.01\,\mathrm{mL\,s^{-1}}$, no influence of the vortex pair is evident. At this low flow rate, the apparent Reynolds number calculated from Eq. (1) with the base length of the wedge of 1 mm, the cross section of the channel in which the vortices are formed of $8.5\,\mathrm{mm^2}$ and the kinematic viscosity at $25\,^\circ\mathrm{C}$ of $0.894 \times 10^{-6}\,\mathrm{m^2\,s^{-1}}$, is 1.3 and results in creeping flow, while no vortex pair is formed. The slightly higher temperature of the farthest heater 3.3 mm behind the bluff body is caused by a longer heating time used for this heater, which is heated first. At a flow rate of $0.04\,\mathrm{mL\,s^{-1}}$ a small influence of the vortices can be seen and the apparent heater temperatures are lower due to the increased cooling by forced convection. For higher flow rates the temperature curves show distinct maxima, indicating the length of the vortex pair where the local flow velocity is nearly zero. As expected, the length of the vortex pair increases with increasing flow rate. The higher temperatures observed close to the bluff body can be explained by a shadowing effect of the bluff body and a resulting lower flow velocity close to zero that is also shown in the simulations (cf. Fig. 9).

Figure 14 shows the measured temperature increase for three volume fractions of methanol in water at constant flow rates of 0.01 and $0.09\,\mathrm{mL\,s^{-1}}$. At the lowest flow rate, the larger temperature increase observed for higher fractions of methanol is due to the reduced thermal effusivity of the mix-

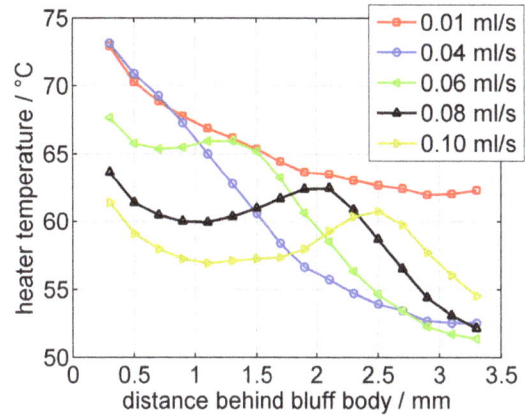

Figure 13. Measured temperature increase (averaged over the last 2 s of every heat pulse) of the individual heaters of the array for different flow rates of water.

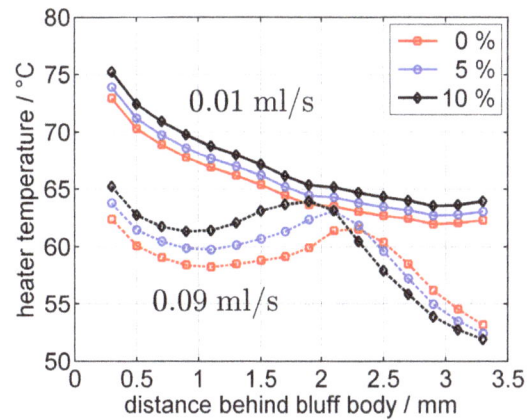

Figure 14. Measured temperature increase (averaged over the last 2 s of every heat pulse) of the individual heaters of the array for different volume fractions of methanol in water for constant flow rates of 0.01 and $0.09\,\mathrm{mL\,s^{-1}}$.

ture. At a flow rate of $0.09\,\mathrm{mL\,s^{-1}}$ the influence of the vortices is visible. With increasing methanol fraction the viscosity of the mixture also increases, resulting in a shortening of the vortex pair, as evidenced by the measurement data.

Figure 15 shows the measured sensitivity of the heater temperatures vs. the methanol fraction for flow rates of 0.06 and $0.09\,\mathrm{mL\,s^{-1}}$. Evidently, the position showing the highest sensitivity shifts to the right with increasing flow rate and also depends on the concentration range. In Fig. 16 the sensitivity calculated for a change from 0 to 5 % of methanol in water is shown for different flow rates between 0.01 and $0.10\,\mathrm{mL\,s^{-1}}$. At a very low flow rate, the sensitivity is nearly independent of the position and corresponds to the sensitivity for undisturbed flow of approx. $0.2\,\mathrm{K\,\%^{-1}}$, which is comparable to earlier results using similar microheaters in a stopped fluid (Schmitt et al., 2013a). With increasing flow rate, the sensitivity initially decreases for all heaters. When

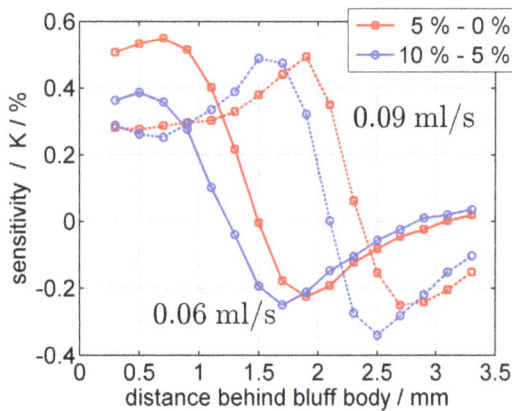

Figure 15. Measured sensitivity (defined as change in heater temperature divided by change in methanol fraction) of the individual heaters of the array for two different flow rates of 0.06 and $0.09\,\mathrm{mL\,s^{-1}}$. Two sensitivity curves are shown for a change in the methanol fraction from 0 to 5 % and from 5 to 10 %, respectively.

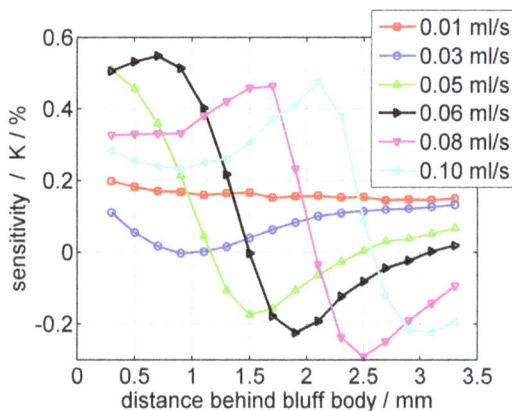

Figure 16. Measured sensitivity (defined as change in heater temperature divided by change in methanol fraction) of the individual heaters of the array for different flow rates.

the vortex pair is formed at flow rates of $0.05\,\mathrm{mL\,min^{-1}}$ and higher, the maximum sensitivity is much higher compared to undisturbed flow. The required number of heaters can be reduced in order to obtain a simpler layout without a significant loss of sensitivity. With only four heaters at distances of 0.5, 1.0, 1.5 and 2.0 mm behind the bluff body and for flow rates between 0.5 and $1.0\,\mathrm{mL\,min^{-1}}$, a sensitivity of at least $0.04\,\mathrm{K\,\%^{-1}}$ is achieved, i.e. twice as high as without the vortex pair.

In Fig. 17 the measured length of the vortex pair is shown, based on the position of the heater in the array showing the temperature maximum. Here, no interpolation between two heaters was used, which could improve the determination of the vortex pair length. In addition, a linear fit of the curve representing 5 % of methanol in water is shown. The expected linear correlation between the vortex length and the flow rate

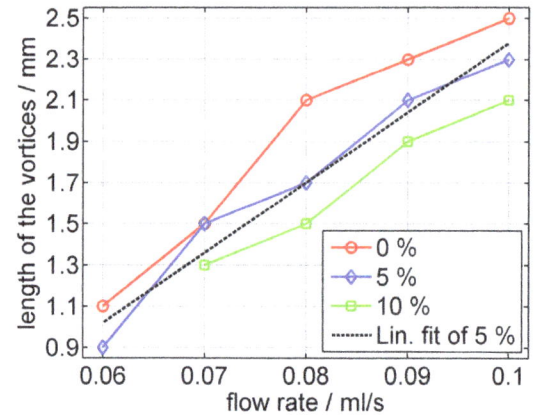

Figure 17. Length of the stationary vortex pair for different flow rates evaluated for three volume fractions of methanol in water.

is evident with deviations from the ideal linear dependence that are less than the spatial resolution of the heaters.

When interpolating the curves shown in Fig. 14, the length of the vortex pair decreases from approx. 2.22 to 1.90 mm when the methanol fraction increases from 0 to 10 %, resulting in a sensitivity of $32\,\mu\mathrm{m}\,\%^{-1}$. However, when regarding the measurement data, it is evident that due to the flat top of the local maximum of the curves, exact determination of the vortex pair length is very difficult. Regarding the influence of the flow rate on the length of the vortices for water and assuming a linear correlation, an increase from 1.19 mm at $0.06\,\mathrm{mL\,s^{-1}}$ to 2.47 mm at $0.1\,\mathrm{mL\,s^{-1}}$ is observed, resulting in a sensitivity of $32\,\mathrm{mm\,(mL\,s^{-1})^{-1}}$. In a previous work, Schmitt et al. (2013a) showed that using the proposed time-of-flight measurement principle for determination of the flow rate achieved a measurement uncertainty of less than $1.34\,\mu\mathrm{L\,s^{-1}}$ in the flow rate range between 0.027 and $0.055\,\mathrm{mL\,s^{-1}}$ which is relevant for DMFCs. Using error propagation, this results in a measurement uncertainty of the volume fraction of approx. 1.34 % due to the uncertainty of the flow rate. The measurement uncertainty for determining the mixture ratio neglecting the influence of the fluid flow is expected to be in the range of 0.5 %, as has been achieved in previous experiments without the use of a bluff body (Schmitt et al., 2013a). In addition, the viscosity of the fluid is strongly influenced by temperature, resulting in a decrease of approx. $2\,\%\,\mathrm{K^{-1}}$ at room temperature. Thus, the accuracy achieved for the temperature measurement of the heaters and temperature sensors of approx. 0.1 K results in an additional measurement uncertainty of the viscosity and hence the mixture ratio of 0.2 %. In total, the expected overall uncertainty is approx. 2 % if both flow rate and reference temperature have to be determined simultaneously.

It is evident that the required measurement time for the serial heating of the heater array presented here is far too long for most applications. An alternative and more feasible approach is to first determine the apparent flow rate using the time-of-flight principle, followed by the evaluation of the temperature increase of the heater in the array that has the highest sensitivity to the methanol fraction for the determined flow rate. Considering a flow rate of $0.09 \, \text{mL s}^{-1}$ as presented in Fig. 14, the heater at a distance of 1.7 mm behind the bluff body shows the highest sensitivity vs. a change in the methanol fraction of approx. $0.5 \, \text{K} \, \%^{-1}$. This represents an increase in sensitivity by a factor of 2.5 compared to measurements based on thermal measurements alone, without using vortices as reported by Schmitt et al. (2013a).

Comparing the simulations and experiments, the measured length of the vortex pair is approx. 30 % shorter than was simulated. The observed reduced length is attributed to the simulated simplified geometry, such as the triangular bluff body used in the experimental setup having rounded edges due to the fabrication process using a 3-D printer. Considering the rounded edges in the simulations, the length of the vortex pair is reduced by about 10 % at a flow rate of $0.07 \, \text{mL s}^{-1}$. With increasing flow rate, the influence of the rounding decreases.

First preliminary evaluations of the reproducibility of the measurements resulted in uncertainties of $\pm 0.25 \, \text{K}$. In combination with the sensitivity of $0.5 \, \text{K} \, \%^{-1}$ this results in a measurement accuracy of 0.5 % for the methanol volume fraction.

6 Conclusion

A novel measurement principle to determine the mixture ratio of binary fluid mixtures based on stationary vortices forming a wake behind a bluff body has been presented. In finite element simulations, the correlation between the length of the vortex pair and the Reynolds number predicted by analytical models has been confirmed. Experiments have verified the simulated results at least qualitatively. Quantitative differences can be attributed to different reference temperatures, the imperfect geometry of the 3-D printed bluff body and the influence of neighbouring heaters on each other, but further investigations are needed. Evaluating the length of the vortex pair using the proposed heater array directly is difficult because a very high spatial resolution is needed. Since no mathematical model has yet been developed to describe the local temperature distribution behind the bluff body, no physically inspired fit of the measured temperature distribution is possible to increase the resolution. In addition, this would require the measurement of multiple heaters, which requires a longer measurement time. Alternatively, when the flow rate is known, the heater providing the highest sensitivity to the mixture ratio can be evaluated, thus reducing the required measurement time to 3 s, or even less. When doing so, the resolution is evaluated to be 0.5 % if flow rate and reference temperature are known. If both have to be determined additionally, a lower resolution of approx. 2 % is achieved. Furthermore, the sensor can be simplified significantly by re-

ducing the number of heaters without a significant loss of sensitivity.

Assuming a critical Reynolds number of 38 according to Zielinska and Wesfreid (1995), the maximum flow rate for the generation of Föppl vortices is approx. $0.29 \, \text{mL s}^{-1}$ at 25 °C. In this case the sensitivity vs. the methanol concentration would be greatly reduced if only one heater without vortices were used, as proposed by Schmitt et al. (2013a). Thus, the applicable flow rate range can be extended with the new concept based on Föppl vortices.

Due to the required vortex pair, the response time of the novel measurement principle after a change in the methanol fraction has to be evaluated in further experiments to determine the achievable measurement rate. Furthermore, vortices forming in the corners of the insert (cf. Fig. 8) could cause carry-over effects. Here, an improvement of the sensor chamber geometry is probably required. Furthermore, adhesion of air bubbles in the flow channel can cause disturbances of the fluid flow and has to be prevented. Finally, vibrations are expected to occur in the intended application of DMFCs as well as in automotive applications. Their influence on the measurement principle also has to be investigated in the future.

Acknowledgements. Funding of parts of this research by the German Federal Ministry of Education and Research (BMBF) in the framework of the project InMischung (support code: 16SV5394) in the Microsystems Technology program is gratefully acknowledged.

References

Albright, P. S. and Gosting, L. J.: Dielectric Constants of the Methanol-Water System from 5 to 55 °, J. Am. Chem. Soc., 68, 1061–1063, doi:10.1021/ja01210a043, 1946.

Bates, O. K., Hazzard, G., and Palmer, G.: Thermal Conductivity of Liquids, Ind. Eng. Chem. Anal. Ed., 10, 314–318, doi:10.1021/ac50122a006, 1938.

Dyer, C. K.: Fuel cells for portable applications, J. Power Sources, 106, 31–34, doi:10.1016/S0378-7753(01)01069-2, 2002.

Föppl, L.: Wirbelbewegung hinter einem Kreiszylinder, Verl. d. K. B. Akad. d. Wiss., München, 1913.

Hucho, W. H.: Aerodynamik der stumpfen Körper, 2. Auflage, Vieweg + Teubner, Springer, Wiesbaden, Germany, 2011.

Kiefer, C., Schmitt, B., and Schütze, A.: Verfahren und Vorrichtung zur Bestimmung der Viskosität einer in einem Strömungskanal strömenden Flüssigkeit, unexamined patent application, application number DE 10 2013 019 872 A1, 2015.

Lienhard, J. H.: Synopsis of Lift, Drag, and Vortex Frequency Data for Rigid Circular Cylinders, Bulletin 300, Technical Extension Service, Washington State University, USA, 1966.

Mikhail, S. Z. and Kimel, W. R.: Densities and Viscosities of Methanol-Water Mixtures, J. Chem. Eng. Data, 6, 533–537, doi:10.1021/je60011a015, 1961.

Ren, X., Zelenay, P., Thomas, S., Davey, J., and Gottesfeld, S.: Recent advances in direct methanol fuel cells at Los Alamos national laboratory, J. Power Sources, 86, 111–116, doi:10.1016/S0378-7753(99)00407-3, 2000.

Schmitt, B. and Schütze, A.: Analytische Beschreibung von Hot Disk-Sensoren zur Bestimmung von Methanolkonzentration und Strömungsgeschwindigkeit für Direktmethanolbrennstoffzellen, Symposium des Arbeitskreises der Hochschullehrer für Messtechnik, Saarbrücken, Germany, 18–20 September 2014, 15–24, doi:10.5162/AHMT2014/1.2, 2014.

Schmitt, B., Kiefer, C., and Schütze, A.: Microthermal sensors for determining fluid composition and flow rate in fluidic systems, in: Proc. SPIE 8763, Grenoble, France, 17 May 2013, 87630O, doi:10.1117/12.2017293, 2013a.

Schmitt, B., Kiefer, C., and Schütze, A.: Neuartiger thermischer Wirbelsensor zur Bestimmung von binären Mischungsverhältnissen in Fluiden, 11. Dresdner Sensorsymposium, Dresden, Germany, 9–11 December 2013, 373–378, doi:10.5162/11dss2013/F16, 2013b.

Shademani, R., Ghadimi, P., Zamanian, R., and Dashtimanesh, A.: Assessment of Air Flow over an Equilateral Triangular Obstacle in a horizontal Channel Using FVM, J. Math. Sci. Appl., 1, 12–16, doi:10.12691/jmsa-1-1-3, 2013.

Thomas M. K. and Mandeville, D. C.: The electrical conductivity of potassium chloride in methyl alcohol-water mixtures, P. Indian Acad. Sci. A, 6, 312–315, doi:10.1007/BF03051254, 1937.

Trautwein, W. P.: DGMK-Research Report 616-1, Deutsche Wissenschaftliche Gesellschaft für Erdöl, Erdgas und Kohle e.V., Hamburg, 2003.

Van Dyke, M.: An Album of Fluid Motion, Fourth Printing, The Parabolic Press, Stanford, California, USA, 1988.

Webster, J. G. (Ed.): The Measurement, Instrumentation, and Sensors Handbook, CRC Press LLC, Boca Ranton, Florida, USA, 1999.

Zielinska, B. J. A. and Wesfreid, J. E.: On the spatial structure of global modes in wake flow, Phys. Fluids, 7, 1418–1424, doi:10.1063/1.868529, 1995.

11

Self-sufficient sensor for oxygen detection in packaging via radio-frequency identification

C. Weigel[1], M. Schneider[1,*], J. Schmitt[1,**], M. Hoffmann[1], S. Kahl[2], and R. Jurisch[2]

[1]Micromechanical Systems Group, IMN MacroNano®, Technische Universität Ilmenau, Ilmenau, Germany
[2]microsensys GmbH, Erfurt, Germany
[*]now at: Robert Bosch GmbH, Reutlingen, Germany
[**]now at: Sonceboz Automotive SA, Sonceboz, Switzerland

Correspondence to: C. Weigel (christoph.weigel@tu-ilmenau.de)

Abstract. A new disposable radio-frequency identification (RFID) sensor for detecting oxygen in packages with a protective atmosphere is presented. For safety reasons and system costs in consumer packages, no battery or energy harvesting devices can be used. Each part of a package, especially in food packaging, must be completely safe even if it is swallowed. Several materials have been investigated that safely react with oxygen and thus change electrical parameters without the need of an additional energy supply. In particular linseed oil was tested, because it is known to react in oxygen-containing atmosphere from liquid to solid. Linseed oil is used not only as food but also as a key part in ecological paint coatings. A significant relative change of capacity was observed during linseed oil drying, which results in -20% after 5 h and -38% after 30 h at an oxygen concentration of 20.5 and 50 % relative humidity, respectively. Pure unsaturated fatty acids were also tested in an oxygen-containing atmosphere and showed similar behaviour. The reaction speed is partially dependent on the level of unsaturation of fatty acids.

The oxygen sensor is coupled with an RFID front end with an internal charge time measurement unit for capacity determination. The combination of sensor element, sensitive material and RFID allows for biocompatible and save systems that indicate the presence of oxygen within a package.

1 Introduction

Metal oxide gas sensors are widely used for oxygen detection. Semiconducting metal oxide surfaces such as titanium oxide, zinc oxide and others show an oxygen-dependent conductivity (Kreisl, 2006). Electrochemical oxygen sensors for measuring liquids and gases are generally based on the amperometric principle, in which the current of a galvanic cell can be measured. One kind of amperometric oxygen sensor is the Clark electrode with liquid electrolyte, which is the most often used oxygen sensor at room temperature. In this case, the oxygen diffuses through a membrane into the liquid electrolyte and generates a current (Clark and Lyons, 1962; Otto, 2011). Furthermore, amperometric sensors with solid electrolyte, which are based on doped zirconia as an ion-conduction material between two electrodes, are employed as well (Tränkler and Obermeier, 1998; Kamp, 2003). Paramagnetic sensors require a heating wire and a permanent magnet. Inside the sensor, gas circulations occur through heating the gas molecules that cool down on the outer wall. As oxygen is paramagnetic it will be accelerated through magnetic repulsion to the wire and cools it at increased rate. The changing heater resistivity can be measured (Reichl, 1989). All these methods are suitable for continuous measurements of oxygen concentrations. Furthermore, those solutions usually require thermal activation energy (Xu et al., 2000) or an electrical potential difference that complicates the usage in self-sufficient systems. In paramagnetic sensors, temperatures of about 300 °C are needed, and metal oxide sensors based on zinc oxide begin to absorb oxygen at temperatures above 120 °C (Reichl, 1989).

Additionally, those systems are complex, resulting in higher costs. These facts do not allow for the usage of sensing elements in packaging of mass products.

Due to cost and environmental constraints, sensing in end consumer packaging excludes energy sources such as batteries or energy harvesting systems. Additionally, hazardous substances could find a way into the packaged goods when those systems are damaged, e.g. during opening of the package.

Low-cost sensors for food application should be as simple as possible and preferentially set up from packaging or food ingredients. The usage of biocompatible substances is preferred. To read out the status, a wireless transmission via radio-frequency identification (RFID) is a suitable way for checking even larger quantities of packages at once. At least a simple identification of leaking packages is possible. Ideally the sensing should be irreversible at a defined oxygen concentration threshold.

2 Concept of a self-sufficient oxygen sensor

To realise self-sufficient, autonomous sensor elements, sensitive materials with low activation energy are required. Some organic materials show a change of conductivity at room temperature. Polyacetylene, polypropylene and polyphenylen sulfide are listed as important representatives. Most of them are assumed to be toxic or lead to irritation of respiratory tracts. The acceptability with regard to food packages is not without limitations (Sigma-Aldrich, 2013; de Moraes Porto, 2012).

But also unsaturated fatty acids are sensitive to reactions with oxygen. In nature many variations and grades of unsaturation exist, which influence the reaction speed at oxygen contact. Some of them are used not only in food but also for applications such as paints and lacquers, sometimes for centuries.

2.1 Oxidation of fatty acids

Fatty acids are aliphatic monocarboxylic acids that include one carboxyl group $(RC(=O)OH)$ and show a non-cyclic carbon skeleton. Natural acids appear in oils, fats and wax from animals and plants (McNaught, 1997). The classification depends on the binding of carbonic atoms and builds two groups: saturated and unsaturated fatty acids, in which double carbon bindings exists (Chow, 2008). The oxygen attacks preferentially these double bindings and causes internal chemical changes. Therefore, the number of unsaturated bindings is an indication of the reaction rate. Holman and Elmer (1947) showed that by measuring the absorbed amount of oxygen, the reaction kinetics depends on the degree of unsaturation. Acids with more than one unsaturated binding are, for example, linoleic acid (two double bindings), arachidonic acid (four double bindings) and docosahexaenoic acid (six double bindings).

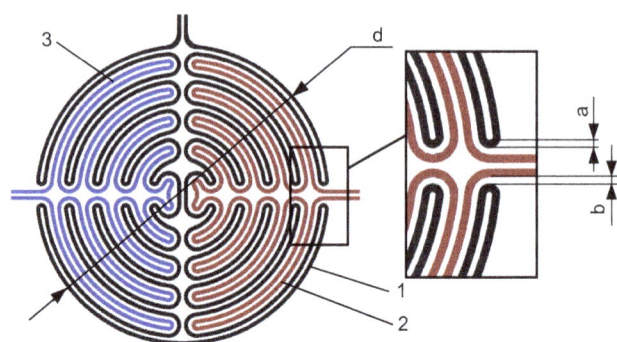

Figure 1. Electrode configuration for measuring capacity and resistivity including the ability to heat the active area (left: outer conductor (1) with a structure diameter (d) to heat and two conductors (2, 3) for referencing and sensing; right: dimensions of circuit path width (a) and distance (b)).

The reaction with oxygen was analysed in a variety of publications. The following description is, for example, given in Kuang Chow (2007):

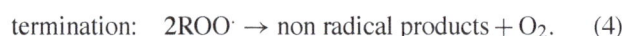

start reaction: $\quad RH \rightarrow R\cdot,$ (1)

propagation: $\quad R\cdot + O_2 \rightarrow ROO\cdot,$ (2)

$\quad\quad\quad\quad\quad ROO\cdot + RH \rightarrow ROOH + R\cdot,$ (3)

termination: $\quad 2ROO\cdot \rightarrow$ non radical products $+ O_2.$ (4)

RH represents multiple unsaturated fatty acid molecules, which lead to a creation of radicals $R\cdot$. By the availability of oxygen, $R\cdot$ reacts quickly and forms the hyperperoxyl radical $ROO\cdot$, which builds the stable hyperperoxide ROOH.

For fast reactions, the percentage of multiple unsaturated radical molecules is essential. Dubois et al. (2008) shows some natural oils divided in three groups of fatty acids. Linseed oil shows a high percentage of multiple unsaturated bindings, which is an indication of short reaction times. Furthermore, after the oxidative polymerisation process, linseed oil approaches a stable condition by forming long-chain molecules. This is contrary to technical oils, which decompose in a reverse direction. During the oxidation, first the volume of linseed oil increases and finally decreases when the polymerisation starts. The reaction time depends on a multitude of ambient conditions such as humidity, temperature and oxygen concentration.

2.2 Sensor element

2.2.1 Design and fabrication of a combined measuring and heating system

The sensor electrodes should be designed for high sensitivity. For this reason, an optimised ratio of large electrode area and small gap size should be obtained for an almost simple

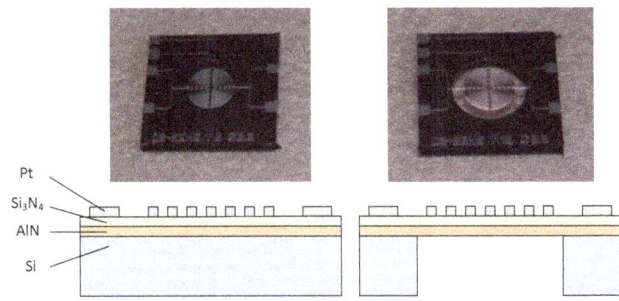

Figure 2. Cross sectional view and fabricated sensors with a diameter of 2.2 mm on bulk silicon (left) and on an AlN membrane (right).

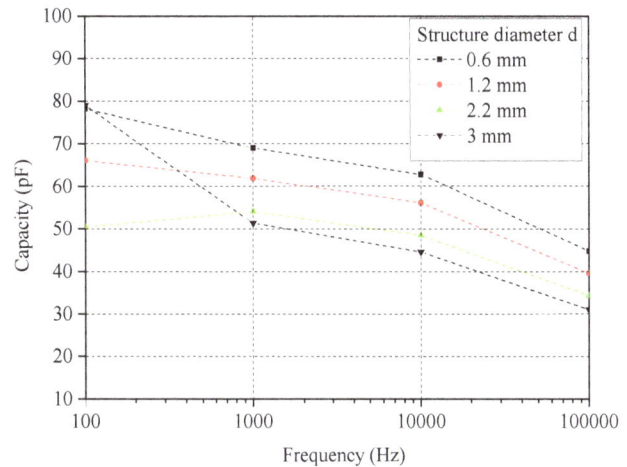

Figure 3. Characterisation of capacity at selected structure diameters (d) with different path widths (a) and distances (b) listed in Table 1 in the frequency range of 100–100 kHz (without sensitive material).

capacity measurement. In order to achieve reliable results, structures for sensing and referencing and the possibility for heating the structures to a reference temperature are included.

The designed interdigital structure is shown in Fig. 1, which has an aspect ratio from circuit path width (a) to distance (b) of 4 : 3, and two separate structures for measuring and referencing including the option to heat the sensor.

This design is able to measure capacity as well as resistivity values in a three-wire configuration. The outer conductor (1) can be used as a heating element, too. This allows for reaching stable and homogenous temperature distribution over the whole structure, if required. Both symmetrical inner conductors can be used for measuring (2) and referencing (3) for the electrical characterisation. The heating and measuring elements are manufactured from a single metallisation layer and allow, subsequently, for a simple fabrication process.

In all, < 100 > silicon wafers (1–20 Ω cm) are used for the demonstrator. The isolation is performed by using a 500 nm thick aluminium nitride (AlN) and an additional 250 nm silicon nitride (Si_3N_4) film. The platinum conductors are fabricated in a lift-off process. Aluminium (Al) conductors are suitable, if heating is not required. A second aluminium layer is applied at the pad areas for improved wire bonding. In principle, the same structure can easily be reproduced on flexible foils with any kind of metallisation process.

Furthermore, sensing devices can be fabricated on membranes by backside deep reactive ion etching (DRIE). This allows for low energy consumption if sensitive materials with higher activation energy require heating. Figure 2 shows devices with and without a membrane.

2.2.2 Characterisation and experiments with sensitive materials for oxygen detection

The initial resistivity and capacity of the conductors were characterized. In Fig. 3, the capacity for different layouts listed in Table 1 at different frequencies is shown. The

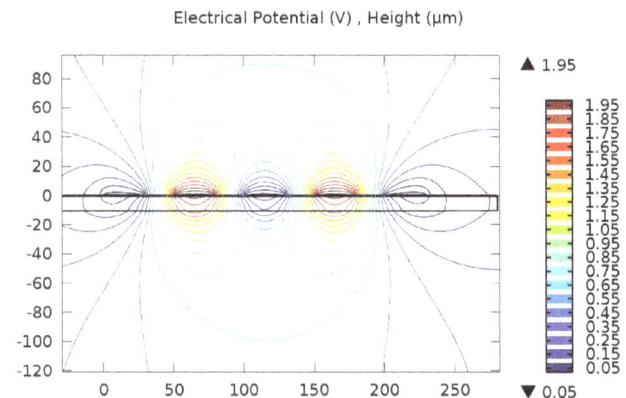

Figure 4. Simplified stray field model of structure with 2.2 mm diameter for simulating the area of high electrical fields (in this case about 50 μm above the electrodes).

ambient gas during this process was nitrogen with a relative dielectric constant $\varepsilon_R = 1$.

The sensitive material is applied onto the sensing structure. The permittivity increases when the fatty acids/oils are introduced. In literature, values between 3.2 and 3.5 are reported for linseed oil. The resulting capacity splits up into the substrate capacity and effect due to the sensitive material. The sensitive material should preferentially cover the volume of the capacitive stray field to achieve reliable changes due to the oxidation. Therefore, COMSOL Multiphysics simulations were performed to determinate the required volume and thickness of the sensitive layer (Fig. 4).

Table 1 shows simulated stray field heights for different layouts that have to be coated with sensitive material. The dispensing of the oils can be done in ambient air and the achieved capacity is measured. The delay before electrical

Table 1. Structure sizes and simulated stray field height of the capacitor.

Structure diameter (d)	Path width (a)	Path distance (b)	Stray field height
0.6 mm	8 µm	5 µm	10 µm
1.2 mm	16 µm	10 µm	20 µm
2.2 mm	30 µm	20 µm	40 µm
3 mm	40 µm	30 µm	60 µm

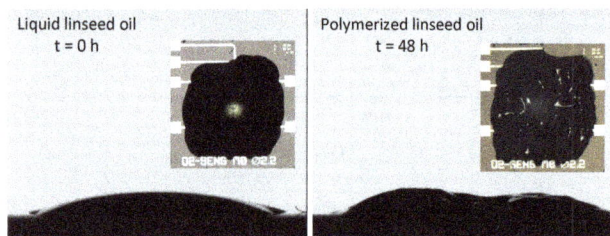

Figure 5. Sensor (structure diameter $d = 2.2$ mm) with liquid linseed oil (left) at the beginning and after polymerisation to linoxyn (right).

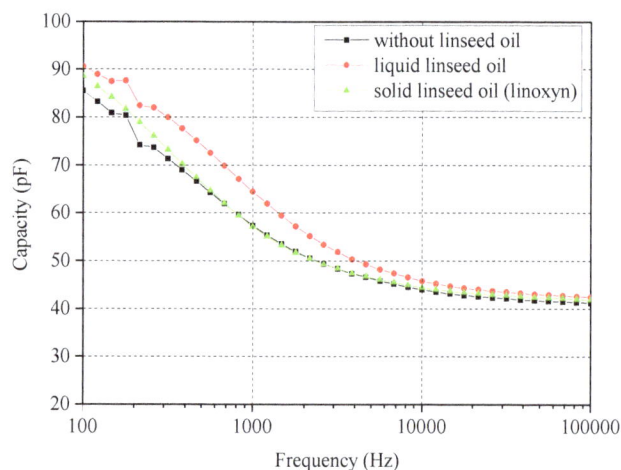

Figure 6. Capacity of a structure with a diameter (d) of 3 mm ($a = 40$ µm, $b = 30$ µm) in a frequency range of 100 Hz to 100 kHz, at conditions with and without solid and liquid linseed oil.

parameters start to drift depending on the fatty acids and it is usually long enough (more than 10 min) to place the sensor into the measuring set-up. Linseed oil is a liquid. In an inert atmosphere, the capacity does not change and it stays liquid. If oxygen is present the unsaturated fatty acids start a reaction and consume oxygen molecules. The permittivity and thus the measured capacity changes during the reaction and equals a stable state after the polymerisation is finished. Figure 5 shows linseed oil before and after the oxidative polymerisation process on a sensor device.

Figure 6 shows the changes of capacity of a structure with 3 mm diameter at ambient conditions (21 % oxygen concentration), 293 K and a relative humidity of 50 % in a frequency range between 100 Hz and 100 kHz. The capacity was determined without, with liquid and with solidified linseed oil in steady state. It can clearly be seen, that the solidified oil has a significantly lower dielectric constant as compared to the liquid state. Due to cross-linking of the fatty acids the reduction of the amount of polar groups causes a reduced dielectric constant.

Further experiments have been performed at a quasi-static state as the measurement principle of the RFID interface uses a similar method. Thus, a voltage ramp is used and the charging current is measured once an hour.

It was confirmed that the reaction speed depends on the humidity and concentration. Therefore, a set-up that allows for varying the concentration of oxygen, humidity and temperature in a measuring chamber is needed to achieve defined conditions. The concentration of oxygen is adjusted by using mass flow controllers for pure nitrogen and synthetic air. Humidity is adjusted by using a bubbler and a humidity sensor for control. Additionally, the chamber

can be heated. The oxygen concentration is measured by using the oxygen sensor Mettler Toledo InPro 6800 G. Furthermore, sensors for pressure and temperature are included and allow for controlling the conditions inside the measurement chamber. Figure 7 shows a simplified scheme of the measurement set-up. Experiments are driven by a LabVIEW program, which allows for the setting and saving of the data. Sensors under test are assembled on a carrier module that connects sensor devices with a reusable RFID system.

In Fig. 8, measured capacity vs. time at oxygen concentrations of 1, 10.25 and 20.5 % are shown at 293 K and 0 % relative humidity. In a second study, the influence of the moisture at constant oxygen concentration is presented (20.5 % oxygen concentration, 293 K). High humidity causes, as expected, an increase of capacity due to the high dielectric constant of water molecules. Due to polymerisation the influence of water decreases and results in a significant decrease of capacity.

Depending on the concentration of oxygen and moisture, the oxidative polymerisation requires a distinct time. Low oxygen concentrations as well as high humidity slow down the reaction.

The reaction can be forced by a higher surface-to-volume ratio as achieved by spin coating or by using materials

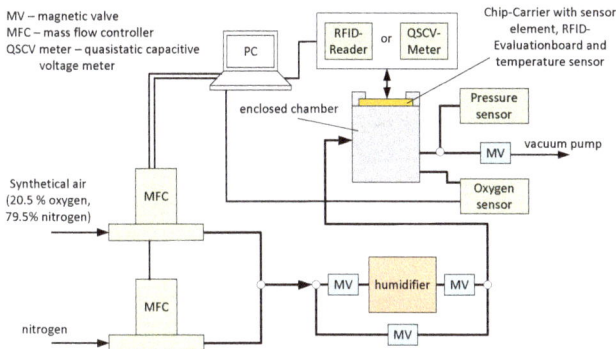

Figure 7. Simplified diagram of the test station with magnetic valves (MV) for flow path selection, mass flow controller (MFC) and sensor element with connected read-out devices via RFID and quasi-static capacitive voltage metre (QSCV).

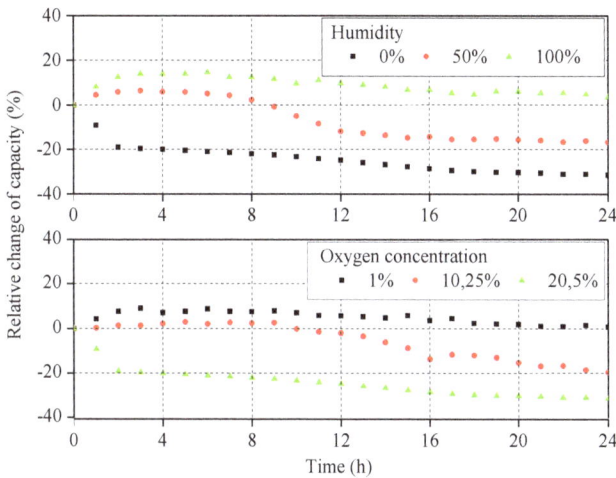

Figure 8. Relative change of capacity by varying the humidity (* high humidity, non-condensing) under constant oxygen concentration at 20.5 % (top) and constant humidity at 0 % at various oxygen concentrations (bottom) for a structure diameter (d) of 3 mm with a path width (a) and a distance (b) of 40 and 30 μm.

with polyunsaturated fatty acids. Also docosahexaenoic acid ethyl ester (DHA-EE), arachidonic acid and linoleic acid that have 6, 4 and 2 double bindings, respectively, have been tested. These are purified ingredients of linseed and tuna oil. Their reaction rate is more reproducible than that of the natural linseed oil with naturally varying contents of fatty acids. Figure 9 illustrates the relative change of capacity in ambient air (21 % oxygen, 293 K, 50 % relative humidity). Polyunsaturated fatty acids such as DHA-EE show a faster reaction as compared to linoleic acid. A volume of about 0.5 μL DHA-EE applied to a 3 mm structure shows a steady-state condition after 25 h. Therefore, a variable sensitivity against oxygen in packages can be achieved.

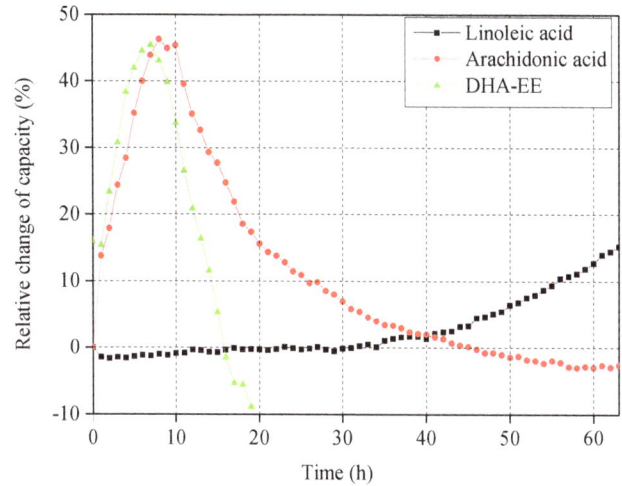

Figure 9. Time-dependent changes of capacity through the oxidation of fatty acids with various unsaturation ions (linoleic acid with 2, arachidonic acid with four and DHA-EE with six unsaturated bindings).

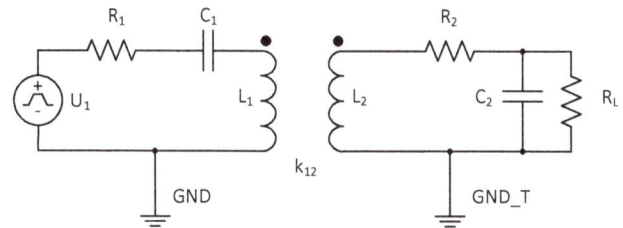

Figure 10. The oscillator circuits of the RFID reader and the transponder with connected sensor element described with the load resistance R_L as well as L_1 and L_2, which represent the transmitter and the transponder antenna.

3 Connection between sensor element and RFID evaluation unit

A suitable interface between the sensor element and RFID is required for a stable energy supply and data communication. Theoretical investigations for the power transmission have been performed. Starting point was the equivalent circuit and the complex equation approach in Finkenzeller (2006). The transmitter is based on a resonant antenna circuit and a resistive load. The sensor element consists of a resonator connected to the sensor, which represents the load. The equivalent circuit is shown in Fig. 10.

The transfer function between the output voltage U_2 and the input U_1 depends on the coupling factor between both oscillator circuits as well as their figure of merit and the load resistance. The load resistance also determines the available current for the sensing element. A load of 1 kΩ corresponds to a current of about 2 mA.

Figure 11. RFID-evaluation boards with front end controller for resistive and capacitive readout (left) and the used M30-Head for connection via USB to a processing unit (right).

3.1 Used RFID hardware

The oxygen sensor is connected to a commercial RFID front end from microsensys GmbH. For the sensor transponder, an interface working at 13.56 MHz was developed. Because of the required power for the transducer and the controller, a high-frequency RFID interface is preferred, which was built according to the ISO14443 type B standard. Therefore, the figure of merit of reader circuit and transponder was adjusted for an optimal bandwidth by decreasing the figure of merit and maximum field strength. The resonance frequency of both oscillating circuits is optimized for the ability to connect other reading systems from microsensys GmbH and for a wide functional range of the system.

Here, the so-called "M30-Head" was chosen and special commands for the transponder were integrated. The protocol is based on the iID® 3000Pro protocol and complements this in order to achieve more flexibility. The sensor adaption is done by the transponder controller. Figure 11 shows the developed RFID transponder for resistive and capacitive read-out and the used M30-Head.

The transponder consists of a commercial RFID front end iID-L® with voltage output, a microcontroller with integrated analogue–digital converter, and a unit to measure capacities. As the interface between front end and controller, a serial bus interface for data exchange is used. The output of the front end supplies a current of 2 mA for the sensor, e.g. for heating. For measuring the signal of the sensor, a high resistance input is used to analyse an analogue voltage as for measuring capacities. The design of the electronic compounds is optimised for a capacitance range between 50 and 90 pF, based on the measurements shown before. Figure 12 shows the block diagram of the used transponder.

3.2 Measuring principle of the interface

For measuring capacitive changes, two approaches can be used:

– A capacity-to-digital converter. Actually, this type is not available for this low voltage and low power application.

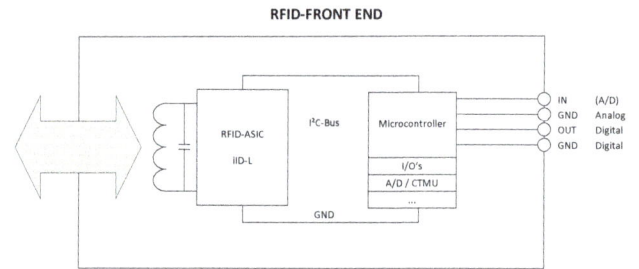

Figure 12. Block diagram of the transponder with integrated units for signal analysis and connections for the sensor element.

– Charge time measurement unit (CTMU), which is already available in the microcontroller. Thereby, a constant, adjustable current is injected for a defined time and the potential is read out as a measure of the capacitance.

The measurement cycle is as follows: the capacitor is discharged before each measurement. The connected capacity C can be calculated from the current I following Eq. (5)

$$C = I \cdot \frac{dU}{dt}. \tag{5}$$

Assuming a starting voltage U of 0 V at a discharged capacitor, correlation Eq. (6) can be applied if the time period t, the current I and the measured voltage U are known.

$$C = \frac{I \cdot t}{U} \tag{6}$$

For final calculation of the capacitance C by using the output signal of the analogue-to-digital converter including internal and parasitical capacitance, the formula Eq. (7) can be assumed:

$$C[\text{pF}] = \frac{I_c \cdot t_c}{\text{ADC} \cdot \frac{U_{cc}}{1024}} - C_{int}. \tag{7}$$

The capacitance C_{int} is constant, which is defined by parasitic capacitance of the electronic circuit. It is empirically determined. I_c is the charging current and t_c the charging time, which are set to 0.55 μA and 200 μs, respectively. "ADC" is the measured digital value. The transponder measures relative to a reference value, which is recorded before starting the measurement. A calibration of the current source and the internal series capacity is not necessary. For the sensor adaption, a test firmware was programmed, which controls the operating voltage. Furthermore, the transponder checks the collected data via RFID–ASIC interface for consistency. After that, the potential and the capacitive value are memorised, handled and the timing is set according to the measured values.

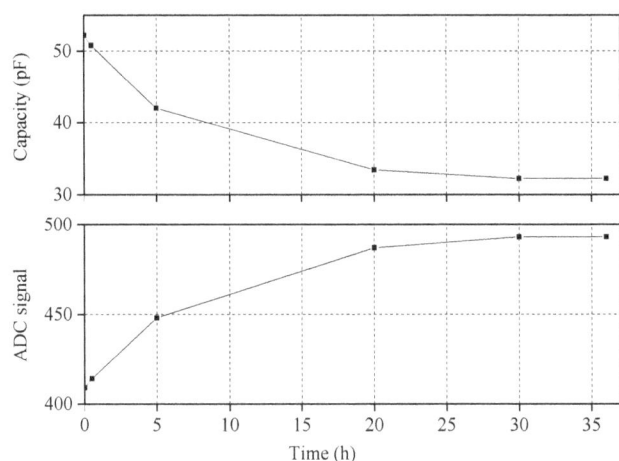

Figure 13. Measured capacity and corresponding ADC values from RFID transponder for a sensor with linseed oil (293 K, 0 % rel. humidity) for a structure diameter of 3 mm.

3.3 Connection between sensor chip and transponder

The sensor was connected to the RFID transponder. First, the value without sensitive layer was read out. The application of the material causes changes in the capacity. Thus, the value of the AD converter decreases. During the reaction time, the ADC value increases. After polymerisation it nearly reaches the value without sensitive material. The tests were performed at different sensor sizes. Larger sensors show higher changes and the evaluation is more accurate. By using Eq. (7), the ADC values can be converted into capacitive values with the typical voltage U_{CC} of 2.35 V. In the following Table 2, data for the investigated test structures are given in ADC and real capacitive values.

Furthermore, the behaviour of linseed oil over time was recorded. After 5 h, the RFID interface shows significant changes. After this time, damages of the package can be detected. By using other fatty acids, a much faster response is possible between the incidence of oxygen and measurable changes. The measured results of linseed oil are shown in Fig. 13.

Hence this demonstrates the applicability of these kind of sensors in combination with an RFID evaluation unit to observe the increase of oxygen in a protective gas atmosphere.

4 Conclusions

A new system concept of self-sufficient oxygen sensors is demonstrated, which records oxygen contamination without external energy. The intended application of the sensor is a threshold statement if unwanted oxygen access has occurred to the atmosphere under test. This is of interest for pharmaceuticals, for food under protective atmosphere and for industrial products such as oxygen-sensitive chemicals.

Table 2. Observed ADC and real capacity values of different chip sizes with liquid and solid linseed oil ($U_{cc} = 2.35$ V).

Membrane size	without		liquid		solid	
	ADC	real	ADC	real	ADC	real
0.6 mm	715	2 pF	700	4 pF	708	2.7 pF
1.2 mm	639	10 pF	616	13 pF	639	10 pF
2.2 mm	469	37 pF	456	40 pF	468	37 pF
3 mm	493	32 pF	409	52 pF	493	32 pF

Therefore, materials which are safely reacting with oxygen were analysed. Unsaturated fatty acids, for example in linseed oil, polymerize in oxygen-containing atmospheres and have been proven to be suitable candidates. The sensitive material was applied to a capacitor and changes its electrical parameter during the polymerisation. Linseed oil shows significant changes of its dielectric constant of about nearly 40 % depending on the ambient conditions. To reach lower response times, fatty acids with more double bonds are suitable at a comparable behaviour than linseed oil. Ambient conditions such as humidity and temperature influence the reaction speed, but do not significantly influence the threshold statement.

External energy is only needed during the readout process of the sensor element. This allows for the integration in packages without batteries or energy harvesting. A wireless readout is usually required as the sensor has to be inside the supervised package. Therefore, an RFID interface was developed that allows for the wireless read-out of the capacitance.

The irreversible chemical reaction guarantees high protection against manipulation. Furthermore, the design is simple and thus it allows for low production costs for mass market applications. A miniaturisation and integration are aspired to allow for the usage in smart transponder housings comparable to D14 from microsensys GmbH. The dimension of such a system are only 14 mm in diameter and 2 mm in height. Also a fabrication on foils can be realised, which would allow for the direct integration in packages.

Acknowledgements. This work was supported by the Federal Ministry of Education and Research (BMBF) by the project "O2-Sens – Niedrigenergie-Sensor zum Nachweis von Sauerstoff in Verpackungen mittels RFID" (contact no. 16SV5277).

The authors thank Jutta Uziel, Birgitt Hartmann, Joachim Döll and Ilona Marquardt from the Technische Universität Ilmenau for supporting the fabrication process and Yahia Cheriguen for his support by creating the stray field simulation models. Furthermore, we thank Ingo Hörselmann from the Technische Universität Ilmenau for his support in electrical measurements.

References

Chow, C. K. (Ed.): Fatty acids in foods and their health implications, 3rd Edn., Food Sci. Technol., CRC Press, Boca Raton, USA, 1281 pp., 2008.

Clark, L. C. and Lyons, C.: Electrode Systems For Continuous Monitoring In Cardiovascular Surgery, Ann. Ny. Acad. Sci., 102, 29–45, doi:10.1111/j.1749-6632.1962.tb13623.x, 1962.

de Moraes Porto, I. C. C.: Polymer Biocompatibility, Polymerization, edited by: de Souza Gomes, A., InTech, doi:10.5772/47786, 2012.

Dubois, V., Breton, S., Linder, M., Fanni, J., and Parmentier, M.: Proposition de classement des sources végétales d'acides gras en fonction de leur profil nutritionnel, OCL, 15, 56–75, doi:10.1051/ocl.2008.0163, 2008.

Finkenzeller, K.: RFID-Handbuch: Grundlagen und praktische Anwendungen induktiver Funkanlagen, Transponder und kontaktloser Chipkarten, 4th Edn., Hanser, München, Wien, 2006.

Holman, R. T. and Elmer, O. C.: The rates of oxidation of unsaturated fatty acids and esters, J. Am. Oil Chem. Soc., 24, 127–129, doi:10.1007/BF02643258, 1947.

Kamp, B.: Beiträge zur Sensorik redox-aktiver Gase, PhD Thesis, University of Stuttgart, Stuttgart, Germany, available at: http://elib.uni-stuttgart.de/opus/volltexte/2003/1241/ (23 October 2014), 2003.

Kreisl, P.: Mikromechanische Plattform für Metalloxid- und MOS-Gassensoren auf SOI-Basis, Berichte aus der Mikrosystemtechnik, Shaker, Aachen, Germany, 168 pp., 2006.

Kuang Chow, C.: Biological Effects of Oxidized Fatty Acids, 3rd. Edn., CRC Press, 855–878, doi:10.1201/9781420006902.ch36, 2007.

McNaught, A. D. and Wilkinson, A.: Compendium of chemical terminology: IUPAC recommendations, 2nd Edn., Blackwell Science, Oxford, UK, Malden, MA, USA, VII, 450 pp., 1997.

Otto, M.: Analytische Chemie, 4th Edn., Bachelor, Wiley-VCH, 2011.

Reichl, H.: Halbleitersensoren: Prinzipien, Entwicklungsstand, Technologien, Anwendungsmöglichkeiten, Kontakt & Studium, 251, Expert Verlag, Renningen, Germany, 1989.

Sigma-Aldrich: Safety Data Sheet Polypyrrole, Version 5.1, available at: http://www.sigmaaldrich.com/MSDS/MSDS/DisplayMSDSPage.do?country=DE&language=EN-generic&productNumber=577030&brand=ALDRICH&PageToGoToURL=%2Fsafety-center.html, 2014.

Tränkler, H.-R. and Obermeier, E.: Sensortechnik, Springer Berlin Heidelberg, Berlin, Heidelberg, Germany, 1116–1119, 1998.

Xu, Y., Zhou, X., and Sorensen, O.: Oxygen sensors based on semiconducting metal oxides: an overview, Sensor. Actuat. B-Chem., 65, 2–4, doi:10.1016/S0925-4005(99)00421-9, 2000.

12

Effect of sintering temperature on adhesion of spray-on piezoelectric transducers

Kyle M. Sinding[1], Alison Orr[2], Luke Breon[3], and Bernhard R. Tittmann[1]

[1]Pennsylvania State University, University Park, Pennsylvania, USA
[2]Misonix Inc. in Farmingdale, New York, USA
[3]Electric Power Research Institute, Charlotte, North Carolina, USA

Correspondence to: Bernhard R. Tittmann (brt4@psu.edu)

Abstract. Conventionally sol-gel spray-on transducers require a high-temperature ($> 700\,°C$) sintering process; however, this process can affect the microstructure of the substrate material. For mechanical elbows and valves utilized for fluid transport in the energy sector, the components are designed to have a specific microstructure, and deviations from these specifications can create weak points in the system. For this reason it is important to investigate how the temperature of the deposition process affects the substrate. This paper investigates the effect of high-temperature and low-temperature ($< 150\,°C$) processing conditions on the surface composition of the substrate. Furthermore, the resultant transducers from high- and low-temperature fabrication processes are compared to determine if a low-temperature processing method is feasible. For these studies a sol-gel spray-on process is employed to deposit piezoelectric ceramics onto a stainless-steel 316L substrate. Energy-dispersive X-ray spectroscopy is utilized to determine the composition of the substrate surface before and after transducer deposition. Results indicate that the high-temperature processing conditions may alter the surface composition of the metal due to a diffusion of the metal into the ceramic, which results in a metal surface that is bonded to the ceramic. Furthermore, it is shown that low-temperature processing of spray-on transducers is a viable method for transducer fabrication where the resultant transducers meet the industry minimum requirement of 30 dB signal-to-noise ratio. In parallel simulation calculations, finite-element method (FEM) studies were performed to model the adhesive strength of the low-temperature processed transducer to the substrate surface. Comparisons between the simulations and experiments suggest that the bond strength is much greater than the commercial gel bonds and closer to hardened epoxy glue bonds. These results indicate that spray-on transducers fabricated under low-temperature processing conditions are a viable solution for leave-in-place monitoring of structures.

1 Introduction

1.1 Goals and objectives

The goal of this work was to do research and development of the fabrication and attachment of leave-in-place ultrasonic spray-on transducers to metallic tubular structures. In particular, the objective was to investigate the effect of the spray-on transducer deposition process on the bond adhesion strength to metal alloy substrates when deposited at relatively low temperatures. A further objective was to compare the bond strength between the spray-on transducer and commercial adhesives such as those used for in-field conventional ul-

trasonic non-destructive evaluation (UNDE). The motivation for this investigation was that permanently installed sensors utilizing this work could offer advantages in implementation over periodically applied conventional UNDE for piping systems in energy production facilities, such as secondary systems in nuclear reactors, where elevated temperatures and low-level radiation limit the presence of personnel.

The following steps were taking to achieve this goal:

1. The physical bond between a high weight percent of bismuth titanate (BiTi) material and a metal substrate fabricated at both high and low temperatures was examined in order to show that the signal-to-noise ratio (SNR) is

independent of the fabrication temperature between 100 and 700 °C.

2. A baseline was set up in order to characterize the spray-on transducer's bond strength. For this a high weight percent of BiTi material (BT200) was obtained and bonded to a metal substrate with a commercial adhesive. This combination was modeled and simulated for various commercial adhesives.

3. Finally, having shown that the fabrication temperature does not play a major role, several transducers were fabricated at low sintering temperatures. This procedure was followed by ultrasonic pulse-echo measurements, and the data were compared to the modeling results to estimate Young's modulus of the bond between the spray-on transducer and the substrate.

1.2 Background

The recent advances in spray-on transducer technology provide a possible avenue toward these objectives. However, conventional spray-on transducers require a high-temperature (> 700 °C) sintering process, which can have an effect on the microstructure of the substrate material. For mechanical elbows and valves utilized for fluid transport in the energy sector, the components are designed to have a specific microstructure, and deviations from these specifications can create weak points in the system. For this reason it is important to investigate how the temperature of the deposition process affects the substrate. This paper investigates the effects of both high-temperature and low-temperature (< 150 °C) processing conditions on the surface composition of the substrate. Furthermore, the resultant transducers from high- and low-temperature fabrication processes are compared to determine if a low-temperature processing method is feasible. For these studies a sol-gel spray-on process is employed to deposit piezoelectric ceramics onto a stainless-steel 316L substrate.

Results indicate that the high-temperature processing conditions may alter the surface composition of the metal due to a diffusion of the metal into the ceramic, which results in a surface of the metal that is bonded to the ceramic. Furthermore, it is shown that low-temperature processing of spray-on transducers is a viable method for transducer fabrication where the resultant transducers meet the industry minimum requirement of 30 dB signal-to-noise ratio. Parallel simulation calculations by finite-element fethod (FEM) were performed to model the adhesive strength of the low-temperature processed transducer to the substrate surface. Comparisons between the simulations and experiments suggest that the bond strength is much greater than the commercial gel bonds and closer to hardened epoxy glue bonds. These results indicate that spray-on transducers fabricated under low-temperature processing conditions are a viable solution for leave-in-place monitoring of structures.

1.3 Recent developments

Here a technology is reviewed that shows promise in allowing ultrasonic transducers to be sprayed directly onto a component and left in place for permanent monitoring. Initially developed by Kobayashi, the method for fabricating such transducers utilizes a powder and a solution combined into a slurry and sprayed onto a substrate (Kobayashi et al., 2000; Kobayashi and Yen, 2004). The ceramic is then heated to drive out the binding compounds. A final heat treatment process is applied to the film to densify the microstructure. Then the ceramic is "electroded" (given a metallic electrode on the surface) and electrically poled to create a transducer. There are several advantages to this technique, including, but not limited to, a transducer that does not require a bonding medium, repeatable measurements at a constant position of the transducer, structural health monitoring of complex geometry systems, and high-temperature monitoring capabilities. Selected works of Kobayashi that deal with the concept of the composite spray-on transducer are given in Kobayashi and Yen (2004), Kobayashi et al. (2004, 2006, 2007), and Barrow et al. (1995). The spray-on process was a new deposition technique for thick film deposition of sol-gel ceramics such as the ones characterized by Barrow et al. (1995, 1997). Searfass et al. (2012) investigated high-weigth-percent BiTi and bismuth titanate–lithium niobate composite transducers (Searfass et al. 2010a, b, 2016; Sinding, 2014). Sinding further pursued the concept of composite spray-on transducers. He utilized a micromechanics analysis code by the generalized method of cells to predict for various weight percentages of BiTi the d_{33} of a transducer fabricated through the conventional spray-on process (Sinding, 2014). The d_{33} is the principal performance-determining parameter for piezoelectric materials and indicates the charge per unit force in the polarization direction of the sample.

While this information is very useful for high-temperature components, it is not always practical or acceptable to implement a high-temperature deposition process for stainless-steel 316L components used in industrial applications. Piping structures installed in energy production facilities are fabricated to specific standards intended to ensure acceptable microstructures (Kalpakjian and Schmid, 2008; Vander Voort, 1991). The high-temperature sintering process that is conventionally carried out at 850 °C for the spray-on deposition process may alter the grain structure of the substrate material (Searfass et al., 2010a). Changes in the grain structure can occur at temperatures as low as approximately 550 °C for stainless steel 316L (American Society for Metals, 1977). When the time–temperature–precipitation diagram is considered, temperatures above 485 °C may cause a change in the microstructure. The continuous cooling transformation (CCT) diagrams of steels with similar compositions to 316L indicate that there could be a transformation as low as 150 °C upon cooling (Ming and Lu, 2012).

Another important factor to consider when dealing with an interface between two materials is diffusion bonding between the two different surfaces. Diffusion at a metal and dielectric interface is a well-researched phenomenon. The diffusion of a metal into a dielectric can be modeled by Eq. (1) (Sinding et al., 2014):

$$\rho(x,t) = \rho_s \left[1 - \mathrm{erf} \left\{ x / \left(2\sqrt{\mathrm{D}t} \right) \right\} \right], \tag{1}$$

where ρ is the concentration of the metal, ρ_s is the fixed metal concentration at the metal–dielectric interface, x is the distance from that interface, t is the time in seconds, D is the diffusion coefficient, and erf is the error function. It is important to note that D is an exponential function of temperature. This means that a higher processing temperature at the metal–dielectric interface results in a higher diffusivity at the interface.

2 Experimental methods

2.1 Experimental plan

The following steps were taken in the experimental studies.

Step 1 Characterization of the spray-on transducer bond.

    i. Preparation of fabrication procedure for optical observations.

    ii. Preparation of samples for SEM imaging of the interface for higher resolution and contrast.

    iii. Preparation of samples for ultrasonic tests by comparing the SNR values for each of the transducers.

Step 2 Modeling and simulation

    i. Surveying and then selecting a finite-element method.

    ii. Simulating wave propagation in test samples.

    iii. Comparing SNR values for different ultrasonic coupling bonds.

Step 3 Transducer performance comparison

    i. Introducing the fabrication procedure for the BiTi low-temperature transducer

    ii. Showing that the differences in fabrication are acceptable:

      i. still having a high weight percent of BiTi;

      ii. electrode thickness did not matter.

    iii. Establishing the method for making a comparison between a spray-on bond and a commercial adhesive.

2.2 Spray-on transducer fabrication

Spray-on transducers were deposited on stainless-steel 316L cylindrical substrates that were 2.54 cm in diameter and 1.27 cm in thickness. Three variations of the deposition process were used. The only difference between each process is the temperature employed for each heat treatment. For simplicity and clarity when a heat treatment process is discussed here, it will be referred to with a process label of 1, 2, or 3. First the substrate was roughened with 200-grit sandpaper. Then the surface was scrubbed with industrial soap and rinsed with water. The surface was then washed with acetone and dried. Next, isopropyl alcohol was used to wash the surface, which was then dried. The protocol for processes 1 and 2 prescribed that the sample was then allowed to sit at 25 °C for 10 min. Process 3, however, let the sample sit at 400 °C for 10 min. After these intermediate processes, the sol-gel solution was mixed. For the mixture, 2.5 g of lead (II) acetate trihydrate was dissolved into 1.925 mL of acetic acid. This solution was placed on a hotplate at 100 °C for 20 min and then taken off and allowed to cool to room temperature. While this solution was cooling, 1.82 mL of zirconium propoxide was combined with 1.085 mL of titanium isoproxide in a separate vial. Next the two solutions were mixed thoroughly until the mixture was completely transparent to form a sol-gel. The sol-gel was either stored for up to approximately 2 months at 70 °C or was used immediately.

To create the slurry that was sprayed onto the substrate, 2.3 g bismuth titanate was combined with 1.2 mL of the sol-gel. This slurry was mixed with a 25.5 kHz ultrasonic horn to create a homogeneous solution. The solution was then poured into an appropriately sized gravity-feed spray gun that was attached to a pressurized air supply with the regulator set to 20 psi. The slurry was sprayed onto the substrate using a simple side-to-side pivot motion while holding the nozzle of the spray gun approximately 30 cm away from the substrates. The resulting thickness was $\sim 20\,\mu$m and could consistently cover a 60 cm^2 area. After each spray, process 1 samples were heated in a tube furnace at 100 °C for 3 h, process 2 samples were heated in a tube furnace at 150 °C for 3 h, and process 3 samples were heated on a hotplate at 400 °C for 15 min. Each process utilized five spray-on layers. After the final pyrolization step, samples processed with method 3 required one additional heating step, which was carried out with an induction heater to approximately 750 °C and held for 15 min. Each sample was electroded with 300 nm of platinum and poled at 55 kV cm^{-1}.

2.3 Characterization methods for substrate surface effects

To identify the effects of a high-temperature processing method compared to a low-temperature processing method on a stainless-steel 316L cylinder, processing methods 1 and 3 were compared. In order to identify any surface changes to the substrate material, the composition of the stainless-steel 316L sample was taken before the deposition process, utilizing energy-dispersive X-ray spectroscopy (EDS). One substrate had a transducer deposited onto it according to process 1, and another substrate had a transducer deposited onto it according to process 3. These samples were neither electroded nor poled. To view the interface between the dielectric spray-on material and the metallic substrate, the sample was sanded on a grinding wheel using 4000-grit sandpaper. The interface was then viewed on a scanning electron microscope (SEM). Once a sufficient image of the interface was collected, the dielectric material was scratched off with a razor, revealing the surface of the substrate. The composition of the surface of the substrate was again determined using EDS. For clarity, process 1 was denoted as the "100 °C tube furnace" sample, process 2 was denoted as the "150 °C tube furnace" sample, and process 3 was denoted as the "induction-sintered" sample (AK Steel Holding Corp., 2007).

2.4 Transducer performance

In order to establish the effectiveness of the low-temperature deposition methods, the resultant transducers were tested. As the deposited piezoelectric was intended to be used as a transducer, the most important parameter to compare was the SNR, which indicated the usability of the transducer; however, material properties also provide insight into the functionality of the transducer. As such, the capacitance, loss tangent, and d_{33} were measured. The d_{33} was measured on a Berlincourt d_{33} meter at 20 Hz. A Stanford LCR meter set to 1.0 V and 1 kHz was used to measure the loss tangent and the capacitance. The transducers that resulted from the two low-temperature (100 and 150 °C) deposition processes were compared to a spray-on transducer that resulted from a high-temperature process such as the one outlined by Searfass (Searfass, 2012), where sintering temperatures can be as high as 850 °C to provide a basis for comparison. The SNR was measured by pulsing the transducer with a Panametrics 5800 pulser–receiver. This device provides a broadband pulse with an energy of 50 uJ and 100 Ω damping. This pulser–receiver was used in the pulse-echo mode (Sinding et al., 2014).

2.5 Substrate effects

Figure 1 depicts the interface between the dielectric and the metallic substrate for the 100 °C tube furnace sample. Although a crack had developed in the ceramic, it was easy to see the interface between the two materials. Figure 1 showed no signs of diffusion at the interface, but diffusion was expected from an atomistic point of view, not a microstructural view. Figure 1 also shows that the interface between the ceramic and the steel was very intimate, suggesting that there was a strong bond between the substrate and the ceramic transducer. Figure 2 depicts the interface between the dielec-

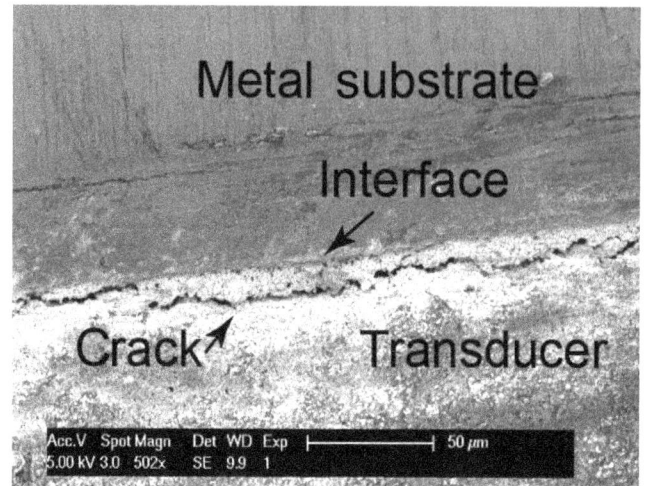

Figure 1. The interface between the ceramic and metal for the 100 °C tube furnace sample.

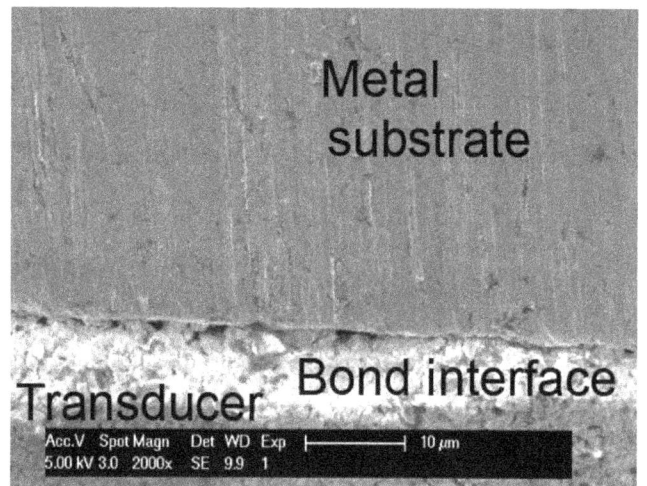

Figure 2. The interface between the ceramic and metal for the induction-sintered sample.

tric transducer and the metallic substrate of the induction-sintered sample. Figure 2 again did not show microscopic diffusion, but this only indicated that the diffusion at the interface was not extremely prevalent and diffusion could have been occurring on a much smaller scale. Noticeable again was that the interface was indicative of a strong bond between the ceramic and the substrate since the two materials seemed to be in contact across most of the interface. If the two materials were not bonded together, the interface would have significantly more pores, such that the two materials would just be touching each other and not bonded together.

Table 1 illustrates the initial weight percent composition of the stainless-steel 316L substrates (Okafor and Nesic, 2007).

A carbide-cutting tool was used to cut the substrates, which seemed to result in extra carbon bonding to the sur-

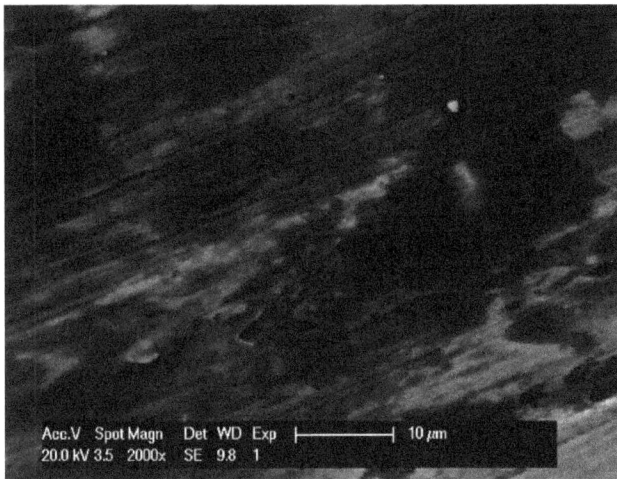

Figure 3. SEM image of the surface of the steel substrate. The lighter regions are less conductive and may indicate that these regions are of higher carbon content due to the use of a carbide-cutting tool to form the surface.

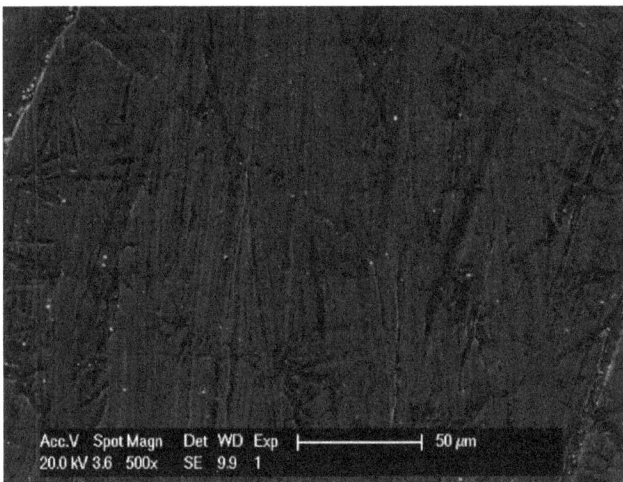

Figure 4. Surface of the steel substrate after the 100 °C tube furnace deposition process.

face. Figure 3 is an SEM image of the surface of the 316L substrate, where the lighter regions indicate that there was some charge buildup – possibly due to higher carbon contents in these regions that lower the electrical conductivity in those regions compared to the rest of the substrate surface. Since the carbon percent was so much higher than its acceptable value, the weight percent of each of the other elements is lower than the actual composition of the 316L substrate. If each weight percent is corrected assuming the weight of carbon is actually 0.3 %, then all of the elements in the substrate fall within the specified values. The specific EDS machine used for these experiments was supposed to produce only a nominal reading within at best approximately ±0.5 %. How-

Table 1. Initial weight percent composition of stainless steel 316L.

Element	Theoretical weight percent	Experimental weight percent
Carbon	<0.03 %	12.5 %
Chromium	16–18.5 %	15.3 %
Nickel	10–14 %	9.3 %
Molybdenum	2–3 %	2.4 %

Table 2. Experimental weight percent composition of stainless steel 316L after transducer fabrication.

Element	100 °C tube furnace	Induction-sintered
Carbon	5.33 %	4.35 %
Chromium	15.63 %	12.2 %
Nickel	9.81 %	6.66 %
Molybdenum	2.48 %	0.0 %
Manganese	1.85 %	1.42 %
Silicon	0.63 %	0.0 %
Phosphorous	0.0 %	0.0 %
Sulfur	0.0 %	0.0 %
Bismuth	0.0 %	14.5 %
Lead	0.0 %	8.65 %
Titanium	0.0 %	1.83 %
Zirconium	0.0 %	2.02 %
Oxygen	0.0 %	3.85 %
Iron	64.2 %	44.5 %

ever, the elements with small weight percent contributions may have had a higher percent error than other elements.

Table 2 presents the results of EDS on the substrates after the respective processing methods.

The weight percent of carbon noticeably dropped between the initial EDS experiment and the final EDS experiment. Although no experiments were performed to investigate this observation, it was noted that acetic acid was the main solvent in the sol-gel, and this acid had been documented to corrode carbon steels (Okafor and Nesic, 2007). This type of reaction could have brought some of the carbon atoms into the sol-gel solution, which was then evaporated out of the ceramic. Furthermore, the longer the acetic acid was allowed to sit on the surface, the longer it was able to dissolve the carbon. For the induction-sintered sample the carbon percent reduced significantly, mainly due to the presence of the bismuth and lead atoms. The actual percent reduction of carbon is small. This is consistent with the previous hypothesis that, since the sample was heated to 400 °C immediately after the spraying process, there was only a small amount of time for the acetic acid to react with the substrate (Okafor and Nesic, 2007).

Figure 4 is an SEM image of the surface of the steel substrate after the 100 °C tube furnace deposition process. There are several white specks on the surface of the substrate that

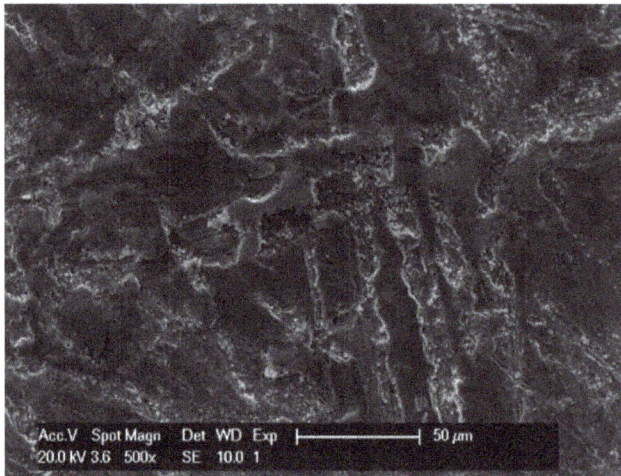

Figure 5. Surface of the steel substrate after the induction-sintering deposition process.

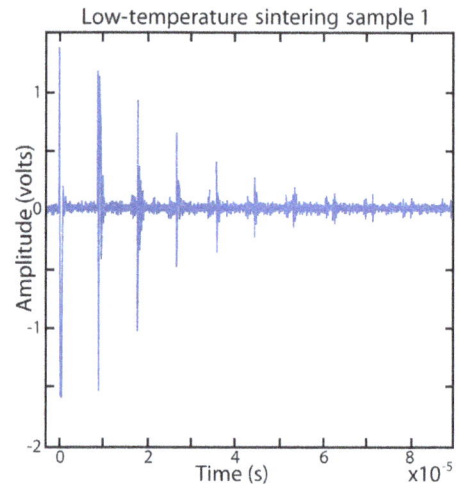

Figure 6. Example of pulse-echo A-scan (relative pulse-echo amplitude vs. time) of low-temperature sintered sample. Notice the large amount of signals produced by the transducer. The receiver gain was at 40 dB. The vertical axis is relative amplitude, which depends on both the transmitter and receiver electronics. The important feature to notice is the high signal-to-noise level.

could be small oxides or particles of ceramic that are left after the transducer is scraped off. These small particles produce results that are within the noise level of the EDS spectrum. Figure 5 is an SEM image of the surface of the steel substrate after the induction-sintered deposition process. This image has significantly more bright areas which could be oxide or ceramic, but the relative area of the bright regions to the total image area is still inconsequential. However, the EDS analysis indicated that the surface contains significant weight percentages of bismuth, lead, oxygen, titanium, and zirconium. These are all elements that are present in the transducer. It is hypothesized that the higher temperature during the deposition process allows the diffusion process to begin. The metal slowly diffuses into the ceramic, and the diffusion occurs enough during the high-temperature deposition process to strongly bond a small amount of the transducer material to the substrate, which cannot be removed by a mechanical scraping.

2.6 Transducer comparison

Table 3 provides a comparison between the transducers fabricated by the 100 °C tube furnace and the 150 °C tube furnace deposition processes and the benchmark high-temperature sintered transducer. It is clear that the conventional transducer generated a signal with a larger SNR, which can be attributed to higher capacitance and d_{33} values. Industry generally uses a cutoff value of 30 dB to determine if a transducer is suitable for non-destructive testing. It is important to note each of the deposition processes produced a transducer that meet the industry requirement. This is important because it indicates that the deposition process can be determined successful by adopting the 30 dB SNR requirement.

As a typical example, the results for a BiTi transducer are shown in Fig. 6. It was fabricated on a 2.54 cm diame-

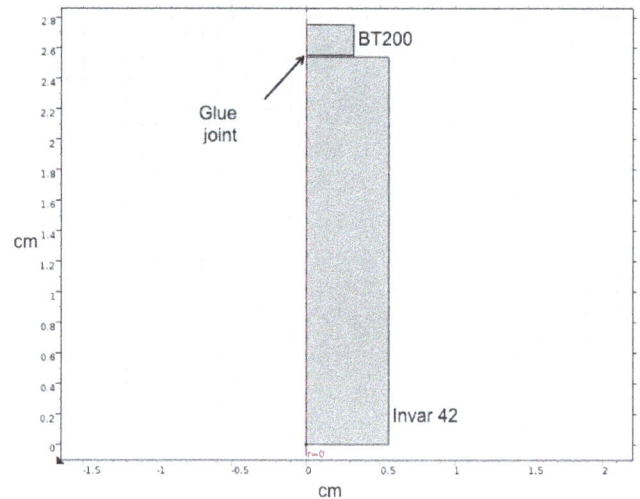

Figure 7. The COMSOL mesh diagram of the system's model.

ter, 2.54 cm long stainless-steel substrate. The substrate was sprayed five times in a row, giving a thickness of about 100 μm. The area of the sprayed piezoelectric is nearly the full top area of the cylinder. The sample was then placed in the furnace at 100 °C for 1 week. After the sample was taken out of the furnace, it was electroded with 100 nm of platinum and poled. The sample was poled with 380 V for 20 min at 98 °C and cooled with 380 V applied. The d_{33} was measured to be 9 pC / N, with a capacitance of 277 pF and a loss tangent of 0.03. After poling, the sample was tested as a longitudinal transducer in pulse-echo mode. The result is shown in Fig. 6.

Table 3. Comparison between spray-on transducers fabricated using high- and low-temperature processes.

Processing temperature	d_{33} (pC/N)	Capacitance (pF)	Tan delta (dB)	Signal-to-noise ratio (°C)
150	12	669	0.035	39.9
100	10	484	0.043	39.0
750	16	500	0.050	34.3

Table 4. Materials used and corresponding Young's modulus values and the resulting peak-to-peak displacement of the wave at the end of the Invar 42 rod from the COMSOL model.

Material	Modulus (Pa)	Displacement amplitude (cm) (pk-to-pk at bottom of rod)
Aremco	2.01E+11	6.009E−07
Silver	8.30E+10	6.025E−07
Glue	7.65E+10	6.076E−07
Epoxy	3.50E+09	4.061E−07
Polypropylene	9.00E+08	1.629E−07
Rubber	4.00E+06	4.061E−08
Commercial gel	1.89E+02	2.507E−12

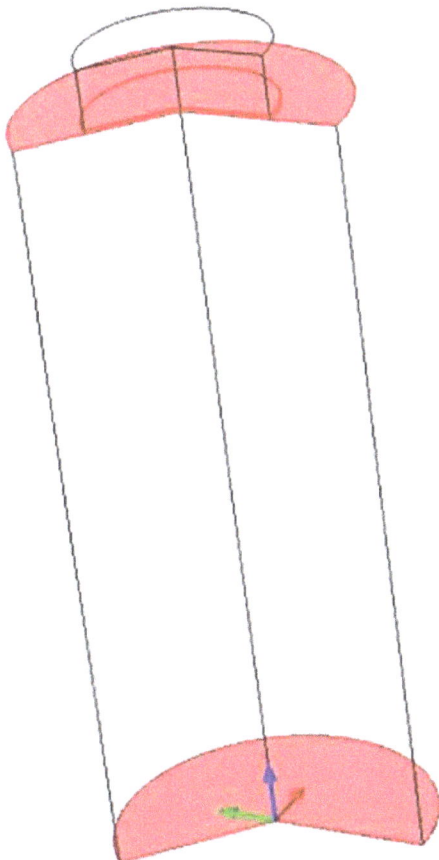

Figure 9. A zoomed-in view of the two planes used at the top of the rod. One plane was at the top of the Invar 42 rod, and the other was at the bottom of the piezoelectric wafer.

3 The bond strength simulations with finite-element methods

3.1 General description

All of the simulations presented in this work used COMSOL Version 4.4, which is the most recent version as of September 2014. The commands and parameters of the model were inserted into the "Model Builder" tree. For this work, the input frequency was the only parameter needed.

3.2 Modeling system–bismuth titanate transducer and Invar rod with glue joint

A model of the system was made to replicate the system used in the experiments (Orr, 2014). The model was built in two dimensions to simplify the model and decrease the computation time. The BT200 piezoelectric ceramic bismuth titanate was a 2 mm thick, 6.0 mm diameter wafer (supplied by TRS Ceramics Corp.) Invar was chosen as the material for the cylindrical rods because the thermal expansion coefficients are comparable to those of ceramics. Invar 42 is a 41 % nickel–iron controlled expansion alloy which has been used in a wide variety of glass-to-metal sealing applications. The Invar 42 rod was modeled as a 2.54 cm long rod with a radius of approximately 0.635 cm. The acoustic coupling bond was modeled as a thin plate approximately 100 μm thick, with the same radius as the BT200. The model is shown in Fig. 7. For each geometry used, the material properties needed to be defined. For the Invar 42, the density given by the company was used, but Young's modulus and the Poisson ratio were

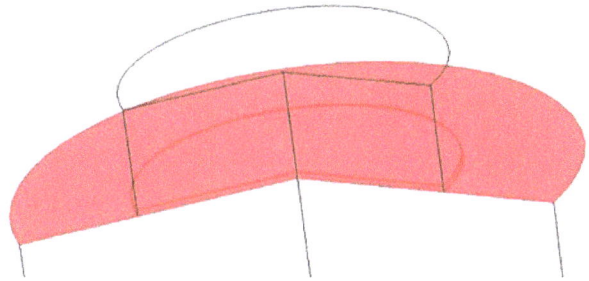

Figure 8. The three planes used to find the average displacement over the area of each vs. time.

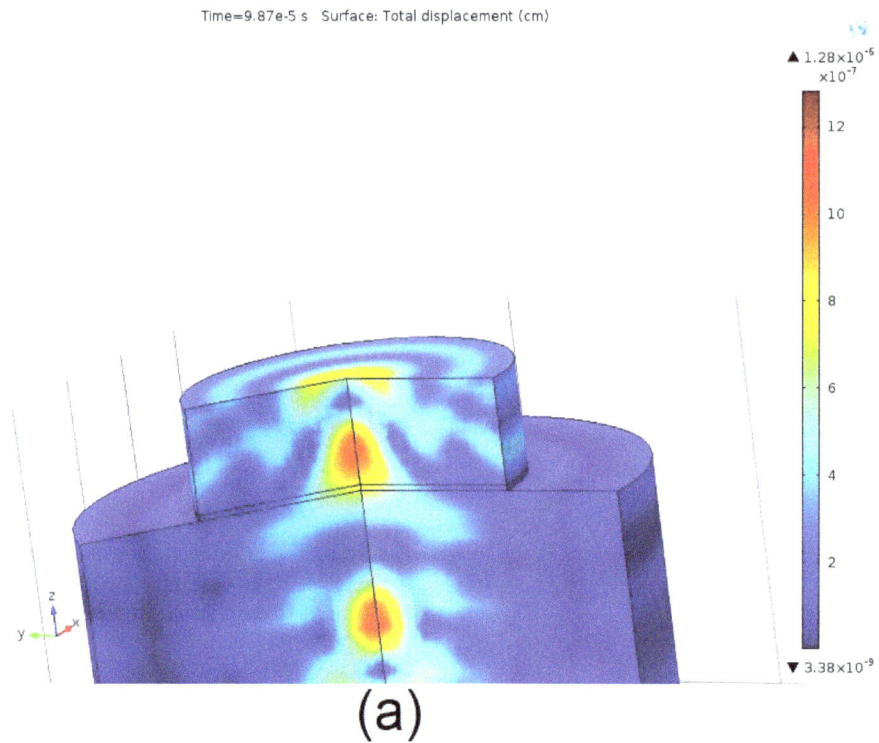

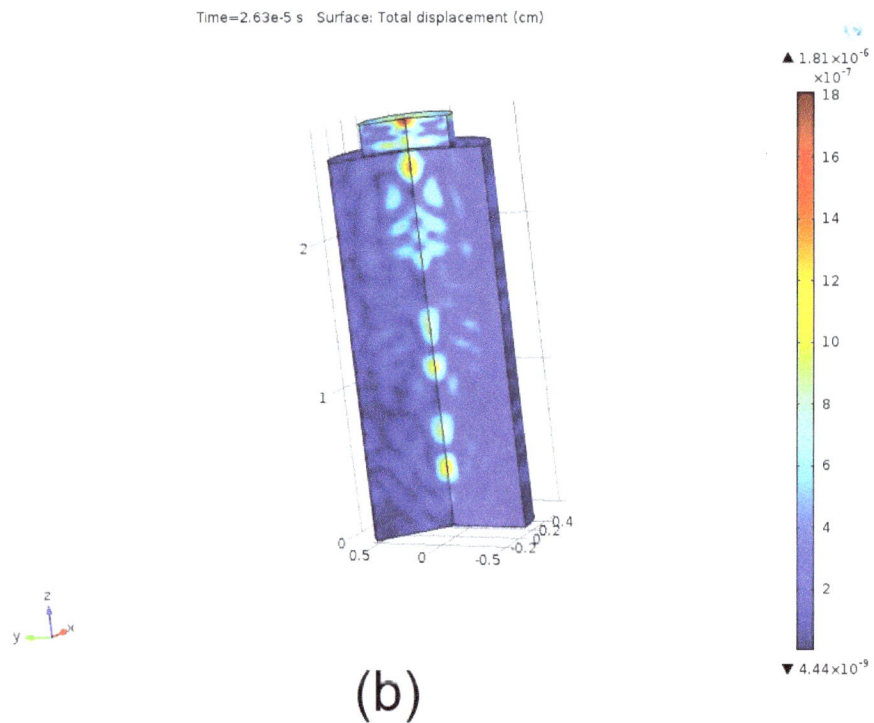

Figure 10. Example of the COMSOL displacement results for the Aremco glue used as the acoustic couplant excited at 1.1 MHz at time $= 26.3\,\mu s$ $(4.10a)$ and time $= 98.7\,\mu s$. Notice how most of the energy is able to travel through the coupling layer, and there is dispersion of energy within the rod. Note that all amplitudes are in centimeters.

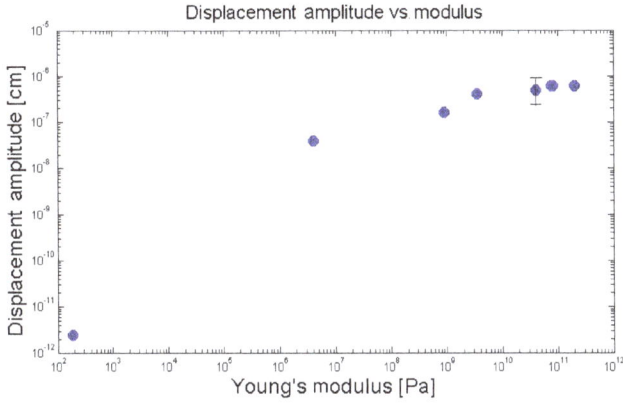

Figure 11. Average peak-to-peak displacement at the bottom of the rod as a function of Young's modulus of the bond materials as simulated with the COMSOL FEM model. The numerical values for various types of bonds are given in Table 4. The point with the error bars is placed by fitting the displacement amplitude given the estimated modulus of the spray-on low-temperature bond. Bond strength appears to lie between the commercial glue and the epoxy bonds.

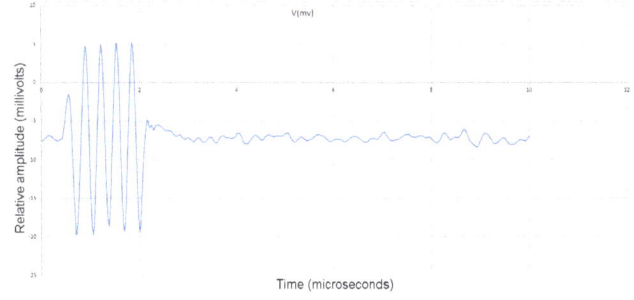

Figure 12. The graph shows the relative amplitude (millivolts) as a function of time (microseconds) as viewed on an oscilloscope with input from a laser interferometer (Polytec Model OFV505) for the out-of-plane displacement of a cylinder surface. The acoustic source is a spray-on bismuth titanate transducer with a 100 V excitation signal on the opposite end of the cylinder. The cylinder is a metal alloy (Invar 42) with dimensions 1.27 cm in diameter and 2.54 cm in length. The estimated signal-to-noise ratio is 28 dB.

3.3 Simulation results

A range of stiffness (modulus) values were used for the bond layer to see how it affected wave propagation in the rod and the sensitivity of the transducer. The materials chosen were the approximate values of the Aremco paste, solid silver, commercial glue (Gorilla), epoxy, polypropylene, rubber (polyisobutylene), and commercial ultrasonic gel (Ultragel II from Sonotech Inc.). With each coupling material, the model was solved with a linear solver with the fundamental frequency of 1.1 MHz.

For each coupling material, three planes were cut to find the average displacement over the area of the plane, as shown in Figs. 8 and 9. The three planes used were the bottom of the Invar 42 rod, the top of the Invar 42 rod, and the bottom of the piezoelectric wafer. The plane at the bottom of the rod was chosen because it shows how much energy arrives at the bottom of the rod through the coupling layer. The top of the rod was chosen because it shows the propagation of the wave down the rod and also shows the returned wave when reflected from the bottom of the rod. The bottom of the piezoelectric wafer was chosen because it shows the launched wave, as well as what is received by the piezoelectric wafer. The plots of the resulting displacements as a function of time are shown in Fig. 10.

estimated based on the literature values for Invar. The bond layer was varied with stiffness, as described below.

In benchtop experiments it was shown that BT200 has similar properties to PZT-5H, so, as a first approximation, PZT-5H was used for all the material properties used in the simulation.

The boundary condition used on the outer surface was a free boundary condition. The same condition was used on the bottom of the Invar 42 rod. The left side of the entire system has an axial symmetry boundary condition, which is why the radius of each material was used. Again, this was done to expedite the calculation time and to simplify the model. Additionally, since a voltage needed to be applied to the BT200 transducer, the top of the transducer has an electric potential boundary condition, while the bottom of the BT200 wafer was defined as the ground.

The BT200 was modeled using wave propagation in the rod, the transducer, and the very thin coupling joint. The equation used to resolve the mesh was the following:

$$\text{mesh size} = \frac{c_0}{f_0 u 5}. \qquad (2)$$

This equation indicated that the maximum size of each element in the mesh is about one-fifth the size of the wavelength associated with the frequency. In this case, c_0 is the speed of sound in the material. Since three different materials were used in this model, each material had a different mesh size. The c_0 was the speed of sound associated with the material being meshed. For the thin coupling layer, a finer mesh was needed since the joint was so small. The equation used was the following:

$$\text{mesh size} = \frac{c_0}{f_0 u 10}, \qquad (3)$$

4 Comparison between COMSOL and experimental results

The boundary layer between the piezoelectric and the Invar 42 rod were varied to understand the sensitivity of the transducer and the behavior of the wave propagation as the stiffness of the bond changes. Seven different materials were chosen to have a large range of materials and stiffnesses. The average peak-to-peak displacement at the bottom of the rod was calculated, and the values are shown in Table 4, whereas a plot of the simulation results are shown in Fig. 11.

The experimental data from the sol-gel fabricated transducers also needed to be compared to the COMSOL model results. Since the stiffness of the boundary layer between the spray-on transducer and the Invar 42 substrate is unknown, it needed to be approximated. Since the bismuth titanate has an oxygen layer on each side of its structure, it was assumed that most of the bonds with the Invar 42 substrate are either nickel oxide or iron oxide.

An average of the two corresponding Young's moduli (Ming and Lu, 2012) were used to approximate the strength of the bond between the composite sol-gel wafer and the Invar 42 rod. Based on this assumption, the sol-gel results fall at about $0.4 \pm 0.1 \times 10^{11}$ Pa. This places the displacement between epoxy and commercial glue (see Table 4). Measurements of out-of-plane displacement with a laser interferometer (Polytec Model OFV505) gave signals with high signal-to-noise ratio (28 dB) as shown in Fig. 12.

The preliminary conclusion is that the sol-gel spray-on transducers follow the same trend as the theoretical results, showing that, as the stiffness of the bond between the transducer and the substrate increases, the transmitted acoustic energy increases. These results show that, when sprayed onto a substrate, the sol-gel transducer boundary condition appears to be comparable to a commercial glue, which allows for greater transmission of energy than a commercial gel couplant.

5 Conclusion

Spray-on transducers were deposited on stainless-steel 316L substrates. The substrates were analyzed to determine the surface effects of a low-temperature and a high-temperature deposition process. It was found that higher-temperature deposition processes promoted a small amount of diffusion from the 316L into the sol-gel, which created a strong bond between the ceramic and substrate. The low-temperature deposition process did not develop as strong an adhesion, probably due to a lesser amount of energy applied to the system. Also transducers fabricated through three different deposition processes were compared, and it was found that low-temperature processing of spray-on transducers produced films with reduced properties compared to the higher-temperature processing method, but the resulting devices still meet the relevant minimum standards of signal-to-noise ratio.

The finite-element method COMSOL was used to see how the boundary layer between the piezoelectric and an Invar 42 cylindrical substrate was varied to understand the sensitivity of the transducer and the behavior of the wave propagation as the bond stiffness changes. Seven different bonding materials were chosen to have a large range of materials and stiffness. The average peak-to-peak displacement at the bottom of the cylindrical rod was measured and plotted against the Young's modulus value; the values show that as the stiffness (Young's modulus) increases, the average displacement amplitude at the bottom of the rod increases as well. Simulations based on an assumed metal oxide layer between the spray-on transducer and the metal substrate show that the bond strength of the sol-gel transducer appears to be between commercial glue and epoxy.

References

AK Steel Holding Corp.: 3116/316L Stainless Steel Data Sheet, UNS S31600 and UNS S31603, AK Steel, 9227 Centre Pointe Drive, West Chester, OH 45069, 2007.

American Society for Metals: Atlas of Isothermal Transformation and Cooling Transformation Diagrams, American Society for Metals, Metals Park, Ohio, 1977.

Barrow, D. A., Petroff, T. E., and Sayer, M.: Thick Ceramic Coatings Using a Sol Gel Based Ceramic-Ceramic 0-3 Composite, Surface and Coatings Technology, 76–77, 113–118, 1995.

Barrow, D. A., Petroff, T. E., Tandon, R. P., and Sayer, M.: Characterization of Thick Lead Zirconate Titanate Films Fabricated Using a New Sol Gel Based Process, J. Appl. Phys., 22 81, 876–881, 1997.

Chen, W.-K.: Linear Networks and Systems, Belmont, CA, Wadsworth, 123–135, 1993.

Kalpakjian, S. and Schmid, S. R.: Manufacturing Processes for Engineering Materials, Upper Saddle River, NJ, Pearson Education, 2008.

Kobayashi, M. and Jen, C.-K.: Piezoelectric Thick Bismuth Titanate/Lead Zirconate Titanate Composite Film Transducers for Smart NDE of Metals, VTT. Symp., 13, 951–956, 2004.

Kobayashi, M., Olding, T. R., Zou, L., Sayer, M., Jen, C.-K., and Rehmen, A. U.: Piezoelectric Thick Film Ultrasonic Transducers Fabricated by a Spray Technique, in: Proc. IEEE Ultrasonics Symposium, 985–989, 2000.

Kobayashi, M., Jen, C.-K., Ono, Y., and Krüger, S.: Lead-Free Thick Piezoelectric Films as Miniature High Temperature Ultrasonic Transducer, in: Proc. IEEE Ultrasonic Symposium, 13, 910–913, 2004,

Kobayashi, M., Ono, Y., Jen, C.-K., and Cheng, C.-C.: High-Temperature Piezoelectric Film Ultrasonic Transducers by a Sol-Gel Spray Technique and Their Application to Process Monitoring of Polymer Injection Molding, IEEE Sensors, 6, 55–62, 2006.

Kobayashi, M., Jen, C.-K., Nagata, H., Hiruma, Y., Tokutsu, T., and Takenaka, T.: Integrated Ultrasonic Transducers Above 500 C, in: Proc. IEEE Ultrasonics Symposium, 10, 953–956, 2007.

Ming, H. and Lu, T.-M.: Metal-dielectric Interfaces in Gigascale Electronics: Thermal and Electrical Stability, Springer, New York, 2012.

Okafor, P. C. and Nesic, S.: Effect of Acetic Acid on Corrosion of Carbon Steel in Vapor-Water Two-Phase Horizontal Flow, Chem. Eng. Commun., 25, 141–157, 2007.

Orr, A.: A Bond Stiffness Study of Sol-Gel Spray-on Transducers, MS thesis, The Pennsylvania State University, Pennsylvania, 2014.

Searfass, C.: Fabrication and Characterization of Bismuth Titanate Thick Films Fabricated Using a Spray-On Technique for High Temperature Ultrasonic Non-Destructive Evaluation, PhD thesis, The Pennsylvania State University, Pennsylvania, 2012.

Searfass, C. T., Baba, A., Agrawal, D. K., and Tittmann, B. R.: Fabrication and Testing of Microwave Sintered Sol-gel Spray-on Bismuth Titanate-Lithium Niobate Based Piezoelectric Composites For Use as a High Temperature (> 500 C) Ultrasonics Transducer, Review of Quantitative Nondestructive Evaluation, 1, 1035–1042, 2010a.

Searfass, C. T., Tittmann, B. R., and Agrawal, D. K.: Sol-gel Deposited Thick Film Bismuth Titanate Based Transducer Achieves Operation of 600 C, Rev. Prog. Q., 2, 1751–1758, 2010b.

Searfass, C. T., Pheil, C., Sinding, K., Tittmann, B. R., and Agrawal, D. K.: A Study of Bismuth Titanate (Bi4Ti3O12) Fabricated Using Spray-on Deposition Technique and Microwave Sintering For Use as a High Temperature Ultrasonic Transducer, IEEE T. Ultrason. Ferr., 63, 139–146, doi:10.1109/TUFFC2015.2501241, 2016.

Sinding, K. M.: The Effect Of Weight Percent On The Properties Of Ultrasonic Transducers Fabricated Though A Sol-Gel Deposition Process, dissertation, Pennsylvania State University, University Park, 2014.

Sinding, K., Orr, A., Breon, L., and Tittmann, B. R.: Sol-Gel Spray Technology, Electric Power Research Institute, Charlotte, NC, 2014.

Vander Voort, G. F.: Atlas of Time-Temperature Diagrams for Nonferrous Alloys, ASM International, United States, 474 pp., 1991.

13

State determination of catalytic converters based on an ultra-wideband communication system

I. Motroniuk, R. Stöber, and G. Fischerauer

Bayreuth Engine Research Center (BERC), Universität Bayreuth, Bayreuth, Germany

Correspondence to: I. Motroniuk (mrt@uni-bayreuth.de)

Abstract. A novel microwave-based approach for monitoring the state of aftertreatment systems such as diesel particulate filters (DPFs), three-way catalytic converters (TWCs), and selective catalytic reduction (SCR) catalysts is proposed. The volume inside the metallic housing of the DPF, TWC, or SCR is considered as a wireless communication channel between two terminals of a communication system. It is shown that, depending on the transmission channel characteristics, the properties of the catalyst, such as the catalyst state, can be inferred. This is done by means of an ultra-wideband (UWB) measurement and the subsequent evaluation and processing of the waveform in the time and frequency domains.

1 Introduction

Modern regulations for exhaust gas components like Euro 5 and Euro 6 require new methods both for the emissions control and for the monitoring and operation of the emission control systems in the diesel and gasoline engines.

During diesel engine operation, soot is captured in the pores of the DPFs (diesel particulate filters) or the SCR (selective catalytic reduction) catalysts. The ceramic filters must be periodically regenerated to avoid blocking. Hence, knowledge about the soot load is useful for the engine control in a number of ways (Feulner et al., 2013; Rose and Boger, 2009; Ochs et al., 2010; Sappok et al., 2010).

For gasoline engines, a TWC (three-way catalytic converter) is used to keep the polluting emissions low. The TWC stores and releases oxygen depending on the operating conditions. As in the DPF case, it would be of great advantage from the control engineering point of view to know the oxygen loading. However, it is the state of the art that the loading state cannot be measured directly but is derived indirectly from the oxygen concentration in the exhaust gas stream before and after the catalytic converter. The oxygen-measuring elements (so-called lambda probes) are implemented for the loading state estimation (Twigg, 2007).

There are a few known approaches for the direct measurement of the desired internal parameters (Zimmermann, 2007; Reiß et al., 2009). These and other similar ones also based on physical sensors can deliver only local information about the catalyst. However, one can learn nothing about the global state, such as the overall loading degree. Additionally, these approaches suffer from the complexities involving the mounting of the sensors in the catalyst and the communication with them.

Since the catalytic converter is typically located in a metallic housing, it and its housing together act as a filled cavity resonator. The resonances are influenced by the changes of material parameters in the catalyst so that the catalyst state can be determined from the appropriate signal characteristics of the resonances. The feasibility of this approach is well documented in the literature (Fischerauer et al., 2008, 2010b; Eichelbaum et al., 2012; Moos et al., 2013). Typically, to observe the cavity resonances, one or two simple thin probe feeds (short stubs) are connected to an automatic vector network analyzer (VNA) via coaxial lines. The changes in the resonant cavity are observed via the scattering matrix parameters S_{ij} measured by the VNA in the laboratory conditions.

It has also been shown that the microwave-based direct observation of TWCs enables superior engine control strategies compared to the indirect observation by way of lambda probes. This superiority consists, for instance, in lower emission levels (Schödel et al., 2014a, b).

The described laboratory equipment cannot be used in the field, i.e., inside a vehicle or in the manufacturing plant, owing to space and cost reasons. For the practical application, it should be replaced by smaller and cheaper solutions.

For solving this miniaturizing task, we have described a novel approach towards the monitoring of exhaust gas aftertreatment systems such as DPFs, SCRs, and TWCs (Motroniuk et al., 2014). It was proposed to consider the interior of the catalyst metal housing as a wireless communication channel. The effect of the catalyst state on such communication parameters as the bit error rate (BER), the packet error rate (PER), the ratio of the energy per bit to the noise power spectral density E_b/N_0 and the data receive rate (Rx data rate) was demonstrated by way of an example. The described approach involves a small-size and cost-effective modular architecture and has the potential for field application, unlike approaches based on laboratory equipment such as automatic vector network analyzers.

In a recent conference contribution, we proposed a modification of our novel approach. Instead of measuring certain communication channel parameters (such as BER, PER, E_b/N_0 levels, and Rx data rate), an ultra-wideband (UWB) pulse is to be transmitted through the catalyst and the changes suffered by it on its way through the catalyst are to be evaluated by a receiving device (Motroniuk et al., 2015). In other words, the propagation medium characteristics and the catalyst state are to be estimated from the UWB waveform changes.

In this contribution, we consider an extension of the UWB approach for the catalyst state determination. It is based on the subsequent evaluation and post-processing of the UWB waveforms in the time and frequency domains.

The rest of the paper is organized as follows. In Sect. 2, a novel wireless approach is presented, followed by a description of a hardware implementation and of an experimental verification with DPF in the time and frequency domains in Sects. 3 and 4, respectively. The conclusions are presented in Sect. 5.

2 Novel wireless approach

It is proposed to consider the interior of the catalyst housing as a communication channel between two terminals of a UWB wireless communication system. The most important parameters during the data transmission are the characteristics of the communication channel (in our case the catalyst). From the wireless channel characteristics, the properties of the catalytic converter such as the catalyst oxygen concentration or soot loading state can be determined. The latter assertion about the measurability of the soot loading state has been demonstrated in Motroniuk et al. (2014) and will be further corroborated in the remainder of this contribution. The former assertion about the measurability of the oxygen concentration follows from the fact that the electrical properties of catalysts depend heavily on their oxidization state; hence, this state can be inferred from the transmission or re-

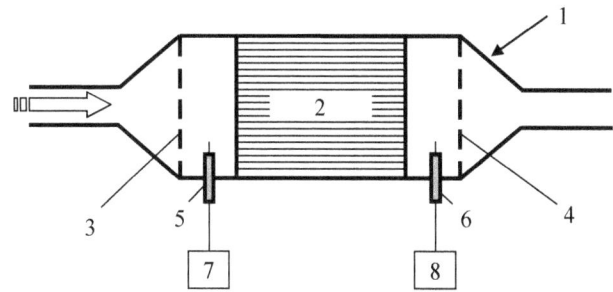

Figure 1. Basic structure of the proposed measurement system.

flection details of electromagnetic microwaves (Fischerauer et al., 2008; Eichelbaum et al., 2012).

The basic structure of the proposed measurement system is shown in Fig. 1. The system consists of a conductive (usually metallic) housing of the catalyst (1) and the catalyst itself (2). Two gas-permeable conductive grids (3, 4) which limit the resonating cavity are of advantage but not absolutely necessary. The grids are effectively short-circuiting the electric field, thus precisely defining the resonating cavity. Two thin-probe feeds (short stubs 5, 6) serve to connect the cavity with the system environment, i.e., with communication end devices (7, 8), by coaxial cables, which are used for signal transmission and reception. The short stubs play the same role for the communication system as the antennas in the common message transmission and the catalyst in its housing plays the same role as the wireless communication channel for the ordinary message transmission.

The signal transmitted through the wireless channel inside the catalyst suffers from such effects as attenuation, fading, multipath propagation, scattering, etc. Note, however, that the sender and receiver in a common wireless transmission link are located in the far field of each other whereas, in the present context, they are located in the near field. This, together with the resonance phenomena in the catalyst-filled cavity, introduces new propagation-path effects compared to wireless transmission links. The characteristics of the received information transferred from the transmitter to the receiver depend on the propagation medium. The effect on the signal is similar to the effect of buildings, trees, autos, rain, humidity, etc. in common wireless communication links: the channel characteristics influence the signal received, either in the time or frequency domain. Here, the closed-space propagation through the soot-loaded or otherwise chemically altered catalyst medium affects the received signal waveform.

In this paper, we specifically focus on a UWB communication system. Rather than measuring individual channel parameters, such as BER, PER, E_b/N_0 levels, or the Rx data rate (Motroniuk et al., 2014), or than measuring the S-parameters of a microwave two-port (Fischerauer et al., 2008), it is the integral UWB pulse waveform transmitted through the device under test (catalyst) which is considered. It is postulated that physical changes of the catalyst material

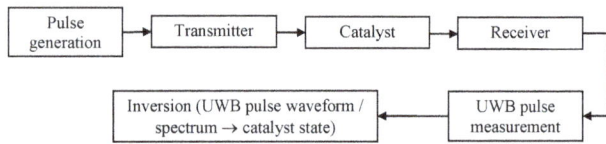

Figure 2. Functional diagram of the proposed measurement system.

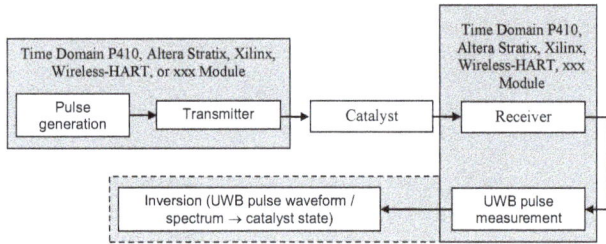

Figure 3. Exemplary hardware implementation of the proposed measurement system.

caused by the absorption or desorption of soot or chemical species are mirrored in the waveform received, or its spectral representation.

The goal, of course, is to estimate the internal state of a catalyst by this approach in a manner substantially simpler than possible with the system architectures demonstrated so far. This is achieved because the measurement can be performed by the existing hardware (wireless communication modules) and software (signal processing techniques for the wireless communication systems).

The functional diagram of such a system is shown in Fig. 2. A pulse generator generates the UWB waveform, which is then transmitted through the device under test (catalyst) and distorted in this process. On the receiver side, the UWB pulse waveform is measured to indirectly estimate the catalyst state.

3 Hardware implementation

Figure 3 depicts some possible ways to implement the architecture proposed in actual hardware. The UWB pulse generation and the transmission are accomplished by a communication module (we have used a Time Domain PulsON P410 module), and the receiver is implemented by another communication module (also PulsON P410). After receiving the pulse modulated by the catalyst properties, it is a matter of experience (calibration, simulation results, etc.) to infer the catalyst state from the waveform distortion.

Devices such as Altera Stratix, Xilinx Virtex, Xilinx Spartan, ZigBee, WirelessHART, Nanonet, PulsON 410, or any other including (but not limited to) field-programmable gate-array (FPGA)-based integrated circuits and devices could be used for building an appropriate communication system for the intended application purpose. The catalyst state can be

Table 1. Soot load of the measured DPFs (Fischerauer et al., 2010a).

DPF number	Soot load (g)	Soot load per DPF volume (g L^{-1})
1	0	0
2	4.6	2.0
3	13.6	5.8

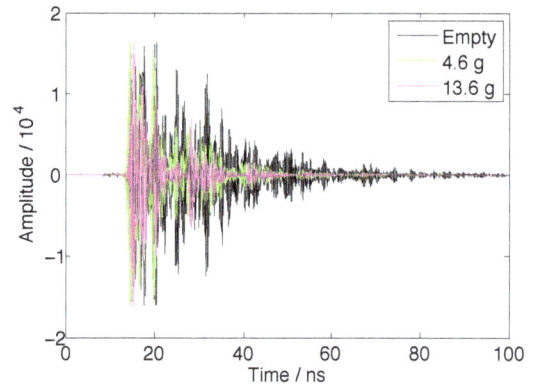

Figure 4. UWB waveforms for three DPFs loaded with different amounts of soot.

estimated by a post-processing of the received UWB waveform with the numerical computing environments and programming languages (such as MATLAB and C).

4 Experimental verification with DPF

4.1 Validation in the time domain

The validity of the approach was tested with DPFs loaded with different amounts of soot. The aim of the tests was to compare the effect of the soot load on the communication channel and, as a result, on the received UWB waveforms in order to prove the validity of the, as we believe, novel approach towards the monitoring and estimation of catalyst states.

One of the DPFs was clean (not used, with 0 g of soot), another one was loaded with 4.6 g of soot, and the third one was loaded with 13.6 g of soot. The filters used and their soot loading are listed in Table 1. For details of the catalyst geometry, the reader is referred to Fischerauer et al. (2010a).

The UWB pulse waveforms recorded during the communication between two commercial PulsON P410 modules via the interior of the DPF-loaded metal housing look substantially different for the various DPFs (Fig. 4). In particular, multiple-transit signals inside the catalyst housing (occurring after 18 ns in Fig. 4) are more attenuated the higher the soot load is. This is explained by the fact that the conducting soot introduces losses into the system. Thus, a wave packet is attenuated each time it traverses the soot-loaded DPF, and

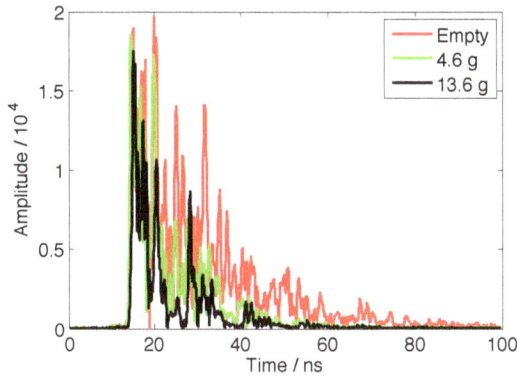

Figure 5. Envelopes of the time-domain responses from Fig. 4.

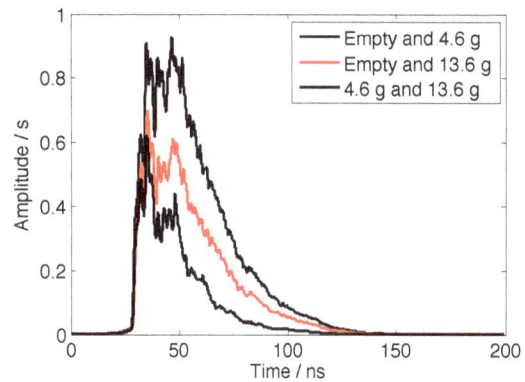

Figure 6. Convolution of the envelopes of the time-domain responses (Fig. 5).

this attenuates multiple-transit signals more than the single-transit (direct input-to-output) signal. In Fig. 4 the vertical scale represents A–D (analog-to-digital) counts proportional to volts as output by the analog-to-digital converter (ADC) in the UWB chip.

The envelopes $e(t)$ of the time-domain responses shown in Fig. 4 have been computed as the magnitude of the analytic signal constructed from the original time response $u(t)$ by adding to it its Hilbert transform (Marko, 1982):

$$e(t) = |u(t) + j\hat{u}(t)| \text{ with } \hat{u}(t) = \frac{1}{\pi} \int_{-\infty}^{\infty} \frac{u(\tau)}{t - \tau} d\tau. \tag{1}$$

Of course, all signals are discrete in time so that Eq. (1) had to be evaluated in the form

$$e(k\Delta t) = |u(k\Delta t) + j\,\hat{u}(k\Delta t)| \tag{1a}$$

$$\text{with } \hat{u}(k\Delta t) \approx \frac{1}{\pi} \sum_{\substack{\ell = \ell_{\min}, \\ \ell \neq k}}^{\ell_{\max}} \frac{u(\ell\Delta t)}{k - \ell}.$$

The result is shown in Fig. 5. One clearly observes a substantial difference in the signal envelopes depending on the soot content in the DPF. In particular, this can be seen starting from the second peak at around 16 ns. The soot loading state is not observable via the first peak because of the technical characteristics of the equipment (at the beginning of the received UWB pulse, the Tx pulse coupling through the Tx switch is showing through).

To quantify the differences, we calculated the convolution of the envelope functions for different DPF states, as always defined by

$$c_{ij}(t) = \int_{-\infty}^{+\infty} e_i(\tau)e_j(t - \tau)d\tau \quad \text{or} \tag{2}$$

$$c_{ij}(k\Delta t) \approx \sum_{\ell = \ell_{\min}}^{\ell_{\max}} e_i(\ell\Delta t)e_j((k - \ell)\Delta t)\,\Delta t$$

(Fig. 6; when comparing the y axis scaling of this and other figures to the scalings in Motroniuk et al. (2015), note that discrete convolution, correlation, and Fourier-transform formulas were used in the reference, whereas we have used the continuous-signal formulas in the current contribution. For example, the continuous-signal convolution approximately computed by Eq. (2) and the discrete convolution differ by a factor of Δt – the sampling interval – and this results in different y axis scalings.) The effect of the soot load on the curves is clearly seen. Multiple-transit signals are more attenuated with increasing soot load. The amplitude of the convolved signals at 50 ns is around 13×10^9 (A–D counts) $\times$ 61 ps $(\Delta t) \approx 0.79$ s for the empty/4.6 g case but only 8.9×10^9 (A–D counts) $\times$ 61 ps ≈ 0.54 s for the empty/13.6 g case.

The difference is observable by the width of the convolution curves as well. Measured at an amplitude of 3.28×10^9 (A–D counts) $\times$ 61 ps (Δt), this width amounts to 52, 42, and 27 ns, respectively, for the three cases.

We also calculated the cross-correlation of the envelope functions, given by

$$\Phi_{ij}(t) = \int_{-\infty}^{+\infty} e_i(\tau)e_j(t + \tau)d\tau \quad \text{or} \tag{3}$$

$$\Phi_{ij}(k\Delta t) \approx \sum_{\ell = \ell_{\min}}^{\ell_{\max}} e_i(\ell\Delta t)e_j((k + \ell)\Delta t)\,\Delta t.$$

For $i = j$, this turns into the auto-correlation $\Phi_{ii}(t)$ of the envelope function (Fig. 7). In particular, $\Phi_{ii}(0)$ is proportional to the signal power. Figure 7 clearly reveals that the signal power and the other values of the auto-correlation function decrease with the soot load. This is to be expected as more soot means higher signal attenuation, which in turn means smaller signal amplitudes at the receiver. In numbers, the maximum of the auto-correlation function is $3.2 \times 10^{10} \times 61$ ps $(\Delta t) \approx 1.96$ s, $1.8 \times 10^{10} \times 61$ ps ≈ 1.11 s,

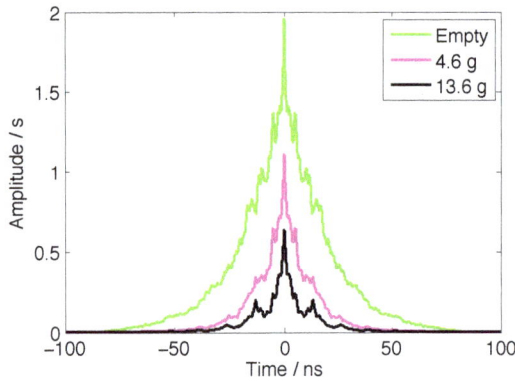

Figure 7. Auto-correlation functions of the envelopes shown in Fig. 5.

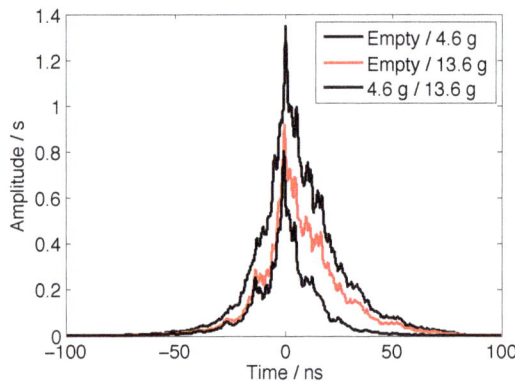

Figure 8. Cross-correlation functions of the envelopes shown in Fig. 5.

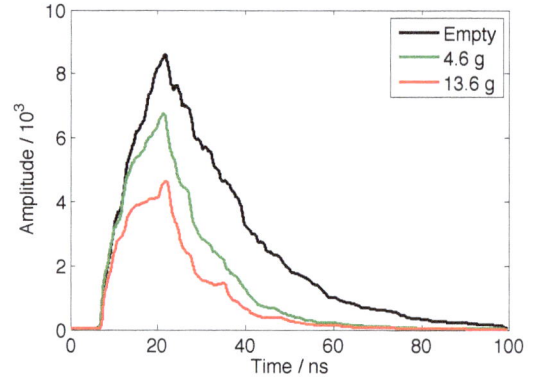

Figure 9. Smoothed envelopes of the time-domain responses from Fig. 4.

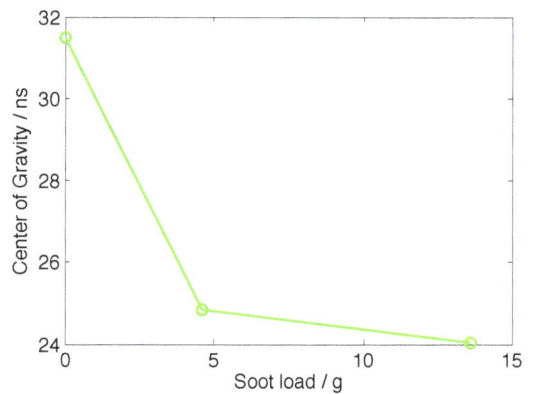

Figure 10. Time coordinate of the center of gravity of the area bounded by the smoothed envelopes and the time axis in Fig. 9.

and $1 \times 10^{10} \times 61$ ps ≈ 0.64 s for the empty DPF, the DPF loaded with 4.6 g soot, and the DPF loaded with 13.6 g, respectively. Again, the differences are also observable via the signal width.

Finally, Fig. 8 shows the cross-correlation functions. The amplitudes and mean values of the peaks are among the parameters suitable for a DPF state estimation algorithm. The maximum amplitude was $2.2 \times 10^{10} \times 61$ ps $(\Delta t) \approx 1.35$ s, $1.5 \times 10^{10} \times 61$ ps ≈ 0.92 s, and $1.3 \times 10^{10} \times 61$ ps ≈ 0.8 s for the cross-correlations of the three envelope signals from Fig. 5. The mean values were $2.8 \times 10^9 \times 61$ ps $(\Delta t) \approx 0.17$ s, $1.8 \times 10^9 \times 61$ ps ≈ 0.11 s, and $1.1 \times 10^9 \times 61$ ps ≈ 0.07 s for the three cases, respectively.

Owing to the rapid amplitude variation of the envelope curves in Fig. 5, it makes sense to first smooth them before further processing (Fig. 9). The amplitude of the smoothed envelopes depends on the soot load – with increasing soot load, the amplitude measured at the receiver becomes lower.

One possible way of processing is to correlate certain features of these smoothed curves with the DPF soot load. This is demonstrated in Fig. 10 for the center of gravity of the ar-

eas bounded by the curves in Fig. 9 and the time axis. It is obvious that this characteristic curve may be used to estimate the soot load from the measured time-domain response of the UWB system. It is also obvious that the sensitivity of this measurement system decreases with increasing soot load.

Along the same line of reasoning, we have computed the characteristic curves linking the maximum and the average values of the smoothed envelopes with the DPF soot load (Fig. 11).

While the overall behavior of the characteristic curves in Figs. 10 and 11 is quite similar, there are differences in the details. It appears as if small soot loads can be better resolved (measured with a higher sensitivity) by way of the center-of-gravity feature (Fig. 10) than by the other features. This would be quite interesting, but, in our opinion, it is too early to draw such conclusions as the number of measurements we have performed so far is much too small to allow for any substantial insight into the statistics of the measurement approach.

Besides, it is also important to note that the curves will look different with different transmitting power because of

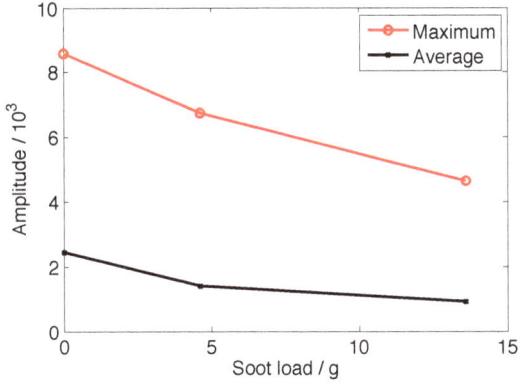

Figure 11. Maximum and average values of the smoothed envelopes (Fig. 9).

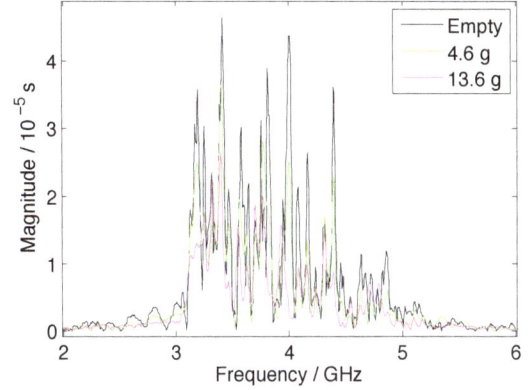

Figure 12. UWB magnitude spectrum for three different DPF loading states corresponding to the time-domain signals in Fig. 4.

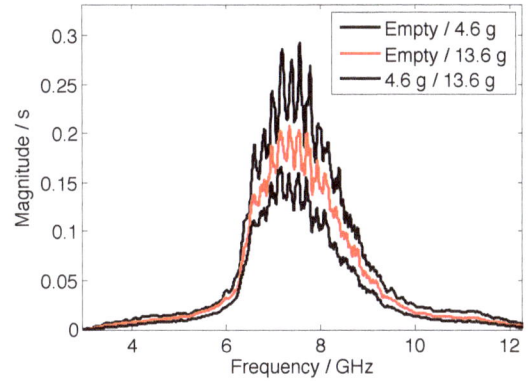

Figure 13. Convolution of the UWB magnitude spectra from Fig. 12.

receiver sensitivity. This, too, will have to be investigated in more detail.

4.2 Validation in the frequency domain

According to the results presented in Sect. 4.1, the state of automotive aftertreatment systems can possibly be inferred from the time-domain characteristics of UWB signals. However, in some cases, it may be easier to characterize and evaluate signals in the frequency domain. This will be considered in the present section.

UWB spectra for the three waveforms from Fig. 4 are shown in Fig. 12 (the spectral range from 3.1 to 5.3 GHz corresponds to the product specifications of the communication units used). One clearly observes substantial differences in the spectral amplitudes between the soot loading degrees, especially at the resonance peaks of 3.18, 3.4, 3.57, 4.0, 4.16, 4.4 GHz, etc. For example, the magnitude of the UWB spectrum at 3.4 GHz amounts to 7.6×10^5 (A–D counts) $\times 61$ ps (Δt) $\approx 4.6 \times 10^{-5}$ s for the empty DPF, to $5.9 \times 10^5 \times 61$ ps $\approx 3.6 \times 10^{-5}$ s for the 4.6 g load, and to $4.3 \times 10^5 \times 61$ ps $\approx 2.6 \times 10^{-5}$ s for the 13.6 g load. Obviously, the signals are more attenuated the higher the soot load is.

The convolution of the UWB magnitude spectra from Fig. 12 with each other is calculated with

$$C_{ij}(f) = \int\limits_{-\infty}^{+\infty} U_i(\varphi)U_j(f-\varphi)\mathrm{d}\varphi \quad \text{or} \quad (4)$$

$$C_{ij}(k\Delta f) \approx \sum_{\ell=\ell_{\min}}^{\ell_{\max}} U_i(\ell\Delta f)U_j((k-\ell)\Delta f)\,\Delta f$$

and plotted in Fig. 13. The maximum magnitude of the convolved signals in the frequency domain is around 0.29 s for the empty/4.6 g case, 0.21 s for the empty/13.6 g case, and only 0.17 s for the 4.6 g/13.6 g case.

Figure 14 shows the auto-correlation functions of the UWB magnitude spectra from Fig. 12, calculated with

$$\Phi_{ij}(f) = \int\limits_{-\infty}^{+\infty} U_i(\varphi)U_j(f+\varphi)\mathrm{d}\varphi \quad \text{or} \quad (5)$$

$$\Phi_{ij}(k\Delta f) \approx \sum_{\ell=\ell_{\min}}^{\ell_{\max}} U_i(\ell\Delta f)U_j((k+\ell)\Delta f)\,\Delta f,$$

and $i = j$. With increasing soot load, the magnitude of these auto-correlation functions decreases. The maximum magnitude observed was 0.49, 0.28, and 0.16 s for the empty DPF, the DPF loaded with 4.6 g soot, and the DPF loaded with 13.6 g, respectively. As in the time-domain case, the difference is also observable via the signal width.

Figure 15 presents the cross-correlation functions of the UWB magnitude spectra, Eq. (5) with $i \neq j$. The maximum magnitudes were 0.35, 0.24, and 0.19 s for the cross-correlations of the three signals obtained with different soot loads.

Furthermore, we have prepared histograms of the spectral magnitudes, measured in A–D counts, for three different

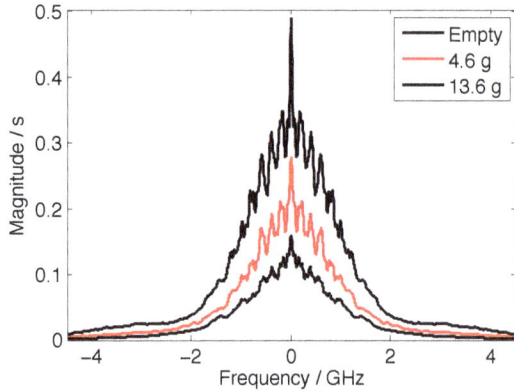

Figure 14. Auto-correlation functions of the UWB magnitude spectra in Fig. 12.

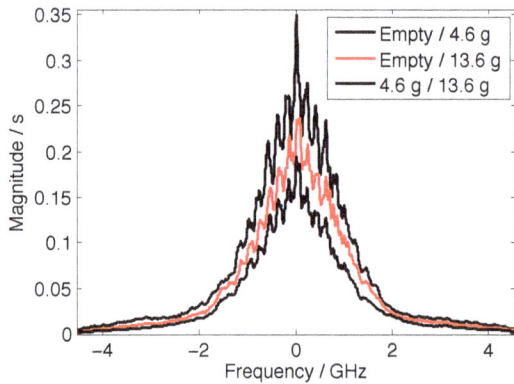

Figure 15. Cross-correlation functions of the UWB magnitude spectra in Fig. 12.

DPF loading states. The results are presented in Fig. 16. The frequency range evaluated was 3.1–5.3 GHz, consistent with the product specifications of the communication units used. The area of every rectangle in the histogram represents the relative frequency of the spectral magnitudes falling into the interval marked by the rectangle basis. The height of the rectangle is the relative frequency density (the relative frequency of the spectral magnitudes associated with the rectangle interval, divided by the width of the rectangle).

An appropriate empirical probability density function (pdf) used to fit the actually observed histogram must be one-sided and supported on the interval $[0, \infty]$ because the spectral magnitudes are non-negative numbers. Possible candidates, among others, are the Pareto distribution and the exponential distribution, both of which are often used in connection with system failure models. Here, we used the exponential distribution

$$f(x;\ \beta) = \begin{cases} \dfrac{1}{\beta} e^{-x/\beta} & \text{for } x > 0 \\ 0 & \text{for } x < 0 \end{cases} . \qquad (6)$$

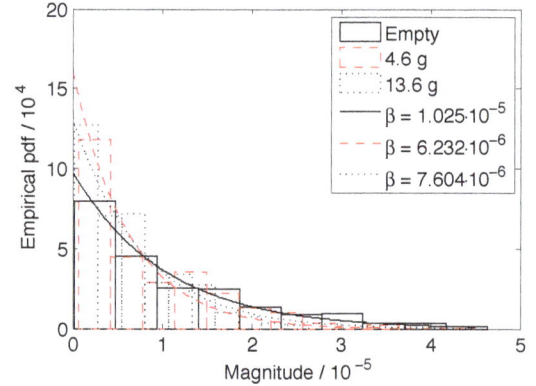

Figure 16. Histogram of the UWB magnitude spectrum for different loading states (bars) and best fits by exponential distributions $p(x) = \beta^{-1} e^{-x/\beta}$ (lines).

The exponential distributions with best-fitting parameters β are also shown in Fig. 16. The widths of the histograms decrease with the soot content or, in other words, the standard deviation of the pdf decreases with the soot content. As before, this is to be expected: more soot means higher attenuation, which in turn results in smaller signal amplitudes at the receiver.

One would expect that more soot load (i.e., higher attenuation) would smooth the spectrum in the sense that the magnitude variations become smaller; this would correspond to a pdf graph with steeper skirts and smaller width. In numbers, this would mean that the pdf parameter β increases with the soot load. The graphs in Fig. 16 support this assumption in principle (compare the 0 and 13.6 g soot cases), but one cannot infer a clear quantitative trend (compare the 4.6 and 13.6 g soot cases). On the other hand, given the small number of cases studied so far (only three different soot load values), there is not enough evidence to rule out the existence of a useful relationship between the soot load and the properties of the pdf describing the signal spectrum picked up by the receiver.

Finally, in analogy with the approach visualized for the time domain in Fig. 11, Fig. 17 depicts the maximum and average values for the magnitude spectrum in the range between 3.1 and 5.1 GHz. As before, the curves look promising in that they are monotonic and would allow for a unique inversion of the form "signal feature → soot load".

5 Conclusions

We have reported on a novel approach for the monitoring of exhaust gas aftertreatment systems, such as DPFs, SCRs, and TWCs, the most attractive feature of which is that it is based on a cost-effective modular architecture. Hence, it can be implemented in situ, e.g., in a car.

In the approach proposed, the interior of the catalyst metal housing serves as a communication channel between two

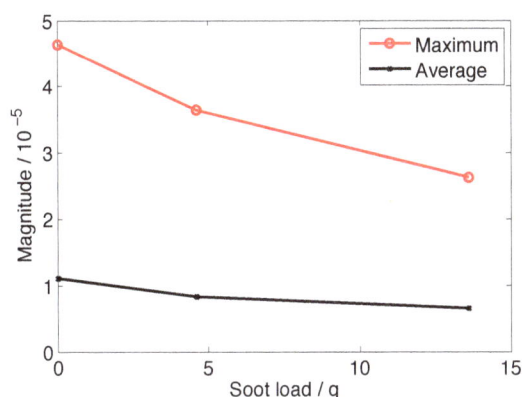

Figure 17. Maximum and average magnitudes of the signal spectrum at the receiver as functions of the soot load (from Fig. 12).

wireless communication nodes. To be more specific, we investigated a UWB-based architecture. It was demonstrated experimentally by way of various soot-loaded DPFs that the catalyst chemistry (i. e., the soot load of the DPF) affects the channel characteristics strongly enough to allow for the identification of the DPF state from the UWB signal details in the time as well as the frequency domains.

In particular, we were able to show that, in principle, certain features of the received time- and frequency-domain signals depend in a manner on the soot load which allows for the inversion of the feature vs. soot load characteristics. Among these features were the position of the center of gravity of the area bounded by the signal envelope and the time axis, the maximal value, the average value of the envelopes in the time domain and the maximum and average values in the spectra.

Acknowledgements. The authors thankfully acknowledge financial support by the German Research Foundation (DFG) under grants Fi 956/3-2 and Fi 956/5-1.

References

Eichelbaum, M., Stößer, R., Karpov, A., Dobner, C.-K., Rosowski, F., Trunschke, A., and Schlögl, R.: The microwave cavity perturbation technique for contact-free and in situ electrical conductivity measurements in catalysis and materials science, Phys. Chem. Chem. Phys., 14, 1302–1312, doi:10.1039/c1cp23462e, 2012.

Feulner, M., Hagen, G., Piontkowski, A., Müller, A., Fischerauer, G., Brüggemann, D., and Moos R.: In-Operation Monitoring of the Soot Load of Diesel Particulate Filters: Initial Tests, Topics in Catalysis, 56, 483–488, doi:10.1007/s11244-013-0002-9, 2013.

Fischerauer, G., Spörl, M., Gollwitzer, A., Wedemann, M., and Moos, R.: Catalyst state observation via the perturbation of a microwave cavity resonator, Frequenz, 62, 180–184, doi:10.1515/freq.2008.62.7-8.180, 2008.

Fischerauer, G., Förster, M., and Moos, R.: Sensing the soot load in automotive diesel particulate filters by microwave methods, Meas. Sci. Technol., 21, 035108, doi:10.1088/0957-0233/21/3/035108, 2010a.

Fischerauer, G., Spörl, M., Reiß, S., and Moos, R.: Microwave-Based Investigation of Electrochemical Processes in Catalysts and Related Systems, Technisches Messen, 77, 419–427, doi:10.1524/teme.2010.0066, 2010b (in German).

Marko, H.: Methoden der Systemtheorie, 2nd Edn., Springer, Berlin, Germany, 1982 (in German).

Moos, R., Beulertz, G., Reiß, S., Hagen, G., Fischerauer, G., Votsmeier, M., and Gieshoff, J.: Overview: Status of the Microwave-Based Automotive Catalyst State Diagnosis, Top. Catal., 56, 358–364, doi:10.1007/s11244-013-9980-x, 2013.

Motroniuk, I., Królak, R., Stöber, R., and Fischerauer, G.: State Observation in Automotive Aftertreatment Systems Based on Wireless Communication Links, in: Proc. 5th IMEKO TC19 Symp. on Environmental Instrumentation and Measurements, Chemnitz, Germany, 23–24 September 2014, 122–126, 2014.

Motroniuk, I., Stöber, R., and Fischerauer, G.: Waveform-Based State Determination for Catalytic Converters, in: Proc. 17th Int'l Conf. on Sensors and Measurement Technology (Sensor 2015), Nuremberg, Germany, 19–21 May 2015, 472–477, 2015.

Ochs, T., Schittenhelm, H., Genssle, A., and Kamp, B.: Particulate Matter Sensor for On Board Diagnostics (OBD) of Diesel Particulate Filters (DPF), SAE Int. J. Fuels Lubr., 3, 61–69, doi:10.4271/2010-01-0307, 2010.

Reiß, S., Wedemann, M., Moos, R., and Rösch, M.: Electrical In Situ Characterization of Three-Way Catalyst Coatings, in: Proc. 8th Int'l Congress on Catalysis and Automotive Pollution Control (CAPOC 8), Brussels, Belgium, 15–17 April 2009, 3, 67–74, 2009.

Rose, D. and Boger, T.: Different Approaches to Soot Estimation as Key Requirement for DPF Applications, SAE Technical Paper, 2009-01-1262, 11 pp., doi:10.4271/2009-01-1262, 2009.

Sappok, A., Bromberg, L., Parks, J., and Prikhodko, V.: Loading and Regeneration Analysis of a Diesel Particulate Filter with a Radio Frequency-Based Sensor, SAE Technical Paper, 2010-01-2126, 14 pp., doi:10.4271/2010-01-2126, 2010.

Schödel, S., Moos, R., Votsmeier, M., and Fischerauer, G.: SI-Engine Control With Microwave-Assisted Direct Observation of Oxygen Storage Level in Three-Way Catalysts, IEEE Trans. CST, 22, 2346–2353, 2014a.

Schödel, S., Votsmeier, M., and Fischerauer, G.: Microwave-assisted oxygen storage level estimation for three-way catalyst control: Model-based development and benchmarking of selected control strategies, Can. J. Chem. Eng., 92, 1597–1606, 2014b.

Twigg, M. V.: Progress and future challenges in controlling automotive exhaust gas emissions, Appl. Catal. B: Environmental, 70, 2–15, doi:10.1016/j.apcatb.2006.02.029, 2007.

Zimmermann, C.: Neuartiger Sensor zur Bestimmung des Zustandes eines NOx-Speicherkatalysators, PhD Thesis, Univ. Bayreuth, Shaker, Aachen, Germany, 2007 (in German).

14

Micro- and nanocoordinate measurements of micro-parts with 3-D tunnelling current probing

A. Schuler, T. Hausotte, and Z. Sun

Institute of Manufacturing Metrology, Friedrich-Alexander-Universität Erlangen-Nürnberg, Naegelsbachstr. 25, 91052 Erlangen, Germany

Correspondence to: A. Schuler (alexander.schuler@fau.de)

Abstract. Measurement tasks of modern micro- and nanometrology are posing a problem for current measurement instruments with decreasing structure sizes and rising aspect ratios. There is an open requirement for nanometre-resolving 3-D capable sensors and corresponding 3-D positioning systems to operate the sensors for 3-D measurements. A 3-D probing system based on electrical interaction is presented which is operated on a nanopositioning system type SIOS NMM-1. Furthermore, we demonstrate the progress and new possibilities for 3-D measurements with the nanopositioning and nanomeasurement machine NMM-1 and also with the application of a rotary kinematic chain. In addition new 3-D measurement routines for the NMM-1, also for micro-tactile probing systems as well as current plans, are shown.

1 Introduction

The progress in micro- and nanometrology was always driven by the developments in manufacturing technology. With newer manufacturing possibilities enabling smaller structures and components with higher complexity, a corresponding improvement in metrology is required (Hansen et al., 2006). The measurement uncertainty has to be 5 to 10 times better than the required manufacturing tolerance (Berndt et al., 1968). This also applies to the uncertainty of the probing sensors for coordinate measuring systems. Modern measurement tasks require a continuous increase of the measurement systems' resolution. Also, the aspect ratio and the three-dimensionality of workpieces are rising with surface angles up to 90°. Examples for such workpieces would be cutting tools for micro-manufacturing. Micro-cutting tools, such as micro-drills, show tapered cutting edges with radii of single micro-metres combined with drill diameters down to $200\,\mu$m. As the shape of the cutting edge is essential for the cutting process and as for micro-tools a homogeneous sharp edge is very difficult to manufacture, the position and shape of the edge has to be measured (Bissacco et al., 2008; Hansen et al., 2006). Other measurement tasks cover micro-components like the micro-gears depicted in Fig. 1. These parts, either manufactured by cutting pro-

cesses or the new technology of LIGA manufacturing show high aspect ratios with poorly accessible corners (Hansen et al., 2006).

The list of challenging components can be extended ranging from optical parts, such as micro-lenses or aspherics, to functional surfaces or micro-fluidic devices (Jiang and Whitehouse, 2012; Weckenmann et al., 2011). Generalizing, a resolution and uncertainty in the nanometre range and full 3-D capability would be required. In contrast current sensors cannot completely meet these requirements, especially the necessary 3-D operation. Optical sensors are limited in their maximum detectable surface angle, usually defined by the numerical aperture of the chosen lens (Rahlves and Seewig, 2009). If the angle is exceeded no surface points (or erroneous surface points) are recorded. Also, widely used tactile profilometers are limited in their maximum angle. As the styli require a certain cone angle of 90 or 60° and a tip radius of 2, 5 or $10\,\mu$m for stability and lateral stiffness, high surface angles lead to a wandering of the assumed contact point with resulting measurement deviation until eventually an unintentional flank collision occurs (Lonardo et al., 1996, ISO 3274:1996). The only sensors fully capable of 3-D measurements are the class of micro-tactile sensors. As the applied probing elements have diameters of $15\,\mu$m and higher,

Figure 1. Gears of a micro-planetary drive. Diameter of the largest gear 4 mm.

the accessibility of small structures and the lateral resolution is limited (UMAP Vision System, 2012; Weckenmann et al., 2004).

Apart from the sensor side progress in the field of co-ordinate measurement systems is also required. To realize the three-dimensional measurement of complex specimen in the nanometre order, the sensor systems (2-D or 3-D probe systems) should be additionally combined with appropriate mover systems (with possibly large movement range) for ma-nipulating and measuring the relative position of the speci-men to the probing tip. To fully utilize high accuracy sensors and to comply with Berndt's statements the positioning sys-tem must also have a resolution of up to sub-nanometre and an uncertainty of only a few nanometres. The classical co-ordinate measuring machines (CMMs) with moving probes, which commonly only fulfil the Abbe comparator principle in the z axis, cannot meet this requirement (Weckenmann et al., 2004). For a three-dimensional measurement with the highest accuracy in the nanometre range, it is necessary to apply the Abbe comparator principle in all three coordinate axes. The current favourable way is to move the measurement object and to keep the probe systems in a fixed position, due to the fact that the workpiece carrier can be extended with a well-known and stable reference surface, e.g. a plane mir-ror, for a displacement measurement system. In this case, the probe system is used as a null indicator. To fulfil the Abbe principle in all axes the probing point lies in the intersection point of the three perpendicular measurement axes (Hausotte, 2011). Some of the mover systems like the positioning stage of the nanomeasuring machine SIOS NMM-1 already fulfil these requirements (Hausotte, 2011; Hausotte et al., 2009; Schott et al., 2013).

As sidewalls of workpieces or other 3-D structures are at present only measureable with tactile sensors, suitable mea-surement machines to operate them are required. In particu-lar, a control system with various scan functions is necessary. A measurement task using a 2-D or 3-D sensitive probing system requires a predefined probing direction. Alternatively, the direction should be calculated in real time during the scan

process for free-form scans and requires corresponding scan routines (Hausotte et al., 2009; Schott et al., 2013).

2 Development of a 3-D nanometre measurement system

To address the needs of modern measurement tasks, research has been conducted at the Friedrich-Alexander-Universität Erlangen-Nürnberg (FAU). Derived from the electrical prob-ing interaction of scanning tunnelling microscopes (STM), a sub-nanometre resolving 3-D sensor was developed and combined with the nanopositioning and nanomeasuring ma-chine NMM-1 to perform measurements on electrically con-ductive workpieces.

Scanning tunnelling microscopes were initially demon-strated by Binnig et al. (1982) and are based on the quantum tunnelling effect. By applying a voltage of a few millivolts between a conductive workpiece and a conductive sharp nee-dle, a current signal of a few nanoamperes can be detected if the gap between both conductive partners is in the nanome-tre range. Due to the quantum tunnelling effect, electrons can pass the isolation barrier between both electrodes and con-tribute to an exponentially increasing current with a decreas-ing distance between the electrodes. The scanning tunnelling microscope easily allows for sub-nanometre resolution in the working direction and also in the lateral direction if the tip is sharp enough. The microscope, as in all scanning probe mi-croscopes, is limited to flat surfaces due to the shape of the tip and the restricted working range of the positioning piezos.

At FAU the working principle was extended to three di-mensions by using, instead of a needle-like tip, a metallic sphere as a sensor electrode allowing a probing on any part of the sphere (Weckenmann and Hoffmann, 2007; Hoffmann and Schuler, 2011). Apart from the gained 3-D capability, the application of probing spheres increases the area of electrical interaction between probe and workpiece, especially when sphere radii in the range from micro-metres to half a mil-limetre are applied. This way with the downturn of lowered lateral resolution, the vertical working range in surface di-rection increases up to 200 nm depending on the alloy of the sphere and workpiece as well as their surface roughness. The sensor characteristics and further aspects of the near-field in-teraction and involved effects were investigated in the works of Hoffmann (2009). The benefits are a force-free probing interaction without the danger of workpiece damage and an isotropic behaviour in all directions. Figure 2 shows the ba-sic application and the resulting exponential characteristic curve if the sphere approaches the surface. Figure 3 displays the schematic of the surrounding set-up and the information flow. A precision voltage source is connected to the sensor electrode enabling the sensor operation. The distance-related current flow over the air gap of some nanometres is measured with a femtoampere amplifier with variable amplification and converted into a voltage signal. Usual amplification factors

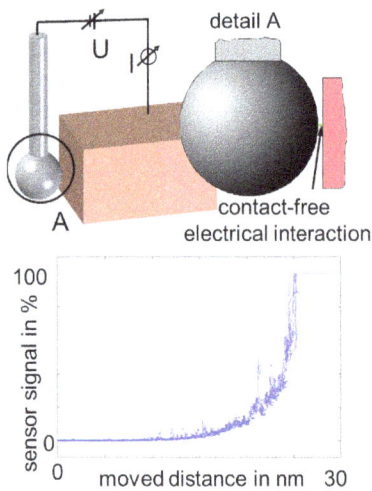

Figure 2. General set-up and characteristic of the electrical 3-D sensor.

range from 0.1 to $10\,V\,nA^{-1}$. As the amplifier's output voltage is limited to $\pm 10\,V$, the amplification factor also influences the measurement range and is to be chosen in accordance with the desired set point. The characteristic in Fig. 2 displays this signal clipping at 100 % sensor signal equivalent to 10 V. The amplified signal is afterwards conditioned to match the input specifications of the connected positioning system, described in the following paragraph.

To make full use of the sensor's capabilities it was integrated in a nanopositioning and nanomeasuring system NMM-1 (Hausotte, 2011; Hausotte et al., 2012; Schott et al., 2013). The NMM-1 features a positioning and measuring volume of $25\,mm \times 25\,mm \times 5\,mm$ and increases the sensor's measurement range equivalently. A corner mirror which is positioned by a three-axis drive system, is used for carrying the object to be measured (see Fig. 4). The accurate reference coordinate system is defined by the corner mirror and the position of the corner mirror is measured with three homodyne interferometers supplied by frequency stabilized lasers (Hausotte et al., 2011). The measuring axes intersect at the point where the sensor tip is situated. This set-up fulfils the Abbe comparator principle in three axes and the probe system acts in this case mostly only as a null indicator. The interferometer measurements are used for a closed-loop position control of the corner mirror. This allows for the compensation of the linear stage's guide path errors. By using two optical autocollimators in x and y axes, angular deviations of the guide system are also measured on the corner mirror. An additional angular control system compensates the angular errors using the four z axis drives. The system provides a reliable measurement resolution of 0.1 nm with the smallest quantisation steps of 0.02 nm. The positioning uncertainty is less than 10 nm for the entire measurement range.

The NMM-1 offers a simple integration of different sensor systems with its set of analogue inputs. The basic integration

procedure is as follows: the voltage outputs of a sensor are recorded during surface approach and third-order polynomials are fitted into the data. The polynomials are applied in the machine's position control circuit to keep the sensor in its chosen working point of the calculated probing interaction signal (e.g. magnitude of probing force vector) (Hausotte, 2011). During a scanning operation with probing in vertical direction, the workpiece is moved (e.g. laterally) and surface height changes result in a change of the sensor signals. The position control circuit continually adjusts the workpiece carrier's height position to keep the configured set point. With the initial control design of the NMM-1 a sensor (1-D) was only used for the height control which limited the application to surfaces with a low curvature. With the recent developments of the NMM-1, a probing and position control via the sensor signal can be performed in all three spatial directions. Resulting possible scan routines are described in Sect. 5. Basically, the position control circuit regulates the distance sensor workpiece in the direction of the probing vector. The vector is either fixed like in the case of an optical point sensor which only operates in beam direction or the vector results from 3-D probing systems like, e.g. a micro-tactile system that outputs its deflection in three spatial directions.

3 Application of the 3-D electrical probing system

The electrical probing system has been used for 2-D as well as 3-D measurement tasks. As mentioned earlier the electrical probing system is capable of probing in any direction because of its spherical probe shape. In comparison to a microtactile probe, which outputs signals for the vector (magnitude and direction) of its deflection, the electrical sensor outputs only a magnitude signal independent of the geometric probing direction. As sophisticated scan commands (described in Sect. 5) require a 3-D probing force vector, mainly simple line scans and single point measurements are possible.

With the electrical sensor basically two different approaches are available. One method is to use a fixed probing vector like the height axis of the NMM-1. This method was used especially with the old control system of the NMM-1 and applied for measurements of flat surfaces and 3-D structures with limited curvature. Figure 5 shows a resulting application where the centre points of two metallic spheres on a ball bar are measured and their 3-D distance is measured.

The spheres' upper hemisphere can be measured nearly to their equators, but with increasing difference between probing vector and surface vector the stability of the position control system drops as the sensor characteristic changes. Furthermore, with the probing vector differing from the surface vector the form of the probing element has to be compensated as the actual probing point wanders from its assumed position.

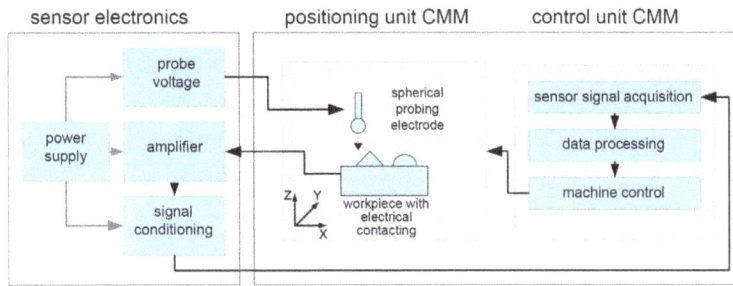

Figure 3. Information flow in the measurement set-up.

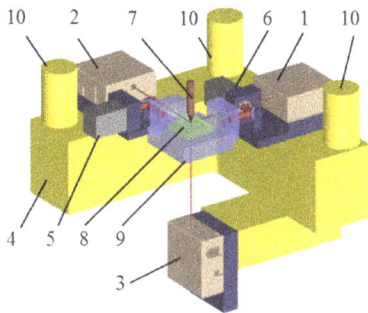

Figure 4. Principle design of the nanomeasuring machine: (1) *x* interferometer, (2) *y* interferometer, (3) *z* interferometer, (4) metrology frame made of Zerodur, (5) roll and yaw angular sensor, (6) pitch and yaw angular sensor, (7) surface-sensing probe, (8) sample, (9) mirror corner, and (10) fixing points for probe system.

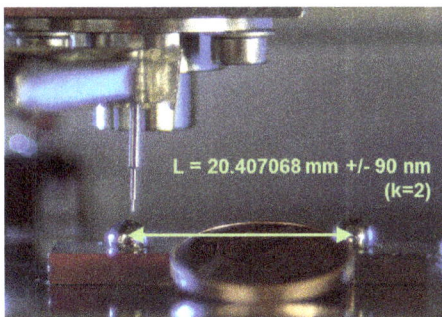

Figure 5. Distance measurement on a ball bar.

Another exemplary 3-D measurement task is displayed with Fig. 6 where the structures on a cutting insert are inspected.

With the newer control system a programmable probing vector is also possible. The probing vector is precalculated to match the local surface angle either derived from a first measurement pass or from a computer aided design (CAD) model. A standard measurement task which was performed with the electrical sensor would be the thickness calibration of a gauge block where the probing is performed in a lateral direction. Another example would be the point-wise mea-

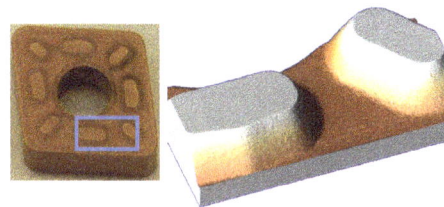

Figure 6. Surface measurement on a cutting insert, area 6 mm × 3 mm.

surement on a sphere where, with the use of a stylus tree (Hoffmann and Schuler, 2011), the full surface, also below the equator, can be measured.

The disadvantage of this 3-D probing system is, similar to micro-tactile systems, the limited lateral resolution on the surface. Only with a very small sphere or very sharp tip can high lateral resolutions or high structure resolutions be achieved, but this again leads to possible shaft collisions and limited 3-D capability.

4 Extension of measurable surface angles

As mentioned in the introduction most sensors show a limited measurable surface angle or at least high measurement deviation for measurements with high inclinations to the normal vector of the surface. Therefore, a method to achieve 3-D capability with 1-D probes was developed at the FAU which can be applied to many sensors. The aim was to increase the effectively measurable surface angle and to reduce measurement deviation by keeping a sensor in its optimal working angle during the measurement. This sensor tilting principle was planned and simulated in theory and the principle was afterwards transferred into a hardware prototype where the suggested principle and its effectiveness were demonstrated (Weckenmann et al., 2012; Schuler, 2013; Schuler et al., 2014). The simulation calculated the occurring measurement deviation of a tactile profilometer on an arbitrary surface and the effect of the tilting was investigated. A main task was the development of the control theory to calculate the sensor's optimal tilt angle at a certain surface position during a surface measurement. Different strategies to calculate

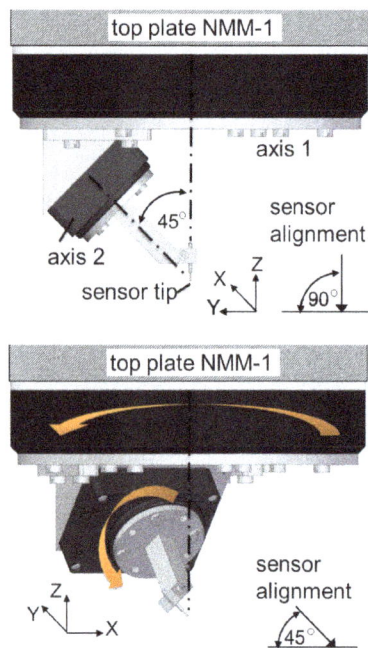

Figure 7. Schematic of the rotary kinematic chain, operated at 90° and 45° sensor alignment.

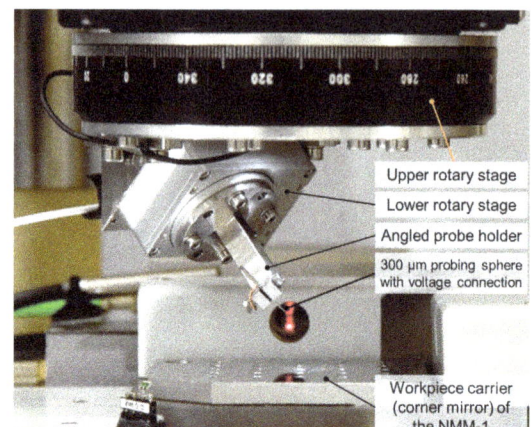

Figure 8. Kinematic chain installed into the NMM-1 with the electrical near-field sensor.

the sequence of tilt angles were investigated. With knowledge based strategies using, e.g. a CAD model, the ideal angles can be calculated beforehand; therefore, with analysing strategies a first scan pass with fixed sensor alignment delivers the calculation basis for a second pass, and with predicting strategies the surface angle is extrapolated during the measurement from already acquired surface points. A series of test cases were simulated and generally a sensor tilt of already 10° could significantly lower the measurement deviation caused by operating the sensor out of its optimal working angle. The simulation results also created the input data for the practical realisation. Different kinematic chains were calculated and compared with the aim that for an effective operation, the probing point of the sensor should not deviate more than ±500 nm after tilting. In result a kinematic system was developed to rotate a 1-D sensor around its probing point in space with a maximum rotation angle of ±90° in 2 degrees of rotation. During a profile measurement this allows for the continuous angular realignment to keep a sensor perpendicular to the workpiece surface (sensor axis parallel to the normal vector of the surface).

As a testing platform for the prototype the NMM-1 was used mounting the rotary kinematic chain above the workpiece carrier. Figure 7 shows a drawing of the developed rotation system and its basic operation principle.

The set-up consists of two stacked rotary tables. The first one with a diameter of 100 mm is a direct-driven stage and the second one with 30 mm diameter uses a self-locking piezo drive. The second axis is mounted below the first axis under an angle of 45° and carries a sensor tip in a sensor

holder again under a 45° angle. In a default orientation the sensor is held perpendicular to the $X-Y$ layer of the NMM-1's workpiece carrier. Both rotation axes are aligned in a way that their centre axes and the sensor probing point virtually intersect, apart from alignment deviations. By turning the lower axis, the sensor revolves around its working point and turns into the $X-Y$ plane of the NMM-1, resulting in rotation components around X and Y. With the additional movement of the upper rotation stage, the X and Y components are translated into each other and a desired combination can be achieved, e.g. a 45° rotation around the y axis in Fig. 7 (lower panel). The controllers of both axes are connected to a host computer and are integrated into the sequence control based on MATLAB which controls the NMM-1 and also realizes the tilting strategy and performs coordinate transformations. To allow for a fast translation of a required Cartesian sensor tilt angle into the corresponding angular positions of the rotary stages, a look up table is also included. Figure 8 displays the finished set-up.

As mentioned before to ensure a feasible operation of this kinematic for micro- and nanometrology, the residual deviation of the sensor's working point after rotation has to be below ±500 nm. As the applied rotation axes can show systematic position errors in the micro-metre range and possible alignment errors also have to be respected, an in situ calibration procedure was created. It relies on the positioning ability of the NMM-1 and uses for best effectiveness the already described electrical 3-D sensor.

The calibration procedure implements the so-called empirical qualification from ISO 10360-5:2010 for articulating probing systems (ISO 10360-5:2010). On the workpiece carrier a calibration sphere is mounted and its centre serves as a reference point being measured under different sensor angles. With the nanometre resolving electrical sensor and its ability to work with spherical probes, the full upper hemisphere of the calibration sphere can be measured, but the

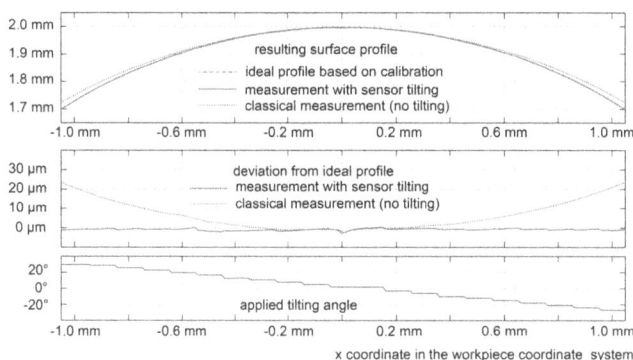

Figure 9. Measurement results with and without sensor tilting on a 4 mm sphere.

procedure can also be used with other sensors, e.g. a tactile profilometer or optical point probe. The resulting centre points allow the calculation of the sensor's probing point in relation to the reference sphere. The movement vectors of the sensor caused by rotation are stored in a compensation field and are compensated by the NMM-1's positioning abilities. With this approach compensating detectable systematic error components the remaining residual position error of the kinematic chain was identified. The range of the sensor position over the calibrated angular range, defined by the non-systematic error components and the non-compensated systematic components, was below 140 nm in the worst Cartesian direction. This parameter is also referred to as P_{LTE} in ISO 10360-5:2010. Compared to the initial requirement of the error range being smaller than $1\,\mu$m (± 500 nm), the realized kinematic chain successfully enables the investigation of the sensor tilting principle in practice.

An exemplary measurement is shown in Fig. 9 comparing measurements on a 4 mm sphere, all conducted with the electrical sensor and a probing tip of $150\,\mu$m curvature radius. The dotted red line in upper part of Fig. 9 shows the results of a regular measurement with the electrical sensor and a vertical sensor alignment. Here the electrical sensor is operated like a profilometer meaning a fixed probing vector and a fixed contact point on the bottom of the tip is assumed. The surface profile is superimposed with the tip shape leading with increasing surface angles to increasing measurement deviation (see middle part of Fig. 9). This convolution can be compensated to a certain degree, especially well on spheres, but the deconvolution is limited on more complex surfaces (Keller, 1991). In comparison, an optical sensor would only be able to record the surface to a certain degree and would get erroneous or no data if the angular limit is exceeded.

With the shown sensor tilting in contrast, the sensor can be kept in its optimal angle, the contact point between tip and surface does not wander and the deviation is avoided. In the middle of the recorded profile in Fig. 9 both curves coincide as the regular measurement operates in its optimal working angle. With increasing distance from the sphere's pole the

deviation increases and the benefit of the tilting principle is visible. Basically with the current kinematic chain, the effectively measurable surface angle of an installed sensor is increased to more than $\pm 90°$.

The electrical near-field 3-D sensor was used for the calibration procedure and to demonstrate the difference between a fixed and a tilted operation. With its 3-D capabilities and the NMM-1's 3-D control unit, a tilting would naturally not be required as the spherical probe can interact in any direction if the probing vectors are chosen appropriately as described in Sect. 3. With a tiltable 1-D probe and very sharp tip a high lateral resolutions or high structure resolutions can be achieved. Current research covers therefore the transfer of this sensor tilting principle to other sensors like optical 1-D sensors.

5 3-D measurement routines of nanocoordinate measurement machines

Apart from developments on the sensor side progress on the side of nanocoordinate measurement machines can also be reported. The current firmware and 3-D position control unit of the NMM-1 provides a complete 3-D control system based on the I++ DME (Inspection plus plus for Dimensional Measurement Equipment) specification for implementing point and scan measurement instructions (Hausotte, 2011; Hausotte et al., 2009; Schott et al., 2013).

Basically the sensors can be operated in an open-loop or a closed-loop mode. Simple open-loop scan commands use only the machine position as the input and the probe is moved along a predefined trajectory across the workpiece to record a profile. If a threshold is specified in this situation where the sensor would exceed its working range, the scan trajectory can be modified during the scan by moving away or towards from the surface. The scanning can also be conducted in a continuous closed-loop operation by maintaining a constant deflection or probing force at all times during the scanning, e.g. along a line, around a circular path or as a helix. Furthermore with a so-called "dodge scan" an open- and a closed-loop scan can be combined. An automatic switching between both the methods is thus feasible in order to improve the measurement speed and accuracy. If the sensor exceeds its working range the position control is switched to closed-loop and evades the obstacle, afterwards it switches back to the open-loop operation. With a two- or three-dimensional sensitive probing system, a measurement operation with the free-form scan is possible. In this case, the probing direction does not have to be predefined. It only depends on the form of the sample surface and is calculated from the deflection vector of the probe.

With different newly implemented scan techniques and various instructions, especially the three-dimensional measurements of micro-structures can be flexibly operated. They are intended to be used with 3-D sensors with a strong focus

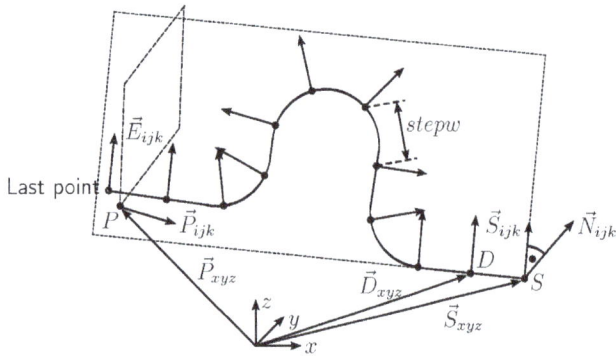

Figure 10. Principle and parameters of free-form scan function "ScanInPlaneEndIsPlane".

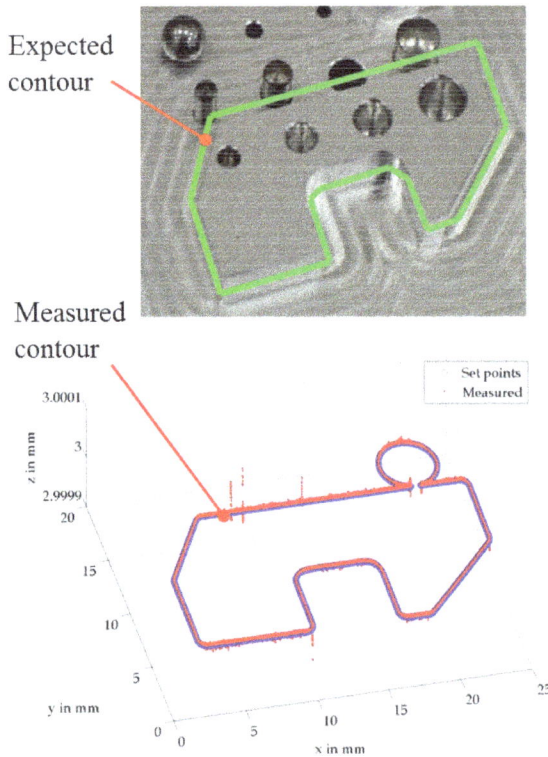

Figure 11. Free-form scan on a sample structure.

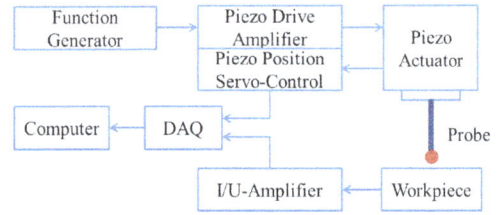

Figure 12. Block diagram indicating key components of the measurement system.

Plane" is shown in Fig. 10. At first a surface approach is executed and the start point S_{xyz} on the surface is probed. The controller attempts to follow the contour in a plane defined by N_{ijk} with given scan speed, and samples the data points with a constant point distance (stepw). The scan direction is calculated from the probing force vector, measured by the tactile probe and its magnitude, and it is calculated with the plane definition. The probing force magnitude is controlled while the deflections and force magnitude are monitored. The scan terminates when the last measured point exceeds the defined end plane.

Figure 11 illustrates a free-form scan of a sample structure using "ScanInPlaneEndIsPlane". A conventional tactile probe with a diameter of 1 mm was used to demonstrate the function. The gap between the biggest bolt and the contour was too slim for the probe sphere to run through. Thus, the controller has chosen another path, around the bolt, to continue the scanning. This prevents expensive probe damage especially on micro-parts where the operator can no longer track the probe system only with his eyes. At each point of the contour the normal vector of the contact surface must be calculated by the NMM-1 controller in order to determine, whether the scan direction should be maintained or with a small change, or even turn sharp to totally another direction.

To benefit from those flexible instructions for free-form scans, it is planned to extend the electrical near-field sensor with capabilities to directly detect its contact direction within an ongoing research project. When the magnitude and the direction of the contact vector are known, the elaborate measurement commands for free-form surfaces can be used.

6   3-D vector extension of the electrical probing system

To address this extension of the electrical probing system's abilities, a superimposed mechanical oscillation is investigated. By using a metallic sphere as a sensor electrode a probing on any part of the sphere is possible and the sensor is hence sensitive in all directions in the measuring space. But differing from the force signal of a tactile probe, the electrical signal is only a scalar, and the direction information has to be gathered otherwise. If a mechanical circular movement is applied the probing direction can be derived from the

on 3-D sensitive tactile micro-probes featuring small probing spheres of down to $15\,\mu m$ diameter. Such micro-probes are commonly designed to detect the probing force as well as to determine the probing direction (Balzer et al., 2011). This type of probe is advantageous for a free-form scan of workpieces with high aspect ratios. In this case the probing vector between probing tip and surface is directly measured and an optimal probing direction can be easily calculated from the detected force vector in real time. It only depends on the form of the sample surface and does not have to be predefined.

As an example for the routines developed for 3-D probing systems, the free-form scan instruction "ScanInPlaneEndIs-

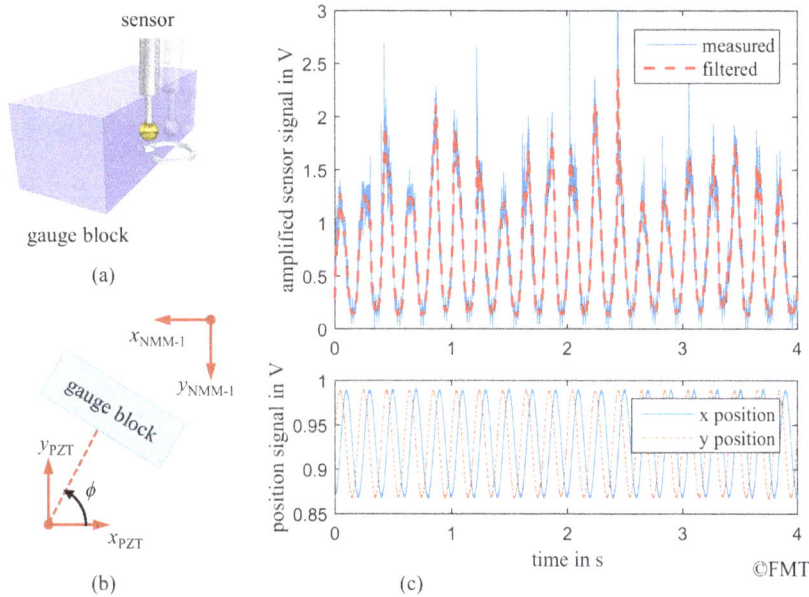

Figure 13. (**a**) Arrangement of the measurement system; (**b**) top view with the coordinate system of the piezo actuator, ϕ: to be determined angle; (**c**) measurement results.

relation between sensor signal and circular sensor position (Ogura and Okazaki, 2002). The frequency of the circular movement and the speed of the vector calculation should be high enough to deliver the NMM-1 a 3-D probing vector during the scanning motion.

Figure 12 shows the configuration of a first measurement set-up. The probe is mounted on a 3-D movable, precise piezo scanner consisting of a function generator, a piezo-drive amplifier, a piezo-position-servo control, and a piezo actuator. As sine and cosine wave signals generated from a function generator are synchronously applied to the actuators, the probe is moving on a circular path of some nanometres. When the path is within the working range of the electrical sensor, a periodic tunnelling current signal will be measured. A data acquisition box (DAQ) can be used to register the data of the actuator position and the amplified current signals synchronously. The NMM-1 is used for a positioning of the workpiece in to the working range of the probe.

As an example a side wall scan of a gauge block with a spherical probe is shown in Fig. 13a. The probe moved circularly in the X–Y plane around the origin of the piezo actuator coordinate system. Depending on the accuracy of the mechanical adjustment/integration of the sensor system, the coordinate system of the actuator is parallel to that of the nanomeasuring machine. The results of the preliminary set-up still indicate drift and vibration effects and the mechanical set-up is therefore being revised. With the planned connection to the NMM-1's position control, the results can be further improved. Nevertheless, with the help of low-pass filtering and Fourier analysis the characteristics of the signals and the resulting probing vector can be determined. The dif-

ference between the phase shifts of the sensor signal and the x position signal is consistent with the angle ϕ, as specified in Fig. 13b. In this condition the probing vector ($\cos \phi$, $\sin \phi$) is consequently identified in the piezo actuator coordinate system.

The further steps involve the set-up of a separate control unit based on a digital signal processor for real-time calculations. This unit will substitute the function generator and the data acquisition. It will be able to create the circular movement path on all three axes and will forward the calculated proving vector to the NMM-1. With the connection to the NMM-1 the position control based on the sensor signal will be realized and the investigations can continue.

7 Conclusions

Driven by modern measurement tasks requiring nanometre resolving sensors with 3-D capability and appropriate nanopositioning systems with suitable scan routines, developments to address these problems have been demonstrated.

To meet these requirements a working principle known from scanning probe microscopes was adapted realising sub-nanometre sensor resolution with a contact free 3-D probing capability. In combination with a nanopositioning system type NMM-1, an operation, such as a 1-D sensor or a 3-D sensor, is possible allowing even the probing of sidewalls and undercuts. Current developments focus on directly detecting the contact vector between sensor and workpiece so more elaborate scan routines can be applied. For this aim research is being conducted to superimpose a circular movement to a

3-D capable probing element and to derive the contact vector from the relation between the circular position and signal.

Apart from the electrical sensor another approach to gain 3-D ability was demonstrated based on the tracking of a 1-D sensor's operation angle during a measurement. A rotary kinematic chain turns a sensor around its working point and keeps its orientation perpendicular to the surface. This effectively reduces measurement deviation on sloped surfaces and extends the accessible surface angle to $\pm 90°$ in 2 degrees of freedom. Paired with an online compensation of systematic position errors the principle was successfully brought to application.

On the side of 3-D positioning systems progress also is reported with the new control system and scan routines of the nanopositioning and nanomeasurement system NMM-1. The routines allow for free-form scans on 3-D objects enabling the operator to make full use of 3-D sensors. Due to the decreased visibility of micro-parts for the operator and the always prevailing danger of damaging a micro-probe elaborate, scan routines are necessary.

Acknowledgements. The authors wish to thank all those colleagues at the FAU and the Ilmenau University of Technology who have contributed to these developments. The authors also thank the German Research Foundation (Deutsche Forschungsgemeinschaft, DFG). Section 4 is based on the research grant WE 918/32-1 conducted at the chair QFM (FAU) under the leadership of Albert Weckenmann. The current development of the electrical sensor is supported by research grant HA 5915/4-1.

References

Balzer, F. G., Hausotte, T., Dorozhovets, N., Manske, E. and Jäger, G.: Tactile 3-D microprobe system with exchangeable styli, Meas. Sci. Technol., 22, 094018, doi:10.1088/0957-0233/22/9/094018, 2011.

Berndt, G., Hultzsch, E., and Winhold, H.: Funktionstoleranz und Meßunsicherheit, Wissenschaftliche Zeitschrift der Technischen Universität Dresden, 17, 465–471, 1968.

Binnig, G., Rohrer, H., Gerber, C., and Weibel, E.: Surface Studies by Scanning Tunneling Microscopy, Phys. Rev. Lett., 49, 57–61, 1982.

Bissacco, G., Hansen, H. N., and Slunksy, J.: Modelling the cutting edge radius size effect for force prediction in micro milling, CIRP Ann. Manuf. Techn., 57, 113–116, 2008.

Hansen, H. N., Carneiro, K., Haitjema, H., and De Chiffre, L.: Dimensional micro and nanometrology, CIRP Ann.-Manuf. Techn., 55, 721–743, 2006.

Hausotte, T.: Nanopositionier- und Nanomessmaschinen – Geräte für hochpräzise makro- bis nanoskalige Oberflächen- und Koordinatenmessungen, Ilmenau, Techn. Univ. Habilitationsschrift. Pro Business, ISBN 978-3-86805-948-9, 2011.

Hausotte, T., Percle, B., Manske, E., Füßl, R., and Jäger, G.: Measuring value correction and uncertainty analysis for homodyne interferometers, Meas. Sci. Technol., 22, 094028, doi:10.1088/0957-0233/22/9/094028, 2011.

Hausotte, T., Balzer, F. G., Vorbringer-Dorozhovets, N., and Manske, E.: Surface and coordinate measurements with nanomeasuring machines, Int. J. Nanomanufacturing, 8, 467–483, 2012.

Hausotte, T., Percle, B., and Jäger, G.: Advanced three-dimensional scan methods in the nanopositioning and nanomeasuring machine, Meas. Sci. Technol., 20, 084004, doi:10.1088/0957-0233/20/8/084004, 2009.

Hoffmann, J.: Elektrische Werkstückantastung für Nanometer aufgelöste Oberflächen- und Koordinatenmesstechnik, Berichte aus dem Lehrstuhl Qualitätsmanagement und Fertigungsmesstechnik, 2009/18, Universität Erlangen-Nürnberg, Dissertation, 2009.

Hoffmann, J. and Schuler, A.: Nanometer resolving coordinate metrology using electrical probing, Tech. Mess., 78, 142–149, 2011.

International Organization for Standardization: Geometrical Product Specifications (GPS) – Surface texture: Profile method – Nominal characteristics of contact (stylus) instruments, ISO 3274:1996, 1996.

International Organization for Standardization: Geometrical product specifications (GPS) – acceptance and reverification tests for coordinate measuring machines (CMM): part 5. CMMs using single and multiple stylus contacting probing systems ISO 10360-5:2010, 2010.

Jiang, X. and Whitehouse, D. J.: Technological shifts in surface metrology, CIRP Ann.-Manuf. Techn., 61, 815–836, 2012.

Keller, D.: Reconstruction of STM and AFM images distorted by finite-sized tips, Surf. Sci., 253, 353–364, 1991.

Lonardo P. M., Trumpold, H., and De Chiffre, L.: Progress in 3-D Surface Microtopography Characterization, CIRP Ann.-Manuf. Techn., 45, 589–598, 1996.

Ogura, I. and Okazaki, Y.: Development of micro probe for micro-CMM, Proc. of ASPE 2002, 349–352, 2002.

Rahlves, M. and Seewig, J.: Optisches Messen Technischer Oberflächen – Messprinzipien und Begriffe, BeuthVerlag, Berlin, Germany, 2009.

Schott, W., Dontsov, D., Langlotz, E., and Manske, E.: Im Nanobereich – Multisensorsystem für Präzisionsmessungen und Kalibrierung, QZ Qualität und Zuverlässigkeit, 58, 42–43, 2013.

Schuler, A.: Erweiterung der Einsatzgrenzen von Sensoren für die Mikro- und Nanomesstechnik durch dynamische Sensornachführung unter Anwendung nanometeraufgelöster elektrischer Nahfeldwechselwirkung, Erlangen, Dissertation, Shaker Verlag, ISBN 978-3-8440-2299-5, 2013.

Schuler, A., Weckenmann, A., and Hausotte, T.: Setup and evaluation of a sensor tilting system for dimensional micro- and nanometrology, Meas. Sci. Technol., 25, 064010, doi:10.1088/0957-0233/25/6/064010, 2014.

UMAP Vision System PR 1207(2), datasheet, Mitutoyo, Germany, 10 pp., 2012.

Weckenmann, A., Estler, T., Peggs, G., and McMurtry, D.: Probing systems in dimensional metrology, CIRP Ann.-Manuf. Techn., 53, 657–684, 2004.

Weckenmann, A. and Hoffmann, J.: Long Range 3-D Scanning Tunnelling Microscopy, CIRP Ann.-Manuf. Techn., 56, 525–528, 2007.

Weckenmann, A., Tan, Ö., and Hartmann, W.: Function Oriented Characterization for Surface Metrology, International Journal of Nanomanufacturing, 7, 517–527, 2011.

Weckenmann, A., Schuler, A., and Ngassam, R. J. B.: Enhanced measurement of steep surfaces by slope-adapted sensor tilting, Meas. Sci. Technol., 23, 074007, doi:10.1088/0957-0233/23/7/074007, 2012.

15

On the use of electrochemical multi-sensors in biologically charged media

S. Sachse[1], A. Bockisch[2], U. Enseleit[1], F. Gerlach[1], K. Ahlborn[1], T. Kuhnke[3], U. Rother[3], E. Kielhorn[2], P. Neubauer[2], S. Junne[2], and W. Vonau[1]

[1]Kurt-Schwabe-Institut für Mess- und Sensortechnik e.V. Meinsberg, Waldheim, Germany
[2]Chair of Bioprocess Engineering, Technische Universität Berlin, Berlin, Germany
[3]Rox GmbH, Radeberg, Germany

Correspondence to: S. Sachse (sachse@ksi-meinsberg.de)

Abstract. For the investigation and characterisation of liquid media with microorganisms, electrochemical sensors are typically used. Usually the microorganisms are part of the process or cannot be excluded for different reasons. This paper describes the application of various electrodes, which are partly miniaturised and combined with multi-sensor systems for several applications in processes containing microorganisms. The application in industrial bioprocesses like beer brewing and biogas production, and in paper manufacturing, is described. The performance of the multi-sensor systems, and thus their suitability for a contribution to improved process monitoring, is evaluated. The multi-sensor systems represent an interesting tool to enhance monitoring capacities at installed systems without the necessity for huge port installations and offer the possibility to monitor the spatial distribution of gradients. The developed systems presented here allow location-independent measurements in process plants with a variable positioning of the sensors in the industrial reactors.

1 Introduction

Microorganisms play an important role in many industrial processes. On the one hand, they can have a negative influence on the process in which they occur if their presence leads to a contamination. An example is the paper manufacturing process (Kiuru and Karjalainen, 2011). On the other hand, there are many bioprocesses in which microorganisms play an important role in producing valuable goods, like in beer and biogas fermentation. In such cases, highest yield and efficiency can only be achieved if the microbial potential for a bioproduction process is utilised under optimal conditions. To ensure this, among others, monitoring of the cultivation conditions (physical and chemical parameters) becomes necessary. This monitoring should usually be performed online and in situ in order to measure important parameters at any point of the reactor and to detect changes or disturbances early. Consequently, it allows sufficient time for intervention (Wollenberger et al., 2003).

Microbial growth leading to biofilm formation is a major problem in the paper manufacturing process. In consequence, holes, stains or paper web breaks occur. It is assumed that the formation of biofilms is influenced by different factors. These include the supply of nutrients, the pH value, the redox potential, the oxygen content and the temperature. In order to keep a certain control over the growth of the biofilms, biocides are usually used (Pauly and Dietz, 2005). However, the addition of biocides might influence the paper manufacturing process itself (Kiuru and Karjalainen, 2011). Therefore, the lowest amount suitable for circumventing excessive biofilm formation should be used. Hence, the process control by means of electrochemical sensors is not only used for monitoring the microbial growth, but also for monitoring the impact of the application of biocides.

Chemical sensors for the monitoring of industrial processes, for example in a steam power plant (Li, 2008), for drinking water treatment (Hashimoto, 2013), waste water treatment (Volbeda, 2004) and in desalination plants (Hashimoto, 2015), have already been described by different authors. Detected parameters in these papers are among others the pH value, the conductivity and the oxygen content, whereby multi-parameter measuring systems are not discussed. Analysis systems for the paper manufacturing industry to determine the moisture content or the cationic charge demand are commercially available (Williamson, 2004).

However, in practice only a few sensors are applied directly in the water circuit for monitoring the paper manufacturing process. The most relevant parameters are the temperature, the conductivity and the pH value. Thereby this refers to fixed installed measuring systems (Kiuru, 2011).

To determine the redox potential, electrochemically electrodes based on noble metals like platinum, gold or palladium are normally used. Thereby, cross-sensitivities against e.g. sulfur-containing compounds have to be taken into consideration. Simultaneously it is possible that the metals themselves can react as catalysts and cause undesired reactions in the medium. Therefore, in this application, a novel redox-sensitive glass-based electrode (Gerlach et al., 2015) which does not show the disadvantages of the noble metal electrodes was used.

Bellin et al. (2014) developed an electrochemical sensor for spatially resolved detection of redox-active metabolites which are formed by microbial biofilms. For the investigation of biofilms in the water circuit of paper manufacturing industries, such a sensor could be suitable. However, in this application, the extreme and constantly varied ambient conditions (composition of the water, flow velocities, etc.) which significantly influenced the growth of the biofilms are problematic. Therefore the described novel robust combination of electrochemical probes has been chosen for the measurement.

Furthermore, the contamination of the measuring system in such complex solutions represents a major problem. In the application presented here, the cleaning was realised by an automatic air purge of the sensor surfaces. The efficiency has been proven in a further paper (Gerlach et al., 2015).

The anaerobic yeast and biogas fermentation processes are the two largest bioprocesses with respect to total product turnover rates. Yeast cells are applied for the production of beer, because they conduct the synthesis of ethanol and carbon dioxide well in the absence of oxygen. However, the brewing process is very complex, so that it relies on many parameters. Even oxygen is required in certain amounts for the cellular growth (O'Rourke, 2002a). Simultaneously, oxygen can negatively affect the quality and stability of the beer during manufacturing and filling (Pöschl, 2006). The enzymes of yeast are responsible for the chemical reactions in a brewing process. They operate optimally only in defined pH and temperature ranges (O'Rourke, 2002b).

In brewing reactors the medium is often not mechanically mixed and homogenised. This causes a high risk of inhomogeneities with regard to different chemical parameters. To detect these inhomogeneities, online measuring systems are suitable devices, especially if they are combined with a variable positioning measuring system for the process characterisation. Consequently, spatially resolved measurements are performed to investigate and compare various areas in the brewing reactor. The movable measuring system has to be as little as possible to avoid the mixing of the medium. Accordingly, only miniaturised sensors can be applied to investigate the required parameters. A further advantage of the miniaturised oxygen sensor is the lower oxygen consumption of the micro-cathode of this probe.

The biogas process is complex, too. Due to various consortia of microorganisms and the non-specific substrate, little is known about the detailed interaction of microorganisms, substrate conversion and product yield, especially at process disturbances or when substrate sources are changed. It is still not clear which parameters can contribute to improved online monitoring due to the little experience of detailed monitoring of the liquid phase in industrial biogas plants. Up to now, there have only been few possibilities to monitor the process, so that many biogas plants work suboptimally (Wiese and König, 2008).

For a large number of biogas plants where measuring systems could already be installed, some usable access points for the introduction of the probes into the fermenters are available. However, the sensing devices are limited in number and are produced in unfavourable sizes. Therefore, a miniaturisation of the electrodes is meaningful for the integration into biogas plants in operation to avoid larger interventions.

Hashimoto (2013, 2015) and Volbeda (2004) have described have described sensors for the determination of pH value, conductivity and oxygen concentration in industrial processes, whereby the sensors are applied individually and not in the form of a multi-parameter probe. Furthermore, the sensors have not been miniaturised, which is essential for the use of small reactor accesses in biogas plants.

There are already developments of electrochemical sensor arrays, e.g. for the online determination of temperature, pH value, dO_2 and biomass concentration (van Leeuwen, 2010). However, they are integrated into micro-bioreactors with a volume of about $100\,\mu L$, so that usage in industrial plants is not possible. In addition, Krommenhoek et al. (2008) published results of investigations using microchips with integrated electrochemical sensors to measure the same parameters. These chips were implemented in a well of a 96-microtitre well plate. Betts und Baganz (2006) compared in a review article several micro-bioreactors (for example, based on shake flask, microtitre plate, miniature stirred bioreactor, and stirred tank reactor) with a maximum volume of $500\,mL$ concerning the use of sensors for the determination of pH value, dO_2 and optical density. All developed sensory systems are only suited for small-scale applications. In industrial reactors, e.g. in fermenters of biogas plants, problems have been expected to be caused by pollution or material degradation on the damageable sensor arrays by several substances contained in the media. The results of the studies presented here were carried out in the main digester in a biogas plant in exercise using a miniaturised multi-sensory probe. In addition, the reactions in the hydrolysis basin of this plant were also investigated over months, which is reported elsewhere (Kielhorn et al., 2015).

2   Development and construction of multi-sensor systems

2.1   Multi-sensor measuring system for on-site analysis in the paper industry

An analysis system consisting of a two-part control cabinet (Fig. 1, left) and a multi-sensor probe with a measuring vessel was developed (Fig. 1, middle). The measuring vessel is directly fed with process water from the paper machine. The upper part of the control cabinet includes the components of the electronic and automation equipment. A mini compressor is installed in the lower part. This device produces compressed air to purge the sensitive electrode surfaces by means of nozzles. The compressed air jets are adjusted next to each sensor. The pressure impulses can be activated or deactivated individually. The multi-sensor probe was cleaned manually at the beginning of each measurement. Moreover, the number and duration of the pressure impulses can be limited. The multi-sensor probe includes sensors to measure the dissolved oxygen content, pH value, temperature and redox potential (Table 1; Fig. 1, middle). In addition to a standard platinum electrode, a novel glass-based thick film structure is used for the determination of the redox potential (Fig. 1, right). This electrode currently introduced in Gerlach et al. (2015) is based on electron-conducting glass. Thereby, a number of drawbacks which are connected to the use of noble-metal-based electrodes (especially in biologically charged media) resulting in negative effects on the electrode performance can be avoided. The drawbacks e.g. concern the possible deactivation of the electrode surface under the influence of sulfur-containing compounds and proteins as well as the ability of platinum to catalyse chemical reactions. Additionally, the glass-based structure leads to a lower dependence on the pH value in comparison to the platinum electrode. Furthermore, the analysis system was expanded by an inlet valve upstream from the measuring vessel. As a result, the inlet flow can be stopped in order to measure in a resting solution for a defined period of time.

2.2   Miniaturised multi-sensor probe for use in bioreactors

A multi-sensor for six miniaturised electrodes was designed for biotechnological applications. This device consists of stainless steel and plastic polyetheretherketone (PEEK). These materials are resistant against typical cultivation conditions and media components. The screwable protective hood protects the electrodes against mechanical influences and offers the possibility to change the electrodes separately.

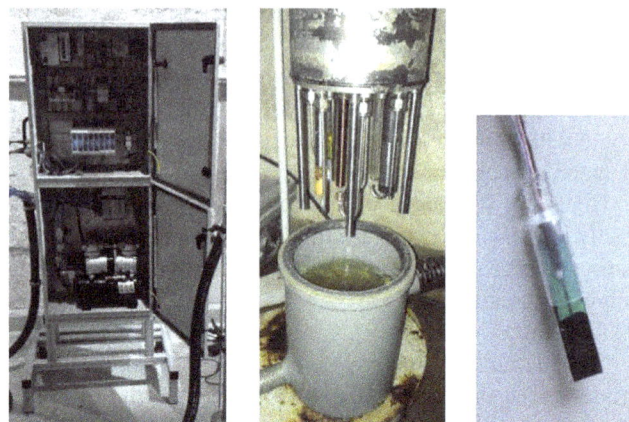

Figure 1. Left: control cabinet of the measuring system for monitoring process water; middle: multi-sensor probe with different electrodes and compressed air jets and the measuring vessel; right: redox glass electrode.

The multi-sensor probes are individually adapted to the different fields of application.

2.2.1   Miniaturised multi-sensor probe for brewing reactors

The developed probe is approximately 140 mm long and it has a diameter of 30 mm (Fig. 2, left). This device contains electrodes for measuring the dissolved carbon dioxide and oxygen, the pH value, the redox potential and the temperature (Fig. 2, right; Table 2). To determine the pH value and the redox potential, a common silver chloride reference electrode is applied. All electrodes were fabricated in-house. The protective hood is equipped with a margin of a few millimetres (Fig. 2, middle) to install a second housing, which includes a pressure sensor.

2.2.2   Miniaturised multi-sensor probe for biogas digesters

The multi-sensor probe (Fig. 3, left), which is used in digesters of biogas plants, has similar dimensions to the previously mentioned one. It consists of the same materials. The stainless steel device includes a temperature probe, a pH electrode, an electrode for the determination of redox potentials and an electrochemical reference electrode (Fig. 3, right; Table 2). Depending on the used measurement device, either one common reference electrode for the determination of pH value and redox potential is applied or two separate reference electrodes are required.

3   Results and discussion

The developed multi-sensor measuring systems are tested in practical applications. Their monitoring abilities in the

Table 1. Overview of the sensors installed in the on-site analysis system.

Sensor/electrode	Design	Measuring principle; manufacturer
Oxygen	Cylindrical; platinum cathode and Ag/AgCl anode; gas-permeable membrane	Amperometric; Sensortechnik Meinsberg GmbH
pH (glass)	Cylindrical; spherical glass membrane; platinum wire with sintered silver chloride body; glass electrode is filled with a KCl-containing electrolyte.	Potentiometric; Kurt-Schwabe-Institut (KSI)
Redox (Pt)	Cylindrical; platinum wire embedded in glass	Potentiometric; KSI
Redox (glass)	Thick film with glass membrane	Potentiometric; KSI
Reference	Cylindrical; Ag/AgCl electrode and KCl gel electrolyte; porous ceramic diaphragm	Reference electrode for potentiometric measurements (pH and redox potential); KSI
Temperature	Cylindrical; platinum (Pt 1000)	Resistance thermometer; KSI

Table 2. Overview of the miniaturised sensors installed in the multi-sensor system.

Sensor/electrode	Design	Measuring principle
Oxygen	Cylindrical (length: 40 mm; diameter: 4 mm); three electrode system (working, counter and reference electrode); micro-cathode (diameter: 30 μm); screwable cup with gas-permeable polypropylene membrane	Amperometric (defined working potential = 800 mV)
pH (glass)	Cylindrical (length: 40 mm; diameter: 3 mm); spherical glass membrane; platinum wire with sintered silver chloride body; glass electrode is filled with KCl-containing electrolyte.	Potentiometric
Redox (Pt)	Cylindrical (length: 40 mm; diameter: 4 mm); platinum wire embedded in glass	Potentiometric
Carbon dioxide	Cylindrical (length: 70 mm; diameter: 7 mm); two electrode system (pH and reference electrode); gas-permeable polymethylpentene membrane; inner pH electrode (glass body) is filled with buffer solution; electrolyte cup contains glycol solution.	Potentiometric (based on pH measurement)
Reference	Cylindrical (length: 40 mm; diameter: 4 mm); Ag/AgCl electrode with KCl inner electrolyte; screwable cup with porous aluminium oxide ceramics diaphragm	Reference electrode for potentiometric measurements (pH and redox potential)
Temperature	Cylindrical (length: 40 mm; diameter: 4 mm); platinum (Pt 1000)	Resistance thermometer

applied environments are evaluated for several weeks and months, respectively.

3.1 Multi-sensor system for application in the paper industry

The multi-sensor measuring system was applied in a paper mill for several months. In the following, two measurement examples of this practical testing are explained.

In Fig. 4, a part of a long-term measurement with the multi-sensor probe in a paper mill is shown. Thereby, the automatic cleaning of the electrodes was performed by air streams every 12 h (3.25, 15.25 h, etc.). The redox potentials, which are measured with a platinum electrode as well as with a glass-based structure, reached a constant level within a few hours. However, the platinum electrode responded to the first pressure impulse (cleaning by air stream) and detected a de-

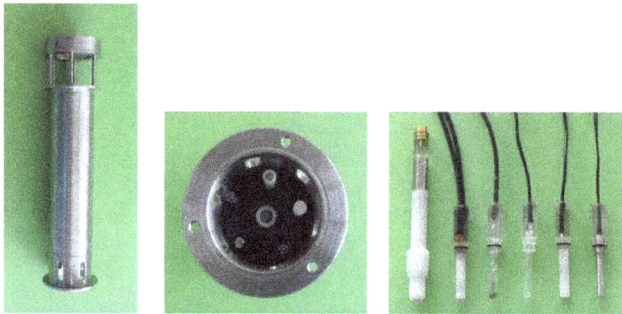

Figure 2. Left: multi-sensor probe; middle: top view of the multi-sensor probe; right: miniaturised electrodes (from left to right: carbon dioxide sensor, oxygen sensor, pH electrode, redox electrode, reference, temperature probe).

crease of 0.035 V. The following purifications caused even greater changes in the redox potential. The curve progression of the redox glass electrode reflected the several air cleaning procedures after the third event only. The measured value started to decrease slightly approximately 4 h before the purification (27.25 h). The base level was reached again within a few minutes after purification. A similar behaviour was observed at the two following cleaning intervals at 39.25 and 51.25 h. It was determined that the decrease in the redox potential curve (glass electrode) prior to the individual purification steps and accordingly the increase in the redox potential after the cleaning became greater over time. The pH value reacted similarly. It showed the same behaviour as the redox glass electrode, starting with the third cleaning interval. This indicates a relevant biofilm formation after approximately 24 h, which covered the sensitive surfaces of the electrodes and the measuring vessel. The steadily increasing changes in the parameters redox potential (glass-based electrode) and pH value over time showed the growing microbial contamination in the measuring vessel. Moreover, the growth of the biofilm on the electrode surfaces as well as in the measuring vessel was faster after each purification step over time. This can be seen in the slow decrease in the measuring values. This drop in the graph started earlier after each cleaning.

In Fig. 5, the redox potential, the oxygen content and the temperature measured in 54 days are shown for a period of 2 days. Thereby, the purification by means of air was automatically performed every 6 h (0.25, 6.25 h, etc.), followed by a 1 h interruption of the inlet flow. The temperature of the process water of the paper machine was about 313 K. This parameter decreased during the 1 h stop, because the solution in the measuring vessel cooled down by 3 to 4 K due to the lower ambient temperature. The oxygen concentration of the solution was varied in the range of 70–90 %. Naturally, the oxygen content decreased significantly during the inlet stop. Additionally, a dependence on the measurement duration was recognisable. The oxygen concentration further decreased with each additional interruption of the in-

let. No oxygen was detectable for a few minutes during the sixth stop. Hereafter, the time in which the oxygen concentration was equal to zero increased clearly. The redox potential (glass-based electrode) showed nearly a constant level (0.14–0.16 V) apart from the inlet stops. However, the decrease in the redox potential became larger during the stationary phase in the course of the observation. The measurement curve of the platinum electrode showed a remarkably different behaviour. The measurements of the redox potential raised up to a maximum value during the first five inlet stops. As expected, the measurement of the redox potential decreased significantly each time the flow through the measuring vessel was recovered. However, in the following three interruptions, only smaller rises could be detected. Furthermore, the increase in the potential began earlier compared to the first inlet resting phases and the decrease was considerably reduced due to the activation of the flow.

The behaviour of the parameters redox potential (glass-based electrode) and oxygen concentration described here in correlation with the purification steps and the interruptions allow the conclusion that by means of both parameters the microbial growth in the measuring vessel can be monitored. If aerobic microorganisms are present in the measurement solution, they consume the oxygen, which is available. This effect was shown by the inlet stops. Simultaneously, changes in the redox potential (glass-based electrode), which also refer to the bacterial growth in the liquid medium, could be detected.

3.2 Miniaturised multi-sensor probe for bioreactors

3.2.1 Brewing reactors

The long-time stability and accuracy of the sensors were tested in several fermentations at the laboratory scale (4 L) before the long-time application at laboratory (12 L) and pilot scale (200 L).

The wort solution that served as an investigation medium consisted of water, various carbohydrates, proteins, minerals and bitter substances. For all processes the brewing was inoculated with bottom fermenting yeast, fermenting at temperatures between 283 and 285 K, and was not stirred. The sensors were calibrated in standard solutions at 285 K. The processes were monitored by the sensor probe during the whole fermentation time. In order to determine the measurement accuracy and deviation over time, the sensors were inserted once per day into the following standard solutions: pH buffers, oxygen-saturated water, autoclaved wort, and standard buffers with different redox potentials (Table 3). In this table the measured deviations after times of 65 and 120 h are also shown. Drifts are given for the pH electrode in pH units, for the oxygen sensor in % air saturation, for the CO_2 sensor in $g\,L^{-1}$ and for the redox sensor in mV which resulted during the operating time. The deviation of the pH measurement after 120 h is much lower than after 65 h. This can be ex-

Figure 3. Left: multi-sensor probe; right: miniaturised electrodes (from left to right: reference, redox, pH electrode, temperature probe).

Table 3. Absolute deviation over fermentation time.

Time frame	pH (in buffer solutions)		dO_2 (%) (in O_2 saturated water)	dCO_2 (gL^{-1}) (in autoclaved wort)	Redox potential (mV) (in standard solutions)	
	pH = 4	pH = 7			250 mV	124 mV
65 h	0.54	0.78	0.37	0.65	2.11	1.11
120 h	0.09	0.27	3.29	0.28	7.61	2.87

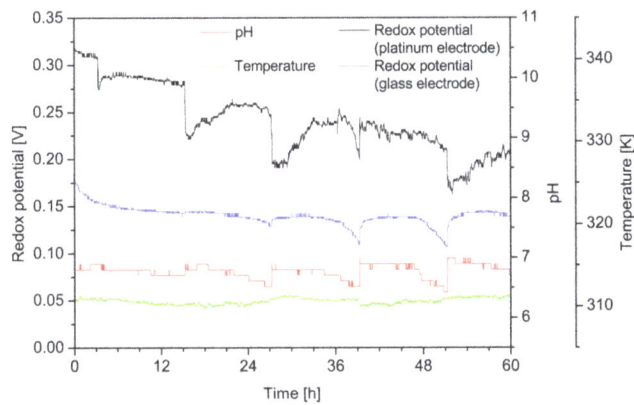

Figure 4. Influence of the air cleaning every 12 h on the parameters redox potential, pH value and temperature during a measurement in the process water of a paper industry.

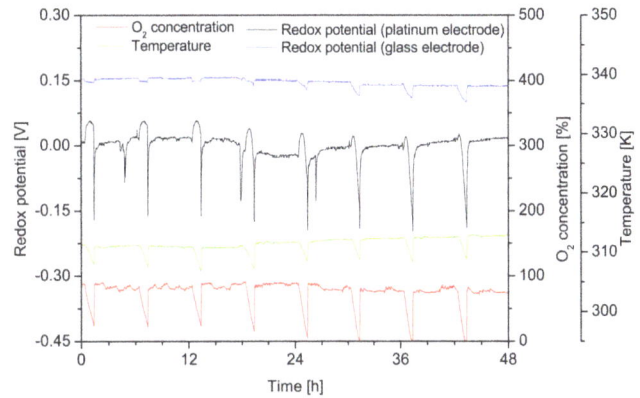

Figure 5. Influence of the air cleaning and the inlet stops (every 6 h) on the parameters redox potential, oxygen content and temperature during a measurement in the process water of a paper industry.

plained by a short-term contamination of the pH chain. During the fermentation this disturbance became detached from the electrode surface. The observed deviations of the redox electrode (less than 8 mV) are almost negligible.

For the estimation of the stability of the pressure sensor, the liquid column of the medium and hence the pressure due to the static head of liquids were used as a reference (data not shown).

The results of two experiments are exemplary displayed in Table 3. The values represent the difference between value measured at the onset and the end of the fermentation in the corresponding reference solution. From these results it can be assumed that the sensor probe can be applied during a brewing process over a fermentation time of 120 h without additional recalibration during the process.

The applicability of the pH sensor could also be proven by the comparison of the *online* measurements with the *offline*

controls (measured by a common glass electrode) during a fermentation process at the pilot scale (Fig. 6). The values correspond to each other over a time of 103 h.

The application of the multi-sensor system in a brewing reactor at the pilot scale (volume = 150 L) for 100 h is shown in Fig. 7. At the start of the measurement (Fig. 8), a rapid decline of the oxygen concentration until a level below 1 mg L^{-1} combined with a decrease in the redox potential could be detected. The reason is the oxygen consumption of the yeast cells. Conversely, the carbon dioxide concentration increased and reached a value of 1.1g L^{-1} after 5 h. The pH value decreased continuously from 5.2 to 4.0 (Fig. 7), because of the carbon dioxide produced by the yeast cells. Carbon dioxide is dissolved in the medium, leading to carbonic acid formation and subsequent decomposition to hydrogen ions. The course of the redox potential (Fig. 7) detected by a platinum electrode presents distinctive changes,

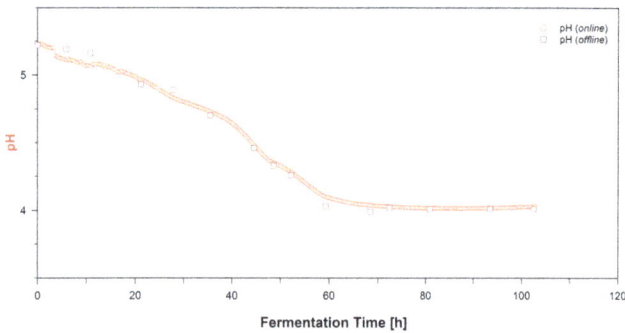

Figure 6. Course of the pH value, monitored online and offline during a fermentation process at the pilot scale (volume: 150 L; 294 K).

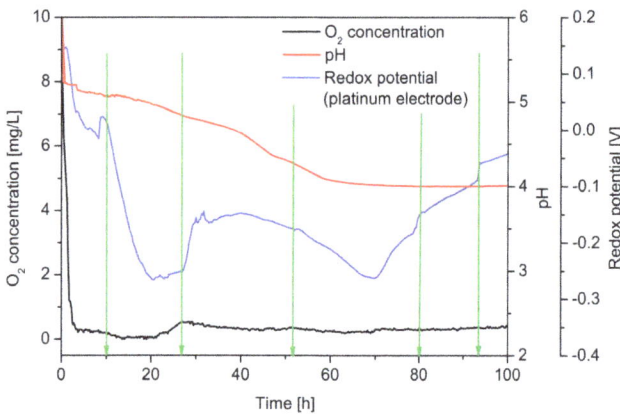

Figure 7. Course of the oxygen concentration, the pH value and the redox potential during a yeast fermentation over 100 h in a laboratory reactor; green arrows: movement of the sensor probe (volume: 150 L; 294 K).

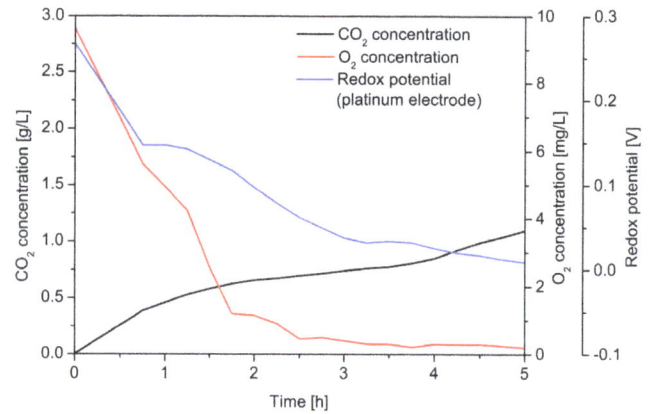

Figure 8. Course of the carbon dioxide and oxygen concentration as well as the redox potential in the start phase of a yeast fermentation in a laboratory reactor (volume: 150 L; 294 K).

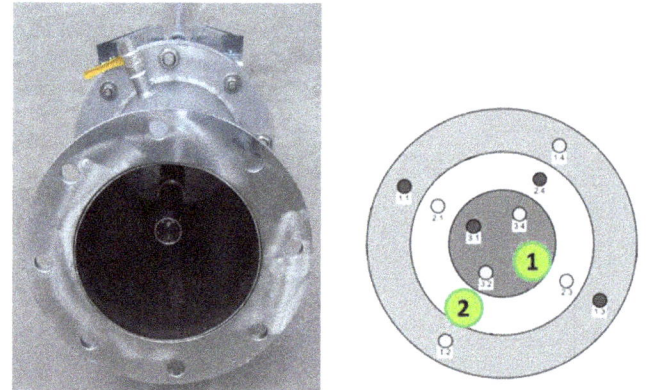

Figure 9. Left: multi-sensor probe in the device to install on a gas-tight lock; right: schematic drawing of the two used fermenter entries by means of gas-tight locks.

which mostly refer to inhomogeneities of the reactor volume or to changes in the position of the multi-sensor probe in the reactor. Some of these events are shown in the courses of the parameters, e.g. after about 10, 27, 52, 80 and 93 h, when sampling was also performed. This was reflected in the redox potential values (Fig. 7, green marking).

3.2.2 Biogas digester

Measurements were performed at a pilot biogas plant. Several gas-tight locks were accessible from the top part of the bioreactor, which was made of concrete. The locks enabled a vertical measurement applying the multi-sensor probe. The measuring unit was attached to a 6 m long guide tube (Fig. 9, left), which could be attached to different sampling spots (Fig. 9, right). Consecutively, the investigations at two measuring positions are described.

At one sampling spot, which is positioned approximately centrally between the outer edge and the middle of the digester, changes in the pH value and temperature are not recognisable at different immersion depths (Fig. 10). Agita-

tion had no noticeable effect on these two parameters. During the measurement, the temperature curve fluctuated only up to 0.1 K. The average temperature was 325 K. The pH value varied in the range from 7.85 to 7.9. Accordingly, at this spot, a good homogeneity could be assumed with regard to the parameters temperature and pH value up to a depth of 3 m. In contrast, by means of the redox potential (platinum electrode), the change in the immersion depth led to a change in the measured redox potential. It varied in the range of 0.475–0.5 V. The immersion depths 1, 2 and 3 m are surveyed twice, whereby at the immersion depths of 1 and 3 m the repeated measurements had similar values.

The investigation of a second sampling spot (data not shown), which is positioned closely to the middle of the digester, has revealed similar pH and temperature values as at the previously described spot. The temperature was 325 K on average. The pH value varied in the range of 7.85–7.89. Again, the redox potential measurements were influenced by the immersion depths in a range between -0.472 and

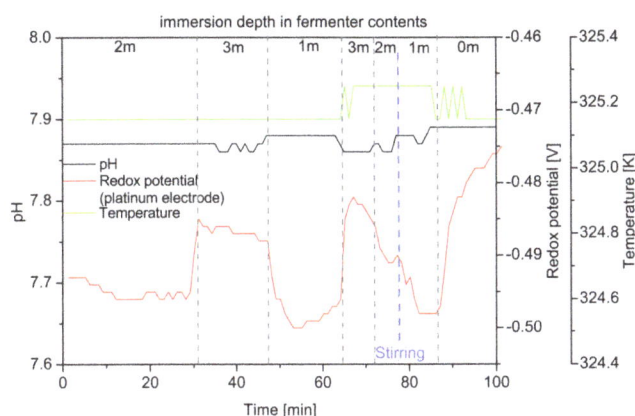

Figure 10. Course of the pH value, the redox potential and the temperature depending on the immersion depth of the measuring probe at test position 2 of the digester.

-0.494 V. Each position of this sampling spot was examined twice, whereby the redox potentials at the immersion depths of 1 and 3 m were nearly reproducible.

After several short-time investigations, the multi-sensor probe was inserted into a second fermenter of the biogas plant for 42 days (data not shown). The installed miniaturised electrodes were calibrated at 323 K before and after the mentioned applications. The testing of the pH electrode against a silver/silver-chloride reference electrode (SSE) showed a negligible change in the sensitivity of 0.3 mV/pH and a decrease in the asymmetry potential of 0.8 mV. The redox electrode was also checked against a SSE, whereby the changes were only several mV. The calibration of both reference electrodes against a SSE resulted in potential increases of 31 and 40 mV, respectively. This corresponds to drifts of 0.7 or 1 mV per day, which were considered mathematically.

In the presented short-time measurements, a dependence of the redox potential on the immersion depth could be detected. Thereby it was remarkable that the present influence by the immersion depth ensured that the redox potential reached strongly negative levels at lower immersion depths.

4 Conclusions

Based on three different examples, we were able to demonstrate that several parameters like pH value, redox potential and temperature are necessary to characterise biological media or solutions, which are contaminated with microorganisms and chemicals. Depending on the application, it is useful to add further appropriate parameters and to choose the dimension of the measuring system. If it is important to disturb the test solution like in brewing reactors as little as possible and to prevent mixing, the present miniaturised multi-sensor probe can be utilised. Their application may also be useful in great fermenters like in biogas plants or brewing reactors, because the reactors normally have only few access

points and these entries are limited due to their size. In many cases, little effort was put into the installation of sensors in the liquid phase; thus, the later integration of sensors during operation requires a certain restriction in dimensions of them. However, if the entrance to the test medium is not limited, the miniaturisation of the measuring system is not necessary. Therefore, greater electrodes, which are more robust and stable over long times due to larger electrolyte volumes, can be used. That is why this system is employed in very demanding conditions, such as the paper manufacturing process. Thereby, huge amounts of process water, which contains different chemical and biocidal substances, are available. Furthermore, it is contaminated with microorganisms, so that the development of an automatic cleaning procedure for the sensors was necessary.

Acknowledgements. Support was given by the Federal Ministry for the Environment, Nature Conservation and Nuclear Safety (03KB059B), the Federal Ministry of Education and Research (03SF0455C), the Federal Ministry of Economics and Technology (KF2218311RH1) and the Investitionsbank Berlin (10033235). Parts of the work were funded by the Kurt-Schwabe-Stipendium. The authors thank Hans Oechsner and Hans-Joachim Nägele from the Universität Hohenheim for the opportunity to apply the multi-sensor system in the pilot biogas plant.

References

Bellin, D. L., Sakhtah, H., Rosenstein, J. K., Levine, P. M., Thimot, J., Emmett, K., Dietrich, L. E. P., and Shepard, K. L.: Integrated circuit-based electrochemical sensor for spatially resolved detection of redox-active metabolites in biofilms, Nature Communications, 5, 3256, doi:10.1038/ncomms4256, 2014.

Betts, J. I. and Baganz, F.: Miniature bioreactors: current practices and future opportunities, Microbial Cell Factories, 5, 21, doi:10.1186/1475-2859-5-21, 2006.

Gerlach, F., Ahlborn, K., Sachse, S., and Vonau, W.: Glass based Electrodes for the Determination of Redox Potentials in Water Samples, Sensor & Transducers, 184, 39–44, 2015.

Hashimoto, R.: Pure by Analysis, Control Engineering Asia, 26–29 March 2013, avaialbe at: http://www.ceasiamag.com/2013/03/pure-by-analysis (last access: 17 September 2015), 2013.

Hashimoto, R.: Improved Conductivity Analysis in Desalination Processes, Water & Wastewater Asia, 40–42, Pablo Publishing Pte Ltd, Singapore, 40–47, January/February 2015.

Kielhorn, E., Sachse, S., Moench-Tegeder, M., Naegele, H.-J., Haelsig, C., Oechsner, H., Vonau, W., and Junne, S.: Multiposition Sensor Technology and Lance-Based Sampling for Improved Monitoring of Liquid Phase in Biogas Processes, Energy and Fuels, 29, 4038–4045, doi:10.1021/ef502816c, 2015.

Kiuru, J.: Interactions of Chemical Variations and Biocide Performance at Paper Machines, Aalto University Publication series, Doctoral Dissertations, 128 pp., ISBN:978-952-60-4454-5, 2011.

Kiuru, J. and Karjalainen, S.: Influence on runnability of paper machines and separate coating lines, ipw – The magazine for the international pulp and paper industry, 10, 11–17, 2011.

Krommenhoek, E. E., van Leeuwen, M., Gardeniers, H., van Gulik, W. M., van den Berg, A., Li, X., Ottens, M., van der Wielen, L. A. M., and Heijnen, J. J.: Lab-scale fermentation tests of microchip with integrated electrochemical sensors for pH, temperature, dissolved oxygen and viable biomass concentration, Biotechnol. Bioeng, 99, 884–892, 2008.

Li, X., Ottens, M., van der Wielen, L. A. M., and Heijnen, J. J.: Lab-Scale Fermentation Tests of Microchip With Integrated Electrochemical Sensors for pH, Temperature, Dissolved Oxygen and Viable Biomass Concentration, Biotechnol. Bioeng., 99, 884–892, 2008.

O'Rourke, T.: The role of oxygen in brewing, The BREWER International, The Institute of Brewing & Distilling, London, 45–47, 2002a.

O'Rourke, T.: The role of pH in brewing, The BREWER International, The Institute of Brewing & Distilling, London, 2, 21–23, 2002b.

Pauly, D. and Dietz, W.: PTS-Forschungsbericht: Vermeidung von Biofilmbildung in Stoff- und Wassersystemen durch optimierte Prozessführung in Altpapier verarbeitenden Produktionsanlagen, 46 pp., 2005.

Pöschl, M.: Bier – eine klare Sache, TUM Mitteilungen, 1, 54–55, 2006.

Wiese, J. and König, R.: Einsatz von Mess- und Automatisierungstechnik auf modernen Biogasanlagen – Ergebnisse großtechnischer Anwendungen, energie I wasser-praxis, 11, 16–21, 2008.

van Leeuwen, M., Krommenhoek, E. E., Heijen, J. J., Gardeniers, H., van der Wielen, L. A. M., and van Gulik, W. M.: Aerobic Batch Cultivation in Micro Bioreactor with Integrated Electrochemical Sensor Array, Biotechnol. Progr., 26, 293–300, 2010.

Volbeda, J.: Double Option – Two ways to measure dissolved oxygen in wastewater applications, Water & Wastes Digest, availabe at: http://www.wwdmag.com/wastewater/double-option (last access: 17 September 2015), 2004.

Williamson, M.: Multi-variable controls, Automation, 2, 4–7, 2004.

Wollenberger, U., Renneberg, R., Bier, F. F., and Scheller, F. W.: Analytische Biochemie, Wiley-VCH, Weinheim, 10–12, 2003.

16

Increasing the sensitivity of electrical impedance to piezoelectric material parameters with non-uniform electrical excitation

K. Kulshreshtha[1], **B. Jurgelucks**[1], **F. Bause**[2], **J. Rautenberg**[2], and **C. Unverzagt**[2]

[1]Institut für Mathematik, Universität Paderborn, Warburger Str. 100, 33098 Paderborn, Germany
[2]Elektrische Messtechnik, Universität Paderborn, Warburger Str. 100, 33098 Paderborn, Germany

Correspondence to: K. Kulshreshtha (kshitij@math.upb.de)

Abstract. To increase the robustness and functionality of piezoceramic ultrasonic sensors, e.g. for flow, material concentration or non-destructive testing, their development is often supported by computer simulations. The results of such finite-element-based simulations are dependent on correct simulation parameters, especially the material data set of the modelled piezoceramic. In recent years several well-known methods for estimation of such parameters have been developed that require knowledge of the sensitivity of a measured behaviour of the material with respect to the parameter set. One such measurable quantity is the electrical impedance of the ceramic. Previous studies for radially symmetric sensors with holohedral electrode setups have shown that the impedance shows little or no sensitivity to certain parameters and simulations reflect this behaviour making parameter estimation difficult. In this paper we have used simulations with special ring-shaped electrode geometry and non-uniform electrical excitation in order to find electrode geometries, with which the computed impedance displays a higher sensitivity to the changes in the parameter set. We find that many such electrode geometries exist in simulations and formulate an optimisation problem to find the local maxima of the sensitivities. Such configurations can be used to conduct experiments and solve the parameter estimation problem more efficiently.

1 Motivation

Piezoelectric effect is the physical phenomenon discovered by Pierre and Jacques Curie in 1880 that is exhibited by several crystalline and synthetic ceramic materials. When voltage is applied across certain surfaces of the solid, it exhibits mechanical strain; conversely, when mechanical stresses are applied, voltage is produced between its surfaces.

Acoustic transducers are required to construct an ultrasonic measurement device of any kind. Transducers based on circular piezoelectric ceramic disks have been manufactured and used in a wide variety of applications.

Ceramics are composed of a large number of randomly oriented crystals. The cumulative properties of each of these crystals together determine the properties of the whole solid. This means that each batch of the ceramic produced will have somewhat different material properties. Additionally, the ma-

terial coefficients are dependent on geometry. If a such a ceramic disk with a specific thickness and diameter is sintered and polarised, the resulting material coefficients are different than a disk of the same material and of different thickness and diameter. Hence, post processing and treatment of the material also influences the material coefficients. Pérez et al. (2010) found the elastic coefficients to vary up to 5 % and the piezoelectric and dielectric coefficients to vary up to 20 %. These variations include measurement uncertainties concerning the determination of the coefficients and also the batch-to-batch and geometry dependencies.

Depending on the application at hand, the full knowledge of the material data set can be crucial to the whole process: for a circular disk with full width electrodes on the top and bottom (see Fig. 1) there is no shear movement; hence, e_{15} is of no importance. Unlike such a disk, in the case of a shear wave transducer the knowledge of e_{15} is however crucial.

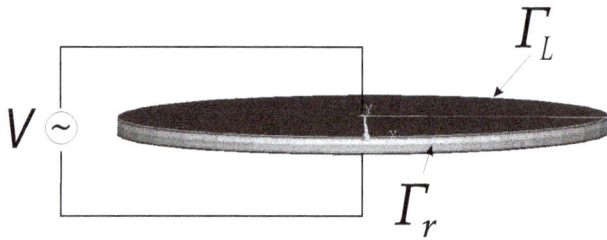

Figure 1. Two-electrode configuration on a circular disk (Lahmer, 2008).

(a)   (b)

(c)

Figure 2. (a) Arrangement of electrodes in thin rings; **(b)** arrangement of electrodes in wide rings; **(c)** schematic diagram of the whole circuit with the piezoceramic shown as a T-Network.

Analytical and numerical modelling of piezoelectricity and the resulting computer simulations (Kocbach, 2000; Unverzagt et al., 2013) have enabled more efficient design and construction of transducers in recent years.

Such numerical modelling requires the precise knowledge of the parameters of the material under analysis. To this end, new methods that utilise the mathematical concept of inverse problems to estimate the material parameters have been designed (Kaltenbacher et al., 2006, 2008).

Previous studies (Rautenberg et al., 2011) have shown that configurations inducing resonance in radial or thickness directions (Lahmer, 2008) are insufficient for parameter estimation, when only the electrical impedance is measured, because the impedance characteristic in the frequency space shows little or no sensitivity to certain material parameters. Especially critical are c_{44}, ε_{11} and e_{15} since in the experiments of Rautenberg et al. (2011) they show no sensitivity to material parameters at all. Low or no sensitivity means that these parameters cannot be estimated using inverse problems from the measurement of impedance characteristics only. Other, more involved measurements have been proposed in Rupitsch et al. (2009); however, they are also much more cost-intensive.

In order to improve the sensitivity of the measured impedance characteristic to the material parameters, we designed a three-electrode configuration with electrodes of various radii as shown in Fig. 2a and b and use simulations in which a non-uniform electrical excitation is applied to the ceramic in order to compute the impedance characteristic and its sensitivity to the material parameters.

Possible applications with non-uniform electrical excitation in piezoelectric ceramics are interdigital transducers (Kirschner, 2010) and annular arrays (Ketterling et al., 2005; Ramli and Nordin, 2011).

In the following sections we first give a short overview of the equations governing our simulation and the excitation of the ceramic. We then define and analyse the sensitivity of the computed impedance to parameters. Finally, we formulate an optimisation problem to find a locally optimal electrode geometry that maximises the sensitivity and show that many such local optima occur.

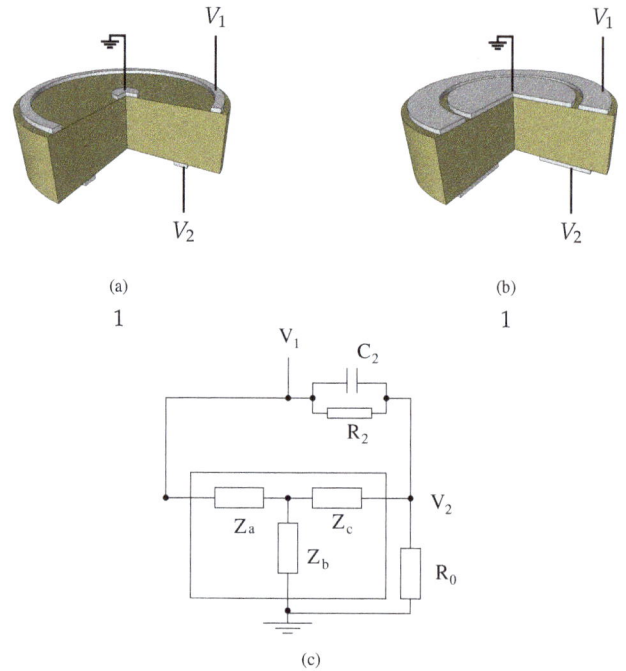

2 Modelling piezoelectricity

In this paper we restrict ourselves to linear piezoelectric effect IEEE Std 176-1987. The equations of linear piezoelectricity in tensor form are given below and form the basis for a finite-element formulation.

$$\sigma = \mathbf{c}S - \mathbf{e}^\top E, \tag{1}$$
$$D = \mathbf{e}S + \boldsymbol{\varepsilon}E, \tag{2}$$

where

- D is the electrical flux density vector

- E is the electrical field vector

- σ is the mechanical stress tensor

- S is the mechanical strain tensor

- $\mathbf{c}$ is elastic modulus tensor

- $\mathbf{e}$ is the piezoelectric coupling tensor

- $\boldsymbol{\varepsilon}$ is the electrical permittivity tensor.

We shall also restrict ourselves to thin circular disks. Therefore, the use of a cylindrical coordinate system is appropriate. We shall also assume that the configuration is rotationally symmetric. Using the notation of Helnwein (2001) and Lahmer (2008) (Voigt notation) in the rotationally symmetric case we can rewrite the tensor notation into matrix formulation:

$$
\mathbf{c} = \begin{pmatrix} c_{11} & c_{13} & 0 & c_{12} \\ c_{13} & c_{33} & 0 & c_{13} \\ 0 & 0 & c_{44} & 0 \\ c_{12} & c_{13} & 0 & c_{11} \end{pmatrix},
$$

$$
\mathbf{e} = \begin{pmatrix} 0 & 0 & e_{15} & 0 \\ e_{13} & e_{33} & 0 & e_{13} \end{pmatrix},
$$

$$
\boldsymbol{\varepsilon} = \begin{pmatrix} \varepsilon_{11} & 0 \\ 0 & \varepsilon_{33} \end{pmatrix}.
$$

This results in the following system of material equations from Eqs. (1) and (2):

$$
\begin{pmatrix} \sigma_{rr} \\ \sigma_{zz} \\ \sigma_{rz} \\ \sigma_{\theta\theta} \\ D_r \\ D_z \end{pmatrix} = \begin{pmatrix} c_{11} & c_{13} & 0 & c_{12} & 0 & -e_{13} \\ c_{13} & c_{33} & 0 & c_{13} & 0 & -e_{33} \\ 0 & 0 & c_{44} & 0 & -e_{15} & 0 \\ c_{12} & c_{13} & 0 & c_{11} & 0 & -e_{13} \\ 0 & 0 & e_{15} & 0 & \varepsilon_{11} & 0 \\ e_{13} & e_{33} & 0 & e_{13} & 0 & \varepsilon_{33} \end{pmatrix} \begin{pmatrix} S_{rr} \\ S_{zz} \\ S_{rz} \\ S_{\theta\theta} \\ E_r \\ E_z \end{pmatrix}. \quad (3)
$$

The above material equation can be extended to a full set of partial differential equations in time and space using Newton's, Gauss' and Faraday's Laws (Meschede and Gerthsen, 2010, Chap. 4 and Chap. 7).

Newton's law of motion (Slaughter, 2002, Cauchy-Navier equation) for the mechanical behaviour is

$$
\mathcal{B}^\top \sigma = \varrho \frac{\partial^2 \mathbf{u}}{\partial t^2}, \quad (4)
$$

where $\mathcal{B}$ is the differential operator relating mechanical strain to mechanical displacement.

$$
\mathcal{B} = \begin{pmatrix} \partial_r & 0 \\ 0 & \partial_z \\ \partial_z & \partial_r \\ \frac{1}{r} & 0 \end{pmatrix}
$$

As ceramic materials are insulators, there is no free charge and Gauss' (flux) law states

$$
\nabla \cdot D = 0. \quad (5)
$$

Neglecting the insignificant changes in magnetic field in this case, Faraday's law states that the electric field is the negative gradient of the electric potential.

$$
E = -\nabla\phi \quad (6)
$$

Additionally, we consider a Rayleigh damping model with positive constant α and β for the energy dissipation and arrive at a system of four partial differential equations in time

and space from Eqs. (3), (4), (5) and (6):

$$
\varrho\ddot{\mathbf{u}} + \alpha\varrho\dot{\mathbf{u}} - \mathcal{B}^\top(\mathbf{c}\mathcal{B}\mathbf{u} + \beta\mathbf{c}\dot{\mathbf{u}} + \mathbf{e}^\top\nabla\phi)
$$
$$
= 0 \text{ in } \Omega, t \in [0, T], \quad (7)
$$
$$
\nabla \cdot (\mathbf{e}\mathcal{B}\mathbf{u} - \boldsymbol{\varepsilon}\nabla\phi) = 0 \text{ in } \Omega. \quad (8)
$$

These equations are sufficient for transient analysis. The electrodes and their excitation occur as boundary conditions, in terms of free charge

$$
q^L = \int_{\Gamma_L} \hat{n} \cdot (\mathbf{e}\mathcal{B}\mathbf{u} - \boldsymbol{\varepsilon}\nabla\phi)\mathrm{d}\Gamma \quad (9)
$$

at the part of the boundary containing the electrode Γ_L with $\hat{n}$ being the normal vector $\hat{n} = (n_r, n_z)^\top$ at the boundary, or the applied potential $\phi^L(t)$ at the electrode L. We call the remaining boundary $\Gamma_r = \Gamma \setminus \Gamma_L$.

The above equations may also be transformed for harmonic analysis into the frequency domain. Application of a Fourier transform changes the unknowns $\mathbf{u}$ and ϕ in Eqs. (7) and (8) to time harmonic complex variables $\hat{\mathbf{u}}$ and $\hat{\phi}$, and the Rayleigh coefficients are scaled according to frequency.

$$
\alpha(\omega) = \alpha_0\omega, \quad \beta(\omega) = \frac{\beta_0}{\omega}
$$

The resulting time harmonic partial differential equations are given as follows:

$$
-\varrho\omega^2\hat{\mathbf{u}} - \frac{1}{1 - \imath\alpha_0}\mathcal{B}^\top\left((1 + \imath\beta_0)\mathbf{c}\mathcal{B}\hat{\mathbf{u}} + \mathbf{e}^\top\nabla\hat{\phi}\right)
$$
$$
= 0 \text{ in } \Omega, \quad (10)
$$
$$
\nabla \cdot \left(\mathbf{e}\mathcal{B}\hat{\mathbf{u}} - \boldsymbol{\varepsilon}\nabla\hat{\phi}\right) = 0 \text{ in } \Omega, \quad (11)
$$

with $\imath$ being the imaginary unit.

This transformation however contains a systematic error as α_0 and β_0 are not constants in practice but change with the central frequency of the Fourier transform.

Weak formulations and finite-element methods for the solution of the above system have been studied in Kocbach (2000) and Lahmer (2008).

3 Electrical excitation

In order to simulate the behaviour of a piezoelectric transducer with finite elements, it is excited by a delta function of the charge or a pulse of potential. This gives the boundary conditions to solve Eqs. (7) and (8).

Electrical current flow at the electrode is the time derivative of the free charge and the impedance between any two electrodes is then defined as the quotient of the potential difference and the current. In the simple case of two electrodes, with a delta function pulse of charge into one electrode with amplitude q_0^L,

$$
q^L(t) = q_0^L\delta(t)
$$

(see Fig. 1), this leads to the straightforward calculation

$$Z_L(\omega_k) = \frac{\hat{\phi}^L(\omega_k)}{q_0^L},$$

where ω_k is some frequency of interest, $\hat{q}_0^L$ denotes the frequency domain charge peak and $\hat{\phi}^L(\omega_k)$ is the calculated response of the potential difference using Eqs. (7) and (8).

In our setup two different electric potentials are applied on two electrodes and the third is grounded. An external circuit (Fig. 2c) is required to achieve this potential difference at the electrodes. The values of the resistors R_0 and R_2 and capacitor C_2 are chosen to maintain $V_1 \geq 2V_2$. This is done by solving the equations of Kirchhoff's laws for the circuit (Meschede and Gerthsen, 2010, Chap. 7). As the circuit representation of the ceramic includes three unknown impedances, the solution of the Kirchhoff laws equations requires three finite-element solutions for Eqs. (7) and (8) using an initial guess for the material parameters and a given electrode geometry (Unverzagt et al., 2015). The chosen values for the external circuit are then kept constant for that particular electrode geometry, while we compute the sensitivity of the total impedance to material parameters and optimise it as discussed in Sect. 5.

4 Sensitivity

For each frequency of interest in the domain, the solution of Eqs. (7) and (8) along with the external circuit results in a total impedance, which can be measured in experiments and compared with the numerical computations in order to estimate the material parameters (Rupitsch and Lerch, 2009; Kaltenbacher et al., 2008). The first step to solve this inverse problem requires the determination of a sensitivity of the numerical solution to variations in the material parameters. This is in fact the derivative of the impedance with respect to the material parameters.

For the numerical solution we use the commercial software package CAPA, which computes the impedance for a given configuration. CAPA is a simulation tool for the numerical solution of electromechanical, coupled field problems and is therefore suitable for the analysis of most mechatronic sensors and actuators such as electromagnetic loudspeakers or piezoelectric transducers.

Besides transient analysis, which is used in this contribution, the harmonic behaviour of the piezoceramic disc can also be calculated. One disadvantage of the transient simulation method is the sole use of the Rayleigh damping model for the energy dissipation processes as mentioned in Sect. 2. This is a rather simple approximation of the damping behaviour in practice. Another disadvantage of precompiled solver packages in general is inflexibility; it is impossible to modify the computation in any way in order to be able to compute more information, like sensitivities.

In contrast to previous studies we look at the real and imaginary parts of the complex impedance separately instead of looking at the magnitude. The formulation in terms of magnitude and phase, although computationally efficient, hides some geometrical structure, which is apparent when looking at the real and imaginary components separately. This is analogous to a polar coordinate system versus a cartesian coordinate system. Figure 3a shows the complex impedance as a function of the frequency in the complex plane, whereas Fig. 3b shows the magnitude of the impedance. The 3-D plot shows more structure and we use the same representation for the sensitivities.

We used finite-difference approximations for the configuration in Fig. 8 for the sensitivity of the impedance w.r.t. the material parameters. Figure 4 shows the sensitivity of the impedance in the frequency domain w.r.t. c_{44}, e_{15} and ε_{11} (occurring in Eq. 3) that are known to show low sensitivity in simpler configurations (Rautenberg et al., 2011).

Using a slightly different geometry (see resulting geometry in Table 1) for the electrodes but the same excitation, the sensitivity curves change both qualitatively and quantitatively (see Fig. 6 and compare with Fig. 4). However, the change is not uniform across the parameters. Hence, there is a need to find the optimal electrode configuration to ensure maximal sensitivity of the impedance to the material parameters.

5 Optimisation

We start by formulating a constrained minimisation problem with the electrode radii as the variables which will then be solved.

The electrode configuration is parametrised using four ring radii $r = (r_1, \ldots, r_{N_r} = r_4)$ and considering the outer radius as constant (see Fig. 5). Let $x = (x_1, \ldots, x_{N_x})$ denote the various material parameters occurring in Eq. (3) and let $\mathcal{F} \subseteq \mathbb{R}$ denote the frequency domain. The impedance $Z(f; x, r)$ can be considered as a function $Z : \mathcal{F} \times \mathbb{R}^{N_x} \times \mathbb{R}^{N_r} \to \mathbb{C}$. For fixed radii parametrisation r and material parameters x the impedance is a function $Z(f; x, r) : \mathcal{F} \to \mathbb{C}$ which maps the argument $f \longmapsto Z(f; x, r)$. This complex-valued function Z is then transformed into a two-dimensional real function $z(\cdot; x, r) : \mathcal{F} \to \mathbb{R}^2$ via the usual Euclidean mapping. We approximate the partial derivative of z towards a change of material parameter x_i using a finite-differences scheme

$$\nabla_{x_i} z(\cdot; x, r) \approx \frac{z(\cdot; x_i + h, r) - z(\cdot; x, r)}{h} \in \{\mathcal{F} \to \mathbb{R}^2\}$$

with $x_i + h := (x_1, \ldots, x_{i-1}, x_i + h, x_{i+1}, \ldots, x_{N_x})$ and for small $h > 0$. Hence, the sensitivity of z towards one specific parameter x_i while remaining in a constant electrode configuration r can be considered as $\left\| \nabla_{x_i} z(\cdot; x, r) \right\|_{L^2(\mathcal{F})}$. We consider the norm of the corresponding function space $L^2(\mathcal{F})$ to measure the sensitivity

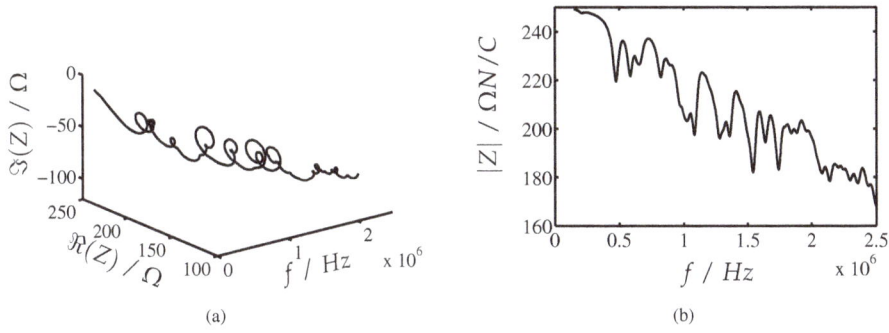

Figure 3. (a) Complex impedance vs. frequency; (b) magnitude of impedance vs. frequency.

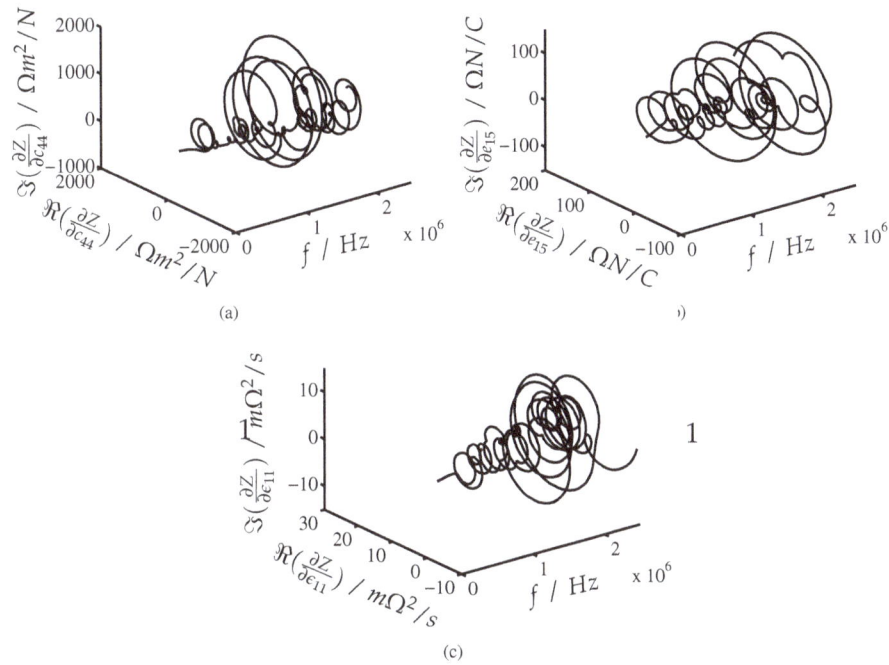

Figure 4. Sensitivity of the impedance for the geometry of the ring electrodes shown in Fig. 8 to various material parameters against frequency w.r.t. (a) c_{44}, (b) e_{15}, and (c) ε_{11}.

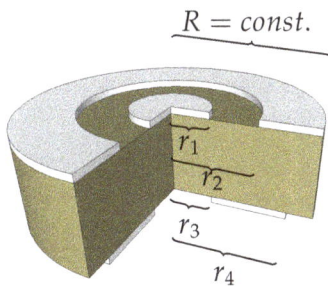

Figure 5. Parametrisation of the ring radii compared with Fig. 4. However, the change is not uniform across the parameters. Hence there is a need to find the optimal electrode configuration to ensure maximal sensitivity of the impedance to the material parameters.

$$\left\| \nabla_{x_i} z(\cdot; \boldsymbol{x}, \boldsymbol{r}) \right\|_{L^2(\mathcal{F})} = \left(\int_{\mathcal{F}} \| \nabla_{x_i} z(f; \boldsymbol{x}, \boldsymbol{r}) \|_2^2 df \right)^{\frac{1}{2}}.$$

Discretising the frequency domain $\mathcal{F}$ equidistantly into intervals $[f_i, f_{i+1}] \subseteq \mathcal{F}$, $1 \le i \le N_f - 1$ and $|f_{i+1} - f_i| = \hat{h}$, $1 \le i \le N_f - 1$ and using the trapezoidal rule leads to the approximation

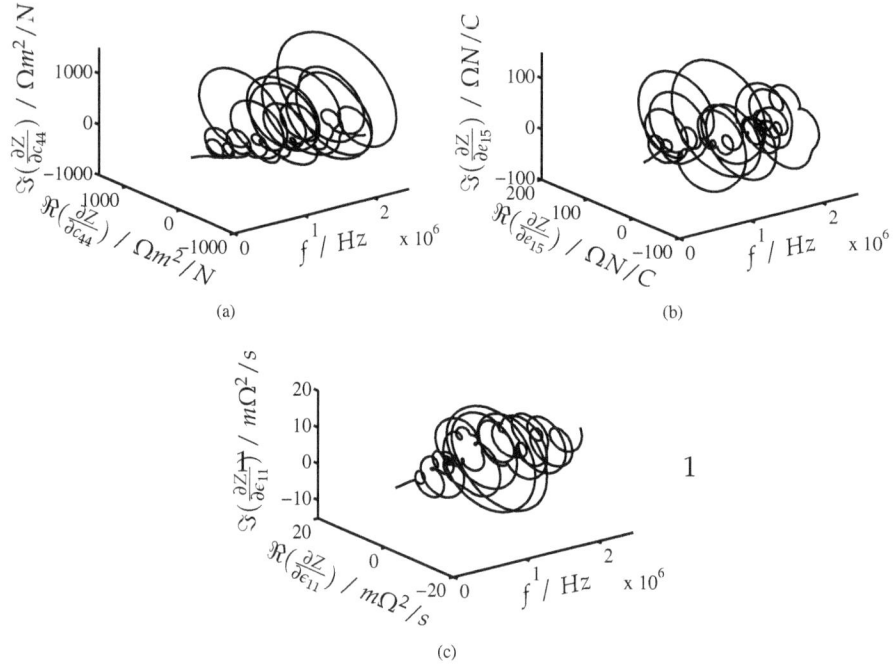

Figure 6. Sensitivity of the impedance for the resulting geometry of the ring electrodes shown in Table 1 to various material parameters against frequency w.r.t. (a) c_{44}, (b) e_{15}, and (c) ε_{11}.

$$\left\| \nabla_{x_i} z(\cdot; \boldsymbol{x}, \boldsymbol{r}) \right\|_{L^2(\mathcal{F})}$$

$$\approx \left(\frac{\hat{h}}{2} \sum_{1 \le i \le N_f - 1} |\nabla_{x_i} z(f_{i+1})|^2 + |\nabla_{x_i} z(f_i)|^2 \right)^{\frac{1}{2}} .$$

Now, we wish to maximise the sensitivities

$$\nabla z(\boldsymbol{r}) = \left[\left\| \nabla_{x_j} z(\cdot; \boldsymbol{x}, \boldsymbol{r}) \right\|_{L^2(\mathcal{F})} \right]_{j=1,\ldots,N_x}$$

with respect to all material parameters x_i.

We have noticed that some parameters are very reactive towards change in geometry and others are not. Due to unevenly distributed sensitivity of the material parameters $\boldsymbol{x}$ towards optimisation, their different orders of magnitudes and the circumstance that the optimisation method used may only minimise an objective function, we introduce a weight matrix $W := \operatorname{diag}(w_1, \ldots, w_{N_x}) \in \mathbb{R}_+^{N_x \times N_x}$ and a scaling matrix $S := \operatorname{diag}(s_1, \ldots, s_{N_x}) \in \mathbb{R}^{N_x \times N_x}$ depending on the orders of magnitude of the initial sensitivity evaluation and reformulate the sensitivity approximation as the minimisation problem:

$$\min_{\boldsymbol{r}} J(\boldsymbol{r}) = \min_{\boldsymbol{r}} \frac{1}{\|W S \nabla z(\boldsymbol{r})\|^2}$$

$$= \min_{\boldsymbol{r} = (r_1, \ldots, r_{N_r})} \frac{1}{\left\| W S \left[\left\| \nabla_{x_j} z(\cdot; \boldsymbol{x}, \boldsymbol{r}) \right\|_{L^2(\mathcal{F})} \right]_{j=1,\ldots,N_x} \right\|_2^2} .$$

The scaling is only used inside the optimiser so that all components of the gradient have a similar scale and has no meaning outside the optimisation routine. For comparison of the results one needs to compare the unscaled objective function with $S = I$, the identity matrix.

5.1 Constraints

In the production process of these piezoceramics, a laser cuts out the electrode rings from a metal plate lying on top of the ceramic. The laser has a width of 0.3 mm; hence, some restrictions to the geometry of the electrodes apply. The minimal distance of two adjacent electrode rings is at least 0.3 mm; also, all radii must be positive and smaller than the constant outer ring radius. These constraints are all linear in the radii and can be reformulated into a vector inequality $A\boldsymbol{r} \le \boldsymbol{b}$ with

Table 1. Uniform weights: improvement of sensitivity of impedance w.r.t material parameters using $W = \text{diag}(1,\ldots,1)$. The sensitivity has increased for all the parameters with exception of e_{15}. These are the best component-wise results we have computed so far. Objective function values are shown unscaled. Figure shows resulting geometry utilising uniform weight matrix in millimetres; $r_1 = 4.01; r_2 = 4.33; r_3 = 2.20; r_4 = 3.82$.

Param.	W	Start	Optimal	Gain ratio
c_{11}	1	2.0770×10^3	6.3875×10^3	3.0753
c_{33}	1	803.3076	2.0959×10^3	2.6091
c_{44}	1	1.9901×10^4	2.0009×10^4	1.0054
c_{12}	1	0.6624	0.8330	1.2575
c_{13}	1	4.4700×10^3	1.2292×10^4	2.7500
ϵ_{33}	1	1.4591	2.1917	1.5021
e_{31}	1	83.2714	254.6764	3.0584
e_{33}	1	62.9616	149.2692	2.3708
ϵ_{11}	1	3.6831	4.6061	1.2506
e_{15}	1	262.1774	246.1737	0.9390
Obj.		2.3748×10^{-9}	1.6756×10^{-9}	0.7056
$\|\nabla Z\|^2$		4.2109×10^8	5.9681×10^8	1.4173

Table 2. Binary weights: improvement of sensitivity of impedance w.r.t. material parameters using $W = \text{diag}(0,0,0,0,0,0,0,0,0,1)$. The sensitivity w.r.t e_{15} has increased. However, the sensitivity toward other parameters has mainly decreased. Objective function values are shown unscaled. Figure shows resulting geometry utilising binary weight matrix in millimetres; $r_1 = 3.4; r_2 = 3.7; r_3 = 1.23; r_4 = 3.54$.

Param.	W	Start	Optimal	Gain ratio
c_{11}	0	2.0770×10^3	1.7465×10^3	0.8409
c_{33}	0	803.3076	670.3565	0.8345
c_{44}	0	1.9901×10^4	1.9167×10^4	0.9631
c_{12}	0	0.6624	0.6789	1.0248
c_{13}	0	4.4700×10^3	3.8569×10^3	0.8628
ϵ_{33}	0	1.4591	1.2413	0.8507
e_{31}	0	83.2714	67.8750	0.8151
e_{33}	0	62.9616	47.5998	0.7560
ϵ_{11}	0	3.6831	4.3195	1.1728
e_{15}	1	262.1774	344.2983	1.3132
Obj.		1.4548×10^{-5}	8.4359×10^{-6}	0.5799
$\|W\nabla Z\|^2$		6.8737×10^4	1.1854×10^5	1.7246

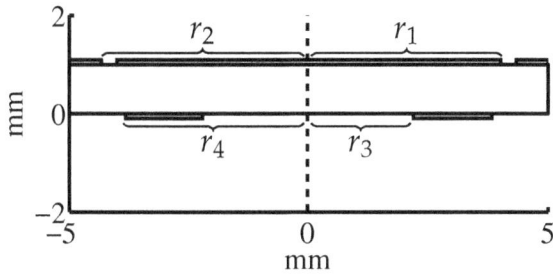

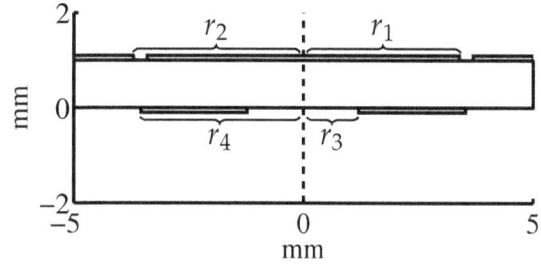

$$A := \begin{bmatrix} 1 & 0 & 0 & 0 \\ 0 & 1 & 0 & 0 \\ 0 & 0 & 1 & 0 \\ 0 & 0 & 0 & 1 \\ -1 & 0 & 0 & 0 \\ 0 & -1 & 0 & 0 \\ 0 & 0 & -1 & 0 \\ 0 & 0 & 0 & -1 \\ 1 & -1 & 0 & 0 \\ 0 & 0 & 1 & -1 \end{bmatrix},$$

$$b := \begin{bmatrix} 4.985 \\ 4.685 \\ 4.985 \\ 4.985 \\ -0.15 \\ 0 \\ -0.15 \\ 0 \\ -0.3 \\ -0.3 \end{bmatrix}.$$

Hence, the resulting minimisation problem is stated as

$$\min_r \frac{1}{\|WS\nabla z(r)\|^2} w.r.t. A r \leq b.$$

5.2 Optimisation method

Since the sensitivity itself is a finite-difference approximation, it makes little sense to use an optimisation method that requires further derivatives. For the optimisation we used Powell's latest derivative-free trust region optimiser LINCOA (LINearly Constrained Optimization Algorithm) (Powell, 2014a, b) for linearly constrained problems.

The LINCOA method is a derivative-free optimisation algorithm for linearly constrained problems written in Fortran by M.J.D. Powell. It is based on Powell's other derivative-free optimisation algorithms with the distinction of incorporating general linear constraints. However, a detailed description of the LINCOA software has not been published yet.

According to Powell (2014a), the LINCOA method uses a quadratic model

$$Q(x) = c + g^T x + \frac{1}{2} x^T H x, x \in \mathbb{R}^n$$

Table 3. Mixed weights: improvement of sensitivity of impedance w.r.t. material parameters using $W = \frac{1}{\sqrt{2}} \cdot \text{diag}(1,\ldots,1,\sqrt{2})$. With exception of c_{44} and e_{15} all partial sensitivities are increased. Objective function values are shown unscaled. Figure shows resulting geometry utilising mixed weight matrix in millimetres; $r_1 = 4.01$; $r_2 = 4.34$; $r_3 = 2.17$; $r_4 = 3.93$.

Param.	W	Start	Optimal	Gain ratio
c_{11}	$\frac{1}{\sqrt{2}}$	2.0770×10^3	6.3742×10^3	3.0689
c_{33}	$\frac{1}{\sqrt{2}}$	803.3076	2.0949×10^3	2.6079
c_{44}	$\frac{1}{\sqrt{2}}$	1.9901×10^4	1.9381×10^4	0.9739
c_{12}	$\frac{1}{\sqrt{2}}$	0.6624	0.8317	1.2555
c_{13}	$\frac{1}{\sqrt{2}}$	4.4700×10^3	1.2346×10^4	2.7619
ϵ_{33}	$\frac{1}{\sqrt{2}}$	1.4591	2.1869	1.4987
e_{31}	$\frac{1}{\sqrt{2}}$	83.2714	254.7393	3.0591
e_{33}	$\frac{1}{\sqrt{2}}$	62.9616	148.4855	2.3583
ϵ_{11}	$\frac{1}{\sqrt{2}}$	3.6831	4.6862	1.2724
e_{15}	1	262.1774	248.5213	0.9479
Obj.		4.7489×10^{-9}	3.4887×10^{-9}	0.7346
$\|W\nabla Z\|^2$		2.1058×10^8	2.8664×10^8	1.3612

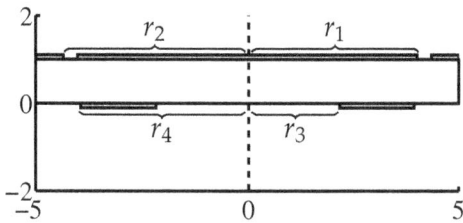

as an approximation to the objective function $J(x) \in \mathbb{R}^n$. However, as no derivatives are available, the quadratic model Q has to be iteratively constructed from function evaluations of J. At the beginning of the optimisation, the user chooses the amount of points m which are further used to interpolate the model Q with $m \in \{n+2,\ldots,\frac{1}{2}(n+1)(n+2)\}$, a typical choice for m being $2n+1$. This leaves $\frac{1}{2}(n+1)(n+2) - m$ degrees of freedom for the choice of Q which are fixed by applying a symmetric Broyden updating method to the model Q. A trust region and an active-set approach incorporating the linear constraints then determine the new points for updating the model function Q, which is iteratively minimised by the process. The trust region size in the process is decreased when certain conditions are fulfilled until ultimately the algorithm halts as the trust region size has reached a user-prescribed lower boundary.

5.3 Results

In the following section we will present results of the sensitivity optimisation. The section is divided into two parts. In the first part we will concentrate on the influence of the weight matrix W on the optimisation. We have tested different weight assignments (uniform, binary and mixed) for W

Table 4. Different starting configurations with uniform weight $W = \text{diag}(1,\ldots,1)$. Initial electrode configurations for case in (a): $r = [0.30.64.44.7]$; for case in (b): $r = [2.02.30.34.7]$; for case in (c): $r = [0.34.682.052.35]$. Successful optimisation of overall and partial sensitivities, however, sensitivity ratio to optimised reference (*SROR) initial point shows that the results are not that good. Objective function values are shown unscaled.

Param.	Gain ratio	SROR*
(a) Obj. func. : 4.0978×10^{-7}. $\|\nabla Z\|^2 = 2.4403 \times 10^6$		
c_{11}	1.2434	0.0849
c_{33}	1.4205	0.1292
c_{44}	5.3202	0.3028
c_{12}	1.0125	0.2835
c_{13}	1.0978	0.0855
ϵ_{33}	1.3629	0.3185
e_{31}	1.1556	0.0947
e_{33}	1.8522	0.1520
ϵ_{11}	2.8385	0.2002
e_{15}	3.2481	0.3084
(b) Obj. func. : 1.5881×10^{-9}. $\|\nabla Z\|^2 = 6.2967 \times 10^8$		
c_{11}	0.9293	0.2234
c_{33}	0.8531	0.3203
c_{44}	1.1749	1.2426
c_{12}	0.5684	1.1232
c_{13}	0.8933	0.2433
ϵ_{33}	0.6659	0.7925
e_{31}	1.0677	0.1759
e_{33}	0.9212	0.2585
ϵ_{11}	1.2864	0.5104
e_{15}	1.4586	0.9625
(c) Obj. func. : 5.7492×10^{-8}. $\|\nabla Z\|^2 = 1.7394 \times 10^7$		
c_{11}	1.1830	0.0068
c_{33}	2.3508	0.0271
c_{44}	4.3539	0.2084
c_{12}	2.6384	0.0748
c_{13}	1.3414	0.0073
ϵ_{33}	2.5290	0.0877
e_{31}	1.2714	0.0109
e_{33}	2.4383	0.0705
ϵ_{11}	2.0128	0.1069
e_{15}	1.8070	0.0977

and we will show the effect it has on the optimisation. In the second part we will demonstrate the influence of the initial starting point, i.e. the initial electrode configurations, on the optimisation procedure.

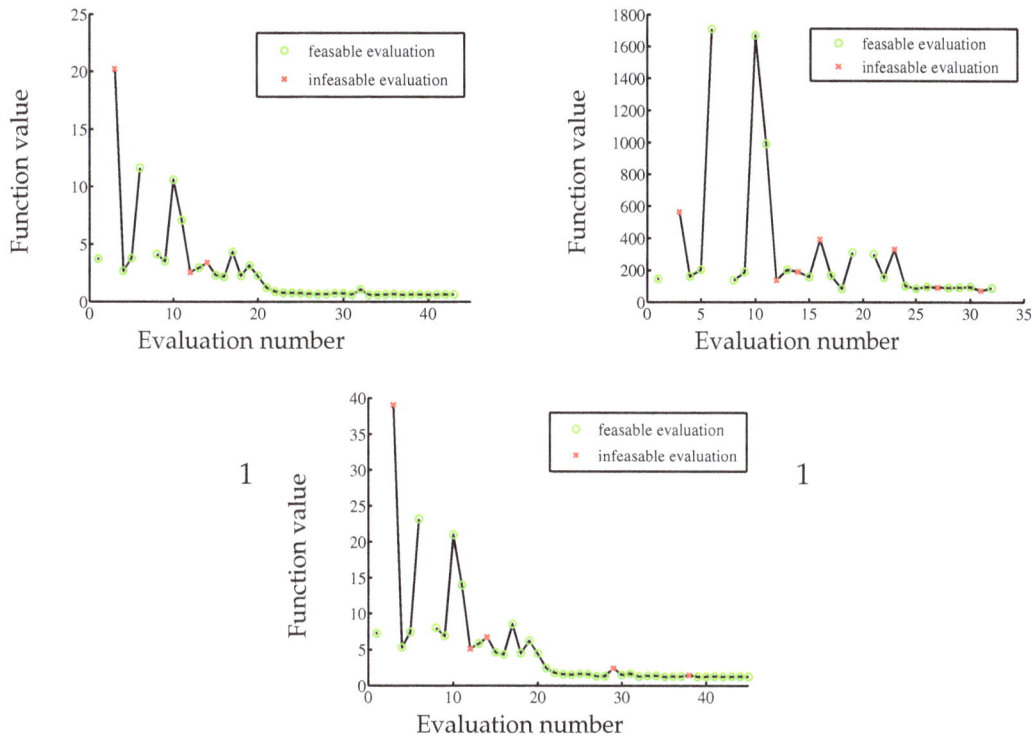

Figure 7. Evaluation history for uniform weights, binary weights, and mixed weights. Initial values are 3.7283, 145.45821 and 7.2703; optimised values are 0.6219, 84.3588 and 1.2397, respectively.

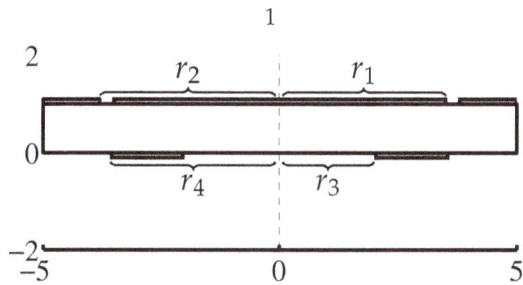

Figure 8. Reference initial geometry in millimetres; $r_1 = 3.5$; $r_2 = 3.8$; $r_3 = 2.05$; $r_4 = 3.55$.

Influence of the weighting matrix

As an initial point for a first optimisation we used the electrode configuration of a piezoceramic (see Fig. 8) we physically possess for measurement purposes. This ceramic configuration has been developed in Unverzagt et al. (2015) using statistical methods. We have made the experience that this configuration has exceptionally good properties with regard to optimisation as opposed to other geometries tested. We shall call this configuration the reference initial point. For the optimisation process we used the uniform weights $W = \mathrm{diag}(1, \ldots, 1)$. A summary of the results can be found in Table 1 along with the final geometry. This shows an overall increase in sensitivity $\|\nabla Z\|^2$ by more than 41 % and some partial sensitivities were increased by up to 307 %. However,

not all partial sensitivities are very reactive towards changes in the electrode configuration, i.e. the sensitivity gain regarding e_{15} is -6.1 %.

As a reaction to this slight partial decrease in sensitivity regarding e_{15} we chose to neglect the partial sensitivities of those parameters which have a positive gain and focus on e_{15} by setting the corresponding weights to 0 and 1, respectively. Through this binary weight-matrix setting, the sensitivity with regard to e_{15} was increased by 31 %. (see Table 2) However, following our expectations, the sensitivities regarding the other parameters have mainly decreased. The evaluation histories can be seen in Fig. 7.

These two resulting geometries combined demonstrate the feasibility of optimising the sensitivity with regard to all parameters, i.e. the two geometries show a combined increase in sensitivity for all parameters.

We have experimented with different mixed weights (see Table 3) with the aim to find a single globally optimal electrode configuration, however, it has not been possible to find such an electrode configuration. Although it is not globally optimal in the sense discussed above, it is still possible to use the configuration from Table 3 for the purpose of solving an inverse problem in the material parameters since each parameter shows some (non-zero) sensitivity. Due to long computational time for each optimisation process we did not try to fine-tune the selection of the weight matrix.

Multiple starting points

To examine the influence of different initial electrode configurations to the optimisation process we have chosen a wide
range of barely feasible configurations (see Tables 4a–c). All
these are optimised using uniform weights and the resulting
optimal is compared to the resulting optimal when starting
from the reference initial point above with uniform weights.
The resulting geometries are all further away from infeasibil-
ity than their initial configuration and locally optimal. These
three cases demonstrate that the problem of identifying a single globally optimal geometry is hard, since there are many
locally optimal configurations.

6 Conclusions and future work

In contrast to Rautenberg et al. (2011) with the two electrode approach we have shown that the sensitivity of the
impedance to various critical material parameters is non-zero
in all our ring electrode configurations with non-uniform excitation. Therefore, parameter estimation techniques as in
Kaltenbacher et al. (2008) can be used with only the measurements of the impedance required. This reduces the cost
of such investigations as the equipment required is compara-
tively cheap.

In order to systematically search for the maximal sensitivity in such a configuration we need to solve an optimisation
problem with the configuration radii as variables. Due to the
inflexibility of the precompiled solver we were forced to use
a derivative-free optimisation algorithm. More efficient algorithms may be used if a more flexible finite-element code
with the possibility of influencing the internal computations
were available, so that one could compute derivatives simultaneously. However the results show clearly that there are
many locally optimal configurations.

We are investigating the possibility of better sensitivity analysis by utilising the simulation software CFS++
(Kaltenbacher, 2010) being developed at the TU Vienna.
Modifications in the software for this purpose are ongoing.
Besides sensitivities, these changes can be used to compute
adjoints and thus solve optimisation problems, including parameter estimation problems. Simulations with CFS++ will
also shed light on the dampening influence of the external
circuit if we increase the number of electrodes, thereby increasing the complexity of the external circuit, in particular
the number of external impedances.

Apart from discrete sensitivities and adjoints, it is also possible to formulate the sensitivity and adjoint equations for
the model in function spaces and discretise these along with
the primal equations and solving them. Another avenue for
future development is to formulate the sensitivity maximi-
sation problem using shape and topology calculus instead of
parametrised rings. This would help generalise the configuration of electrodes to ceramic geometries that are not radially
symmetric.

Acknowledgements. Part of the research presented here was
done under the financial grant of the Research Prize 2012 awarded
by the University of Paderborn to Kshitij Kulshreshtha and
Jens Rautenberg.

The authors are thankful to M. Kaltenbacher for his support
and making the simulation software CFS++ available for future
development.

References

Helnwein, P.: Some remarks on the compressed matrix representa-
tion of symmetric second-order and fourth-order tensors, Computer methods in applied mechanics and engineering, 190, 2753–
2770, 2001.
IEEE Std 176-1987: IEEE Standard on Piezoelectricity, The Insti-
tute of Electrical and Electronic Engineers, Inc., New York, IEEE
Std 176-1987 edn., 1988.
Kaltenbacher, B., Lahmer, T., Mohr, M., and Kaltenbacher, M.:
PDE based determination of piezoelectric material tensors, Eur.
J. Appl. Math., 17, 383–416, doi:10.1017/S0956792506006474,
2006.
Kaltenbacher, M.: Advanced simulation tool for the design of sen-
sors and actuators, Procedia Engineering, 5, 597–600, 2010.
Kaltenbacher, M., Lahmer, T., Leder, E., Kaltenbacher, B., and
Lerch, R.: FEM based determination of real and complex elas-
tic, dielectric and piezoelectric moduli in piezoceramic materials,
IEEE Transactions on Ultrasonics, Ferroelectrics and Frequency
Control, 55, 465–475, 2008.
Ketterling, J. A., Aristizabal, O., Turnbull, D. H., and Lizzi, F. L.:
Design and fabrication of a 40-MHz annular array transducer,
Ultrasonics, Ferroelectrics, and Frequency Control, IEEE Trans-
actions on, 52, 672–681, 2005.
Kirschner, J.: Surface Acoustic Wave Sensors (SAWS): Design
for Application, Micromechanical Systems, available at: http:
//www.jaredkirschner.com/uploads/9/6/1/0/9610588/saws.pdf
(last access: 11 June 2015), 2010.
Kocbach, J.: Finite element modeling of ultrasonic piezoelectric
transducers, PhD thesis, Department of Physics, University of
Bergen, 2000.
Lahmer, T.: Forward and inverse problems in piezoelectricity, PhD
thesis, University of Erlangen-Nuremberg, 2008.
Meschede, D. and Gerthsen, C.: Physik, vol. 24 überarbeitete Auflage, Springer, 2010.
Pérez, N., Andrade, M. A., Buiochi, F., and Adamowski, J. C.:
Identification of elastic, dielectric, and piezoelectric constants
in piezoceramic disks, IEEE T. Ultrason. Ferr., 57, 2772–2783,
2010.
Powell, M.: Derivate Free Optimization, Tech. Rep. 2014:02,
Linköping University, Optimization, http://liu.diva-portal.org/
smash/get/diva2:697412/FULLTEXT03.pdf, 2014a.
Powell, M.: On fast trust region methods for quadratic models
with linear constraints, Tech. rep., Technical Report DAMTP
2014/NA02, Department of Applied Mathematics and Theoretical Physics, University of Cambridge, Silver Street, Cambridge CB3 9EW, http://www.damtp.cam.ac.uk/user/na/NA_
papers/NA2014_02.pdf, 2014b.

Ramli, N. A. and Nordin, A. N.: Design and modeling of MEMS SAW resonator on Lithium Niobate, in: Mechatronics (ICOM), 2011 4th International Conference On, 1–4, IEEE, 2011.

Rautenberg, J., Rupitsch, S., Henning, B., and Lerch, R.: Utilizing an Analytical Approximation for c_{44} to Enhance the Inverse Method for Material Parameter Identification of Piezoceramics, in: 7th International Workshop on Direct and Inverse Problems in Piezoelectricity, 04–07 October 2011, Duisburg, 2011.

Rupitsch, S. J. and Lerch, R.: Inverse method to estimate material parameters for piezoceramic disc actuators, Appl. Phys. A, 97, 735–740, 2009.

Rupitsch, S. J., Wolf, F., Sutor, A., and Lerch, R.: Estimation of material parameters for piezoelectric actuators using electrical and mechanical quantities, in: Ultrasonics Symposium (IUS), 2009 IEEE International, 1–4, IEEE, 2009.

Slaughter, W. S.: Constitutive Equations, in: The Linearized Theory of Elasticity, 193–220, Birkhäuser Boston, 2002.

Unverzagt, C., Rautenberg, J., Henning, B., and Kulshreshtha, K.: Modified Electrode Shape for the Improved Determination of Piezoelectric Material Parameters, in: Proceedings of the 2013 International Congress on Ultrasonics (ICU), edited by: Siong, G. W., Siak Piang, L. and Cheong, K. B., 2013.

Unverzagt, C., Rautenberg, J., and Henning, B.: Sensitivitätssteigerung bei der inversen Materialparameterbestimmung für Piezokeramiken, Tech. Mess., 82, 102–109, doi:10.1515/teme-2014-0008, 2015.

17

High-temperature stable indium oxide photonic crystals: transducer material for optical and resistive gas sensing

Sabrina Amrehn, Xia Wu, and Thorsten Wagner

Department of Chemistry, University of Paderborn, Paderborn, Germany

Correspondence to: Thorsten Wagner (thorsten.wagner@upb.de)

Abstract. Indium oxide (In_2O_3) inverse opal is a promising new transducer material for resistive and optical gas sensors. The periodically ordered and highly accessible pores of the inverse opal allow the design of resistive sensors with characteristics independent of structure limitations, such as diffusion effects or limited conductivity due to constricted crosslinking. Additionally the photonic properties caused by the inverse opal structure can be utilized to read out the sensors' electronical state by optical methods. Typically semiconducting sensors are operated at high temperatures ($> 300\,°C$). To maintain a good thermal stability of the transducer material during operation is a minimum requirement. We present results on the synthesis and investigation of the structural stability of the In_2O_3 inverse opal structure up to a temperature of $550\,°C$ (limit of substrate material). As will be shown, their optical properties are maintained with only slight shifts of the photonic band gaps which can be explained by the results from the structural characterization using X-ray diffraction and electron microscopy combined with optical simulations.

1 Introduction

Indium oxide is not only known as a sensing material for resistive semiconducting gas sensors for the detection of NO_2, O_3, ethanol, hydrogen or CO (Ivanovskaya et al., 2001; Wagner et al., 2011, 2013; Takada et al., 1993; Zheng et al., 2009; Martin et al., 2004; Yamaura et al., 1996). Because of its optical properties it is also utilized for transparent conductors (typically doped with tin, ITO) for electronic components such as flat screen displays, solar cells and LEDs (Lewis and Paine, 2000; Kim et al., 1998).

The combination of these two properties, namely the gas sensitivity and the transparency in the visible regime combined with its high refractive index, makes indium oxide an interesting candidate for building optical transducers for gas sensors. Periodically ordered nanostructures of dielectric materials with a periodicity in the order of the wavelength of interested electromagnetic waves show interesting new, structure related optical properties. Therefore these structures are commonly referred to as photonic crystals (Joannopoulos et al., 1997). The most prominent feature of a photonic crystal is its photonic band gap. Light of certain wavelengths cannot propagate along one (stop band) or all (complete band gap)

directions of the photonic crystal. Photons with the energy within the photonic band gap region are reflected. Therefore, the photonic crystals operating in visible range could show intense color impressions. The position and size of the band gap are determined by structure parameters (e.g., periodicity, symmetry, geometry, filling fraction) and the refractive index contrast between the wall material and the fluid in the pore (Joannopoulos et al., 1997). Variation of one of these properties can be observed by a change in the reflectance spectra.

Photonic crystals can be synthesized with different methods, e.g., electron beam lithography (Cheng and Scherer, 1995), direct laser writing (Deubel et al., 2004) or laser holography (Miklyaev et al., 2003).

In the following, we focus on the inverse opal structure, a specific type of three-dimensional photonic crystals. By utilizing self-assembly of spherical particles (artificial opals) as a template and a consecutive casting step the inverse opal structure offers a relatively simple method of production (Stein et al., 2008) with the drawback of a high amount of lattice defects and macroscopic cracks due to shrinkage. These structural imperfections are crucial for some highly demanding applications such as informatics and telecommunications; but for most sensing applications the reflectance at the photonic band gap of these chemically prepared photonic crystals is high enough to be used as an optical signal for sensing. So the advantage of a fast and cheap synthesis combined with the possibility of scaling up for mass production allows this method to find its own niches of application where extremely high structural quality is not required.

As for sensing applications, a typical sensing mechanism of photonic crystals is to optically read out the change of their reflection spectra resulting from the change of their refractive index contrast between the solid phase and the introduced fluids (Nair and Vijaya, 2010). This mechanism can be used, for example, to detect liquids with different refractive indices (Amrehn et al., 2015). Another typical mechanism is based on the reaction of the detected species with the solid phase itself, inducing a variation of electronic and optical properties of the photonic crystals (Xie et al., 2012).

Besides sensors utilizing the inverse opal structure for optical readout also resistive type sensors take some advantage of the highly accessible pores of the inverse opal. The structure allows the design of sensing layers with characteristics independent of structure limitations such as diffusion effects or limited conductivity due to constricted crosslinking (Scott et al., 2001).

As for some special sensing applications, such as sensors operating at high temperature (above 500 °C), remote sensing using optical readout of photonic crystals is of special interest. Because of remote sensing there is no need for the wire connection, which may break down at high temperature, between the transducer and the optical devices for measuring the signal change. Since the optical signal of a photonic crystal is caused by its structure, any degradation of structure at high temperature can be detected by the optical devices, which is difficult to be realized for sensors based on other porous materials.

Despite of all these advantages for photonic crystals used as sensors at high temperature, the thermal stability of the photonic crystals needs to be tested before they are applied to this new type of sensor concept. In this paper, indium oxide is synthesized in the inverse opal structure and heat treated at different temperatures. Afterwards structural changes and the corresponding optical properties changes are investigated, to evaluate the potential for indium oxide with the inverse opal structure for high temperature sensing applications.

2 Experimental

The indium oxide inverse opal structures in this study are prepared utilizing a modified three-step version (Fig. 1) of a literature-known, template-based synthesis protocol (Stein et al., 2008). In the first step, monodisperse polymethylmethacrylat (PMMA) spheres are synthesized by surfactant-free emulsion polymerization. Controlled deposition of these spheres (second step) leads to artificial opals which are, in

Figure 1. Scheme of the inverse opal synthesis by casting a PMMA opal: the PMMA spheres are deposited onto a glass slide at 60 °C, and then the opal pores are filled with an indium nitrate solution. This composite is dried, the indium nitrate is thermally converted to indium oxide and the PMMA template is removed by combustion.

the third step, used as rigid structure matrices for the casting of the indium oxide inverse opal, which results from thermal conversion of an infiltrated indium precursor species and removal of the matrix.

To investigate the thermal stability of indium oxide in the inverse opal structure, the photonic band gap position (optical spectroscopy), the crystallite size (powder X-ray diffraction) and the structure of the indium oxide films (electron microscopy) are compared before and after heat treatment. The temperature limit of this investigation is given by the glass transition of the glass slides as substrates (550 °C).

2.1 Synthesis

The polymethylmethacrylat (PMMA) spheres were synthesized by surfactant-free emulsion polymerization (Egen and Zentel, 2004), this leads to monodisperse spheres. A total of 400 mL water was heated to reflux in nitrogen atmosphere. After stopping the nitrogen flow, methylmethacrylat (21.3 mL, 0.5 mol L^{-1}, Merck, 99 %) and ethylene glycol dimethacrylat, (0.57 mL, Merck, 97.5 %) as a crosslinking agent, were destabilized by filtrating with Al_2O_3 and added to the flask. The mixture was stirred for 10 min (300 rpm). $K_2S_2O_8$ (27 mg, Bayer) was dissolved in 1 mL water and added. The mixture was stirred for 2 h at 100 °C, then the flask was opened and the mixture cooled to room temperature. By filtering the mixture through a paper filter, large aggregates were removed.

For the opal template structure preparation, microscopy slides were cleaned with ethanol and acetone, then heated up to 60 °C on a heating plate. A total of 40 μL of the PMMA dispersion was placed on the slide with a microliter pipette and the solvent was removed by evaporation at 60 °C.

The In_2O_3 inverse opal films were synthesized by casting the PMMA opal. Therefore, the opal pores were filled with an indium nitrate solution (0.016 g $In(NO_3)_3 \cdot xH_2O$ (Sigma Aldrich, 99.99 %) in 0.043 mL ethanol) by placing 2 μL of this solution on top of the opal film. The resulting composite was dried at room temperature for 24 h and at 60 °C for 72 h. Finally, the indium nitrate was converted to indium oxide in an incineration furnace at 300 °C (2 h; heating rate 0.5 °C min^{-1}). A scheme of the synthesis procedure is shown in Fig. 1.

2.2 Thermal stability tests

The optical and structural change of the indium oxide inverse opal films before and after each heat treatment step were characterized with a modified Fourier transform infrared spectrometer with wavelength extension to the visible regime (FTIR-vis) and with the X-ray diffraction method, respectively. The heat treatment was performed in an incineration furnace under air. The heating rate was $5\,°C\,min^{-1}$, the dwell time was 5 h and the tested temperatures were 350, 400, 450, 500 and $550\,°C$. For each temperature, a fresh sample was prepared to avoid the superposition of time-dependent aging effects. The sample annealed at $550\,°C$ was again tested after characterization at $550\,°C$ for 72 h to investigate the influence of prolonged heating on the stability. Before this additional heating step the sample was cooled down slowly over 24 h to reduce unwanted effects by thermal stress and maintain comparability with the other samples. To gain some information on reproducibility, an additional series of samples was synthesized and annealed at $550\,°C$ for 72 h.

2.3 Characterization

The reflectance spectra were recorded with a Vertex 70 FTIR spectrometer under the Hyperion 1000 light microscope (Bruker) modified with aluminum mirrors and a xenon light source to extend the wavelength range to visible regime (referred to as FTIR-vis). A silicon wafer with known reflectance was used as a reference for normalization to obtain the absolute reflectance of samples. For every sample, four measurement spots were randomly chosen and their reflectance was averaged. Gauss fit was used for determination of the reflectance maximum. The scanning electron microscope (SEM) images were taken with a Zeiss Neon 40. The powder X-ray diffraction (PXRD) patterns were recorded with a D8 Advance (Bruker). The X-rays have a wavelength of 0.154 nm, generated by a CuK_{α}-tube. The measurement range was from 20 to $70°\ 2\theta$ with a step size of $0.02°$ and an integration time of 3 s. The crystallite size is calculated with the Scherrer equation. Therefore the full width at half maximum and the peak maximum of the (222) peak is evaluated with Lorentzian fit; the shape factor is 0.94.

2.4 Simulation

The photonic band structures of both the inverse opal structure and rod structure were calculated using a free software package "MIT Photonic-Bands" (MPB 1.4.2). MPB calculates fully vectorial eigenmodes of Maxwell's equations with periodic boundary conditions by preconditioned conjugate-gradient minimization of the block Rayleigh quotient in a planewave basis (Johnson and Joannopoulos, 2001). The structural model of the inverse opal structure is constructed by placing air spheres in the lattice points of fcc structure and filling the complementary solid phase with In_2O_3 with the refractive index of 1.85 taken from literature (Senthilkumar

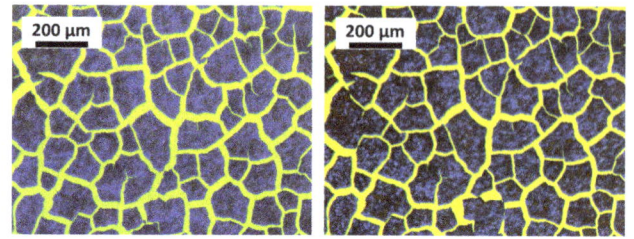

Figure 2. Optical microscope images from an as-synthesized indium oxide film (left) and after annealing at $550\,°C$ for 77 h (double heated for $5+72$ h, see text) (right) at the same location of the same sample.

and Vickraman, 2010). The dispersion and absorption of the light in visible range is assumed to be negligible. The radius of the air sphere is varied to model the structure with different volume fractions of the solid phase varying from 1.8 to 62.0%. The rod structure is still an fcc structure. This structure is constructed by connecting the octahedral and tetrahedral interstitial sites in the conventional unit cell of fcc structure by rods to form eight tetrahedral bonded structures in the unit cell, which corresponds to two tetrahedral boned structures in the fcc primitive cell. The rod is an ideal structural motif to model the materials connecting the interstitial sites, where the detailed surface roughness and topology as observed in SEM is simplified.

These rods are filled with In_2O_3 with the same refractive index as in the inverse opal structure. The radius of the rod is varied to vary the volume fraction of the solid phase from 4.7 to 47.7%. For the simulation of both structural models, the grid resolution is 32 pixels per basis vector in a given direction. The mesh size used to average the refractive index at each grid point is set to be 7. At each k point of high symmetry, 10 bands are calculated. The stop gaps along Γ-L direction, which corresponds to the $\langle 111\rangle$ surface normal of the real photonic crystal, are used to evaluation. The resulting photonic band structure diagram in frequency is calculated into a diagram in wavelength for the specific photonic crystals synthesized here with a lattice constant of 339 nm.

3 Results and discussion

A great advantage for sample evaluation of the here presented inverse opals compared to many other types of nanostructured materials is the structural color in the visible regime due to the photonic stop bands. This allows fast evaluation of the quality of the synthesized product even by naked eye and might be later used as a feature for self-testing a high-temperature sensor device by colorimetry. As can be seen in the optical microscope (Fig. 2), the indium oxide inverse opals do show an intense blue color after synthesis. This color as well as the crack pattern is preserved after the heat treatment which is a first indication for thermal stability of the photonic crystals' framework. A more detailed optical

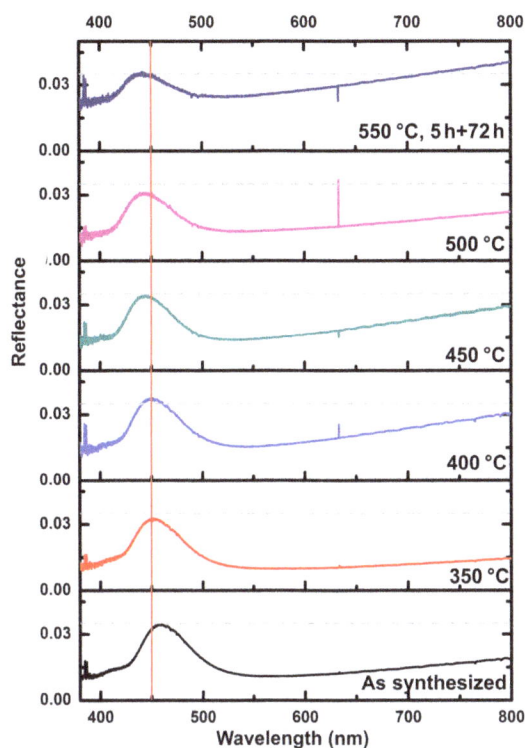

Figure 3. Reflectance spectra of the indium oxide inverse opals: average of the spectra before heat treatment, and spectra after heat treatments at different temperatures. The grey horizontal line marks the maximum reflectance value of 3.5 % after 77 h treatment at 550 °C. The vertical red line at 450 nm serves as a guide for the eye to recognize the shift of the maximum intensity.

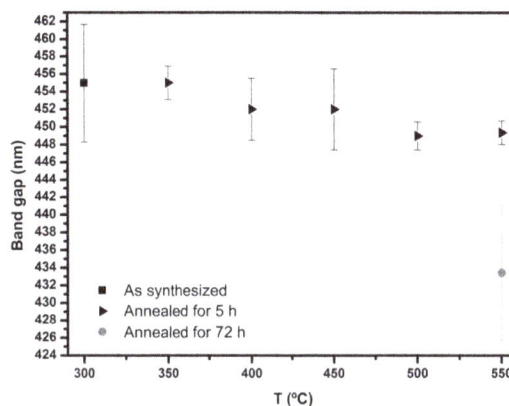

Figure 4. Reflectance maximum after heat treatments at different temperatures. Error bars: standard deviation of the mean value of the measurement of five different spots on each sample. Values were measured from 6 samples for the cases of both as-synthesized and 72 h treatment, and the rest of the values were measured from 2 samples for each case.

characterization of the indium oxide inverse opal films before and after heat treatment (Fig. 3) reveals reflection bands with maxima between 446 and 466 nm (corresponding to blue). Since there is no such reflection observed in nonstructured indium oxide films and the wavelength range corresponds to the results of theoretical simulations (see later) this color is taken as a strong evidence for (i) the successful casting of the inverse opal structure and (ii) the conservation of the framework after heat treatment. Figure 3 also shows the evolution of the maximum reflectance. As can be seen, it is conserved (ca. 3.5 %) throughout the heat treatment. This implies that (i) the size of the ordered regions is conserved, (ii) the position and size of the macroscopic cracks remains constant and (iii) there is no large change of refractive index of the In_2O_3. However, especially after the prolonged 550 °C treatment there is a strong increase in the background reflectance. In principle, the background reflectance is caused by the opal–substrate interface. Therefore we assume that this strong increase is due to changes of the utilized glass substrates which, at 550 °C, is close to the glass transition temperature.

A more detailed analysis of the position of the reflectance maximum (Fig. 4) reveals a shift to shorter wavelengths. Be-

fore the heat treatment of the indium oxide the mean value of the band gap maximum for the different samples is 454 nm. After the heat treatment the reflectance maximum shifts e.g., to 449 nm after the 550 °C treatment for the 5 h treated samples. However, as discussed before, the absolute value of the reflectance of the indium oxide photonic crystal does not decrease. For the samples treated for 72 h, a stronger change in the reflectance peak position is observed. As will be discussed below, this is most likely due to the thinning of the inverse opal framework by diffusion of an indium species into gas phase. For later applications this has to be evaluated more carefully in the context of the targeted working conditions. Especially the oxygen concentration in the surrounding atmosphere will have a strong impact on the diffusion behavior.

Summarizing the results from the optical characterization it can be concluded that indium oxide inverse opals are a thermally stable transducers for high-temperature applications in the range of 550 °C for a limited timespan (72 h) since their optical properties are preserved. For lower temperature as e.g., 500 °C it is assumed that even hundreds of hours can be achieved according to Arrhenius law, since the processes which lead to the degradation of the material are thermally activated.

To further investigate the origin of the observed shift of the reflection bands, a detailed structural analysis utilizing PXRD and SEM was carried out. The XRD results (Fig. 5) show the typical reflection pattern for cubic crystalline indium oxide phase (JCPDS 71-2194). As the heat treatment temperatures increase from 300 to 550 °C, the crystallite size calculated by the Scherrer equation increases from about 11.5 to 20 nm (Fig. 6). Similar increase of crystallite size after heat treatment was also reported for In_2O_3 film deposited

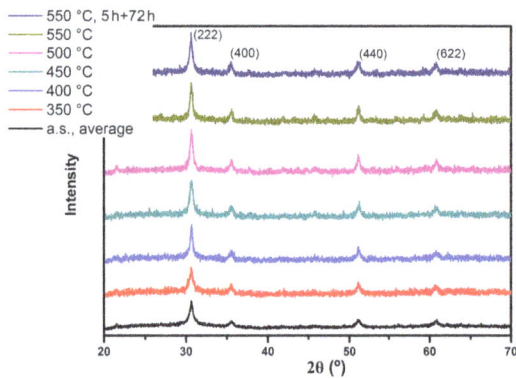

Figure 5. PXRD data: average of the as-synthesized samples and after heat treatments at different temperatures.

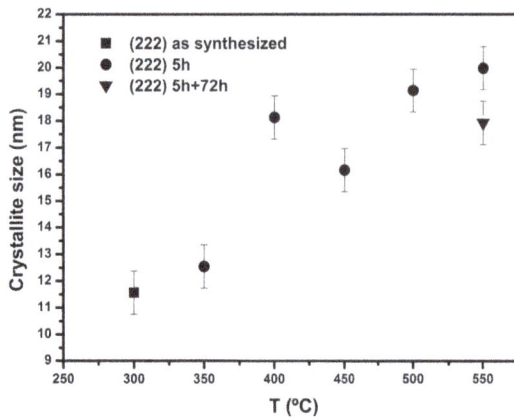

Figure 6. Crystallite sizes derived from the (222) peaks using the Scherrer equation for samples before and after heat treatment. Error bars: standard deviation of the mean value of the crystallite size of 12 different "as-synthesized" samples.

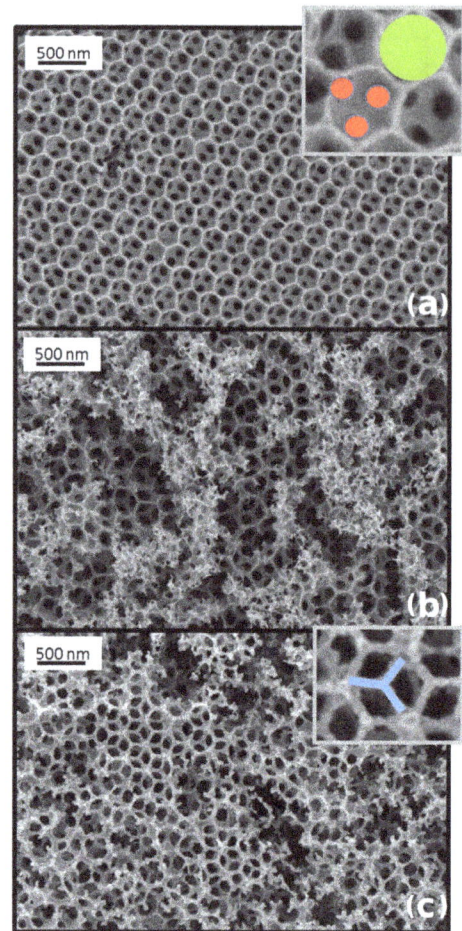

Figure 7. SEM images of the inverse opal structures: as synthesized **(a)**, annealed at 450 °C **(b)** and at 550 °C for $5 + 72$ h **(c)**. Insets show zoomed out region of **(a)** and **(c)** with spherical pore marked in green, windows connecting pores in red and rods forming after 550 °C treatment in blue.

by the electron beam evaporation method (Senthilkumar and Vickraman, 2010).

As the position of the stop band is strongly affected by the periodicity of the inverse opal structure, SEM analysis was carried out. Figure 7 shows the evolution of the structure after different temperature treatments. Contrary to the expected change in periodicity, which was not observed, the periodicity remains stable at about 240 nm (distance between the centers of two neighboring pores), but there are strong variations in the pore shape to be observed. The as-synthesized samples show a typical nanocast inverse opal structure with spherical pores interconnected by circular windows (Fig. 7a, spherical pore marked in green, windows in red). For the 450 °C treated sample the window size is increased (Fig. 7b) and after 550 °C the windows are widened in such a way that the inverse opal framework is formed by a rod-type structure (Fig. 7c).

Compared to the as-synthesized sample, which shows a smooth surface in the SEM, the indium oxide annealed at

550 °C shows thickness variations along the rods. The narrow necking on the rods' surface might be correlated to the thermal grooves typically formed along grain boundaries on the surface of thin polycrystalline materials after annealing (Gottstein, 2004). Mainly driven by the energy reduction of the system via eliminating grain boundaries, the crystallite size (grain size) increases as the heat treatment temperature increases which is shown above in the XRD results. Meanwhile, the margin of the window of the as-synthesized inverse opal structure changes its shape to the rod-type structure driven by the reduction of surface energy. The thermal grooves on the surface of rods form when the average grain size reaches the size of the smallest sample dimension, which is the diameter of these rods here in nanometer range. These grooves lead to a retarding force on the grain boundary migration reducing the grain growth rate, which is beneficial for the thermal stability we are aiming at. These grooves are, in principle, even able to stop grain growth when the grain size

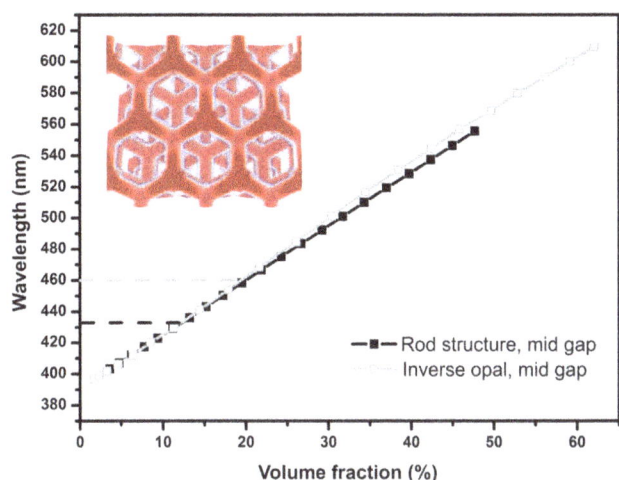

Figure 8. Simulated mid-gap wavelengths for indium oxide in the inverse opal structure and the after annealing rod-type structure (structural model as inset) at different volume fractions of solid phase, the dashed lines show the measured reflectance maxima.

is about twice the sample thickness in the case of a thin sheet sample (Gottstein, 2004). This effect may explain the cease of grain growth after prolonged heat treatment (72 h) at high temperature (550 °C) (Fig. 6). However, to further prove this, prolonged treatments have to be carried out.

We consider Ostwald ripening and surface diffusion to be the dominating mechanisms which might lead to the formation of the new rod-type structure. Ostwald ripening is driven by indium oxide vapor pressure difference at the highly curved window margins (high pressure) and at the less curved surface of rods (low pressure). The indium oxide evaporates from the margin and condenses on the rods. Surface diffusion (mass transfer diffusion) can also occur since the surfaces of as-synthesized inverse opal wall structures most probably contain many surface defects, and therefore allow the formation of mobile adparticles to transfer material.

To test if these structural changes due to heat treatment might be responsible for the observed blueshift of the reflectance maxima, simulations of the photonic band structures of the as-synthesized sphere-type as well as of the rod-type structure have been carried out utilizing MPB (Johnson and Joannopoulos, 2001). Results (Fig. 8) show a band gap of the mid-gap wavelength of 460 nm (the same as the average wavelength of the reflection maxima of the as-synthesized samples) would be opened for the inverse opal structure when the volume fraction of indium oxide is assumed to be 19 %. Given the shrinkage of materials during the casting process, this smaller volume fraction is considered to be a reasonable and close-to-reality volume fraction of the solid phase in the as-synthesized inverse opal structure, instead of 26 % for the ideal case of face-centered cubic close packing. For the very same volume fraction, the simulation re-

sults show that the band gap blueshifts by 2 nm (from 460 to 458 nm) from sphere-type to rod-type structure. This theoretical blueshift qualitatively agrees well with the experimentally observed shift in the same direction, but the amount of the measured shift is larger (21 nm). However, the simulation results show that loss in indium oxide mass also will have a strong impact on the position of the reflectance maximum. A reduction of the solid volume fraction by about 7 %, probably due to evaporation of the indium oxide, can lead to the observed shift to 433 nm.

Summarizing the results of the SEM, XRD characterizations and the photonic band structure simulations it can be concluded that the structural change from sphere-type to rod-type in combination with a 7 % mass loss can explain the observed changes in optical properties. However, this theoretical estimation is only accurate under the assumption that the refractive index of indium oxide remains constant. At the current stage, it is difficult to measure the refractive index change of the inverse opal films as a porous material after the heat treatment. In the literature, it was reported that the refractive index of In_2O_3 film deposited by electron beam evaporation increases as the temperature increases, which may be attributed to the improved crystallinity, and the changes of the packing density and porosity of the film after annealing (Senthilkumar and Vickraman, 2010). Since this synthesis method is completely different from the one used in the present study, it is difficult to estimate whether the similar trend of refractive index change occurs here. In case of increased refractive index after annealing, the expected mass loss which can lead to the observed blueshift should be larger than 7 %.

4   Conclusions

Indium oxide inverse opals show stable optical properties up to the temperature of 550 °C in air for 72 h. The observed structural changes due to heat treatment, such as grain growth and pore shape modifications, occur in the length scale which is 1 order of magnitude smaller than the periodicity of the photonic crystals. The visible light as a sensing signal is not sensitive to these minor structural changes as long as the periodicity and refractive index of the photonic crystals remain stable at high temperature. This inherent optical "inertness" of photonic crystals to structural changes at certain level makes them interesting candidates for a new class of optical gas sensing transducers at high temperatures. Furthermore, it is practically very convenient that the structural color of photonic crystals can be utilized as an indicator for the structural integrity since it can be readout by the naked eye. Not only is the possibility of remote sensing using optical readout attractive for high-temperature application but also for conventional resistive sensing the open inverse opal structure might be of interest since it allows the design of sensors with characteristics independent of structure limitations.

Acknowledgements. We thank the Federal Ministry of Education and Research (BMBF, 13N12969) for financial support.

References

Amrehn, S., Wu, X., Schumacher, C., and Wagner, T.: Photonic crystal-based fluid sensors: Toward practical application, Phys. Status Solidi A, 212, 1266–1272, doi:10.1002/pssa.201431875, 2015.

Cheng, C. C. and Scherer, A.: Fabrication of photonic band-gap crystals, J. Vac. Sci. Technol. B, 13, 2696–2700, doi:10.1116/1.588051, 1995.

Deubel, M., von Freymann, G., Wegener, M., Pereira, S., Busch, K., and Soukoulis, C. M.: Direct laser writing of three-dimensional photonic-crystal templates for telecommunications, Nat. Mater., 3, 444–447, doi:10.1038/nmat1155, 2004.

Egen, M. and Zentel, R.: Surfactant-Free Emulsion Polymerization of Various Methacrylates: Towards Monodisperse Colloids for Polymer Opals, Macromol. Chem. Phys., 205, 1479–1488, doi:10.1002/macp.200400087, 2004.

Gottstein, G.: Physical Foundations of Materials Science, Springer Berlin Heidelberg, Berlin, Heidelberg, ISBN 978-3-662-09291-0, 2004.

Ivanovskaya, M., Gurlo, A., and Bogdanov, P.: Mechanism of O_3 and NO_2 detection and selectivity of In_2O_3 sensors, Proceedings of the 8th International Meeting on Chemical Sensors IMCS-8 – Part 2, 77, 264–267, doi:10.1016/S0925-4005(01)00708-0, 2001.

Joannopoulos, J. D., Villeneuve, P. R., and Fan, S.: Photonic crystals: putting a new twist on light, Nature, 386, 143–149, doi:10.1038/386143a0, 1997.

Johnson, S. and Joannopoulos, J.: Block-iterative frequency-domain methods for Maxwell's equations in a planewave basis, Opt. Express, 8, 173–190, doi:10.1364/OE.8.000173, 2001.

Kim, J. S., Granström, M., Friend, R. H., Johansson, N., Salaneck, W. R., Daik, R., Feast, W. J., and Cacialli, F.: Indium–tin oxide treatments for single- and double-layer polymeric light-emitting diodes: The relation between the anode physical, chemical, and morphological properties and the device performance, J. Appl. Phys., 84, 6859–6870, doi:10.1063/1.368981, 1998.

Lewis, B. G. and Paine, D. C.: Applications and Processing of Transparent Conducting Oxides, MRS Bulletin, 25, 22–27, doi:10.1557/mrs2000.147, 2000.

Martin, L. P., Pham, A.-Q., and Glass, R. S.: Electrochemical hydrogen sensor for safety monitoring, Solid State Ionics, 175, 527–530, doi:10.1016/j.ssi.2004.04.042, 2004.

Miklyaev, Y. V., Meisel, D. C., Blanco, A., von Freymann, G., Busch, K., Koch, W., Enkrich, C., Deubel, M., and Wegener, M.: Three-dimensional face-centered-cubic photonic crystal templates by laser holography: fabrication, optical characterization, and band-structure calculations, Appl. Phys. Lett., 82, 1284–1286, doi:10.1063/1.1557328, 2003.

Nair, R. V. and Vijaya, R.: Photonic crystal sensors: An overview, Prog. Quant. Electron., 34, 89–134, doi:10.1016/j.pquantelec.2010.01.001, 2010.

Scott, R. W. J., Yang, S. M., Chabanis, G., Coombs, N., Williams, D. E., and Ozin, G. A.: Tin Dioxide Opals and Inverted Opals: Near-Ideal Microstructures for Gas Sensors, Adv. Mater., 13, 1468–1472, doi:10.1002/1521-4095(200110)13:19<1468::AID-ADMA1468>3.0.CO;2-O, 2001.

Senthilkumar, V. and Vickraman, P.: Annealing temperature dependent on structural, optical and electrical properties of indium oxide thin films deposited by electron beam evaporation method, Curr. Appl. Phys., 10, 880–885, doi:10.1016/j.cap.2009.10.014, 2010.

Stein, A., Li, F., and Denny, N. R.: Morphological Control in Colloidal Crystal Templating of Inverse Opals, Hierarchical Structures, and Shaped Particles, Chem. Mater., 20, 649–666, doi:10.1021/cm702107n, 2008.

Takada, T., Suzuki, K., and Nakane, M.: Highly sensitive ozone sensor, Sensor. Actuat. B-Chem., 13, 404–407, doi:10.1016/0925-4005(93)85412-4, 1993.

Wagner, T., Hennemann, J., Kohl, C.-D., and Tiemann, M.: Photocatalytic ozone sensor based on mesoporous indium oxide: Influence of the relative humidity on the sensing performance, Proc. 7th International Workshop on Semiconductor Gas Sensors, 520, 918–921, doi:10.1016/j.tsf.2011.04.181, 2011.

Wagner, T., Kohl, C.-D., Malagù, C., Donato, N., Latino, M., Neri, G., and Tiemann, M.: UV light-enhanced NO_2 sensing by mesoporous In_2O_3: Interpretation of results by a new sensing model, Selected Papers from the 14th International Meeting on Chemical Sensors, 187, 488–494, doi:10.1016/j.snb.2013.02.025, 2013.

Xie, Z., Xu, H., Rong, F., Sun, L., Zhang, S., and Gu, Z.-Z.: Hydrogen activity tuning of Pt-doped WO_3 photonic crystal, Thin Solid Films, 520, 4063–4067, doi:10.1016/j.tsf.2012.01.027, 2012.

Yamaura, H., Jinkawa, T., Tamaki, J., Moriya, K., Miura, N., and Yamazoe, N.: Indium oxide-based gas sensor for selective detection of CO, Sensor. Actuat. B-Chem., 36, 325–332, doi:10.1016/S0925-4005(97)80090-1, 1996.

Zheng, W., Lu, X., Wang, W., Li, Z., Zhang, H., Wang, Y., Wang, Z., and Wang, C.: A highly sensitive and fast-responding sensor based on electrospun In_2O_3 nanofibers, Sensor. Actuat. B-Chem., 142, 61–65, doi:10.1016/j.snb.2009.07.031, 2009.

Characterisation of the polarisation state of embedded piezoelectric transducers by thermal waves and thermal pulses

Agnes Eydam, Gunnar Suchaneck, and Gerald Gerlach

Solid State Electronics Laboratory, Technische Universität Dresden, Dresden, Germany

Correspondence to: Agnes Eydam (agnes.eydam@tu-dresden.de)

Abstract. In this work, we apply the thermal wave method and the thermal pulse method for non-destructive characterisation of the polarisation state of embedded piezoelectric transducers. Heating the sample with a square-wave modulated laser beam or a single laser pulse leads to a pyroelectric current recorded in the frequency or time domain, respectively. It carries information about the polarisation state. Analytical and numerical finite element models describe the pyroelectric response of the piezoceramic. Modelling and experimental results are compared for a simple lead–zirconate–titanate (PZT) plate, a low-temperature co-fired ceramics (LTCC)/PZT sensor and actuator, and a macro-fibre composite (MFC) actuator.

1 Introduction

Piezoelectric smart structures are created by embedding piezoelectric transducers into structural components to make them controllable or responsive to their environment. These structures find applications, for instance, for health-monitoring of safety components, for reducing noise emission in automobile engineering, or for damping vibrations. Their mass-production requires control of the polarisation state due to mechanical and thermal loads appearing during device fabrication.

Non-destructive methods for obtaining polarisation profiles rely on an external excitation of the material leading to a local change of material properties (Mellinger et al., 2007). A thermal excitation in terms of thermal waves or thermal pulses gives rise to a pyroelectric current, which carries information on the polarisation profile. In the frequency domain, the laser intensity modulation method (LIMM) is well-established. Thereby temperature oscillations are generated by a periodically modulated laser beam (Lang and Das-Gupta, 1986). When thermal pulses are applied with a pulsed laser, the signal is recorded in the time domain (Collins, 1977). The advantage of the thermal pulse method is a higher pyroelectric signal in a shorter measuring time. In Pham et al. (2009) both methods are used to map the polarisation pro-

files in thin dielectric films with a high resolution. Pham et al. (2009) came to the conclusion that they provide similar results with the thermal pulse method being up to 50 times faster. The use of scanning LIMM to generate a polarisation map of the sample surface is described in more detail by Stewart and Cain (2009). In Stewart and Cain (2015), piezoelectric films of thicknesses down to 100 nm are measured with high-frequency LIMM.

In this work, we present simplified analytical and numerical finite element models to describe the pyroelectric response of the LIMM and the thermal pulse method, and we compare them to experimental results.

2 Theory

The pyroelectric current is described by a fundamental relation:

$$I(t) = \frac{A}{d} \int_0^d p(z) \frac{\partial}{\partial t} \Theta(z, t) \mathrm{d}z, \qquad (1)$$

where A is the heated area, d the piezoelectric film thickness, $p(z)$ the pyroelectric coefficient distribution, and $\Theta = T - T_0$ the temperature difference to the environment.

Assuming a homogeneous polarisation equivalent to $p(z) = p_0 = \text{const.}$, Eq. (1) simplifies to

$$I(t) = \frac{A}{d} p_0 \int\limits_0^d \frac{\partial}{\partial t} \Theta(z,t) \mathrm{d}z = A p_0 \frac{\mathrm{d}\Theta_m}{\mathrm{d}t}, \tag{2}$$

where $\Theta_m(t)$ is the temperature averaged across the sample.

In this work, we use different models to determine the temperature distribution $\Theta(z,t)$ and the resulting pyroelectric current.

2.1 Laser intensity modulation method

In the case of LIMM, we consider a harmonically heated piezoelectric plate exhibiting heat losses to the environment, characterised by a thermal relaxation time τ_{th} (Suchaneck et al., 2012). The steady-state periodic solution of this problem in the form of an infinite series is given by Bauer and Ploss (1990) and Carslaw and Jaeger (1959):

$$\Theta(z,t) = \frac{\Phi_0}{c\rho \cdot d} \left[\frac{\tau_{th}}{1 + i\omega\tau_{th}} + 2 \cdot \sum_{n=1}^{\infty} \cos\left(\frac{n\pi \cdot z}{d}\right) \frac{\tau_d/n^2}{1 + i\omega\tau_d/n^2} \right] \exp(i\omega t), \tag{3}$$

where Φ_0 is the heat flux absorbed by the plate surface, c the specific heat, ρ the density, $\tau_d = d^2/\pi^2 a$ the heat diffusion time, and a the thermal diffusivity of the plate. The pyroelectric response then yields (Bauer and Ploss, 1990)

$$I(\omega) = \frac{\Phi_0 A}{c\rho \cdot d} \left(p_0 \frac{i\omega\tau_{th}}{1 + i\omega\tau_{th}} + \sum_{n=1}^{\infty} p_n \frac{i\omega\tau_d/n^2}{1 + i\omega\tau_d/n^2} \right) \exp(i\omega t), \tag{4}$$

where p_0 is the average pyroelectric coefficient and p_n are its spatially dependent parts.

For a continuous distribution of relaxation times (instead of a single time constant) and a homogeneous polarisation with $p_n = 0$, Eq. (4) results in (Suchaneck et al., 2013)

$$I(\omega) = \frac{\Phi_0 A}{c\rho \cdot d} p_0 \left[1 - \frac{1}{\left[1 + (i\omega\tau_{th})^\alpha\right]^\beta} \right] \exp(i\omega t). \tag{5}$$

This model is equivalent to the Havriliak–Negami function as known from dielectric relaxation with the empirical parameters α and β accounting for the broadness and asymmetry of the distribution function. Two special cases are the Cole–Cole relaxation for $\beta = 1$ and $\alpha < 1$ and the Cole–Davidson relaxation for $\alpha = 1$ and $\beta < 1$.

Depending on the sample design, a superposition of several relaxation processes is also possible.

2.2 Thermal pulse method

2.2.1 Analytical solution

The temperature distribution after heating the piezoelectric plate by a laser pulse can be described by a transient pulse

Table 1. Material properties of the PZT plate (PI Ceramic GmbH, 2015).

Density, $\mathrm{g\,cm^{-3}}$	Thermal conductivity, $\mathrm{W\,m^{-1}\,K^{-1}}$	Specific heat, $\mathrm{J\,kg^{-1}\,K^{-1}}$
7.8	1.1	350

model of Camia (1967). It assumes an infinitely short laser pulse, a thermally isolated top surface and thermal coupling of the backside to an ideal heat sink. The temperature distribution is given by an infinite series of exponential terms:

$$\Theta(\xi,\tau) = \sum_{m=0}^{\infty} 2\cos\left[\frac{(2m+1)\pi}{2}\xi\right] \exp\left[-\frac{(2m+1)^2\pi^2}{4}\tau\right]. \tag{6}$$

$\xi = z/d$ and $\tau = at/d^2$ are dimensionless variables of depth and time, respectively. Inserting Eq. (6) in Eq. (2) yields for the time-dependent pyroelectric current:

$$I(\tau) = -A p_0 \frac{a}{d^2} \sum_{m=0}^{\infty} (2m+1)\pi \cdot (-1)^m \exp\left[-\frac{(2m+1)^2\pi^2}{4}\tau\right]. \tag{7}$$

Another analytical approach is a one-dimensional transient heat transfer model introduced by Bloß et al. (2000). It additionally considers the thermal mass of the electrode. The temperature distribution yields

$$\Theta(\xi,\tau) = \sum_{k=0}^{\infty} 2 \frac{\sin[x_k(1-\xi)]}{(1 + r + r^2 x_k^2)\sin x_k} \exp(-x_k^2\tau), \tag{8}$$

where r is the ratio of the thermal mass of the electrode to that of the pyroelectric element. x_k are the solutions of a transcendent equation, which is solved numerically. The resulting pyroelectric current is given by

$$I(\tau) = -A p_0 \frac{2a}{d^2} \sum_{k=0}^{\infty} 2 \frac{x_k(1 + \cos x_k)}{(1 + r + r^2 x_k^2)\sin x_k} \exp(-x_k^2\tau). \tag{9}$$

2.2.2 Numerical solution

The temperature distribution was also determined by a finite element model (FEM) solved by ANSYS 15.0.

The transient thermal analysis was performed by applying a heat flux in the area of the laser spot on the top of the sample for the duration of the laser pulse ($0.5\,\mu\mathrm{s}$). The backside of the sample was assumed to be an ideal heat sink. The initial temperature of the device was set to $20\,^\circ\mathrm{C}$. The material properties used for modelling are listed in Table 1.

3 Experimental methods

Two different types of embedded piezoelectric transducers were evaluated:

i. A low-temperature co-fired ceramics (LTCC)/lead–zirconate–titanate (PZT) sensor and actuator consisting of an already sintered PZT plate (CeramTec Sonox® P53) with a size of $(25 \times 10 \times 0.2)$ mm³ embedded in the centre of a sintered LTCC module ($(45 \times 20 \times 0.7)$ mm³, Heraeus HeraLock® Tape-HL2000). Sample fabrication is described in detail elsewhere (Flössel et al., 2010). The sample capacitance was 30 nF, the dielectric loss tangent amounted to about 2 % at 10 kHz.

ii. A commercial M-8528-P2 macro-fibre composite (MFC) actuator (Smart Materials, Dresden, Germany) with an overall length of 105 mm, an active length of 85 mm, an active width of 28 mm, a thickness of about 0.3 mm, a sample capacitance of 170 nF, and a dielectric loss tangent of about 5 % at 10 kHz. PZT macro-fibres are embedded in epoxy resin. They are electrically contacted by copper electrode strips and covered by a Kapton film.

In comparison, a non-embedded PZT plate with a size of $(10 \times 7 \times 0.2)$ mm³ covered by a nickel–chromium electrode was analysed.

For LIMM measurements, the samples were periodically heated by an array of six laser diodes or a single laser diode (LCU98A041A, Laser Components GmbH, Olching, Germany) square-wave modulated with frequencies of up to 1 kHz each with a power of 14 mW at a wavelength of 980 nm. The complex pyroelectric current was determined by an impedance/gain-phase analyser (Solartron 1260, Solartron Analytical, Farnborough, UK) with DC coupling. In order to reduce noise, 30 measurement repetitions were used for averaging.

Thermal pulse measurements were carried out by heating the samples with a pulsed laser diode (LC905D1S3J09UA, Laser Components GmbH, Olching, Germany) at a wavelength of 905 nm with a maximal peak power of 75 W, a pulse width of 0.15 or 0.5 µs and a repetition frequency of 1 Hz. The pyroelectric current was transformed to a voltage by a current amplifier (DLPCA-200, Femto Messtechnik GmbH, Berlin, Germany) and the signal was recorded by a Waverunner® Xi-A oscilloscope (LeCroy, Chestnut Ridge, USA).

4 Results and discussion

4.1 Laser intensity modulation method

Figure 1 shows the pyroelectric current spectrum of the non-embedded PZT plate fitted to Eq. (5). The thermal relaxation time amounts to 0.8 s. The fit shows a deviation from an ideal Debye-like model due to a slight time distribution, which is attributed to the impact of the electrodes and the electrical contact by a wire. The spectrum is best described by a Cole–Davidson relaxation with $\beta = 0.9$. The PZT plate shows a homogeneous polarisation distribution. The obtained

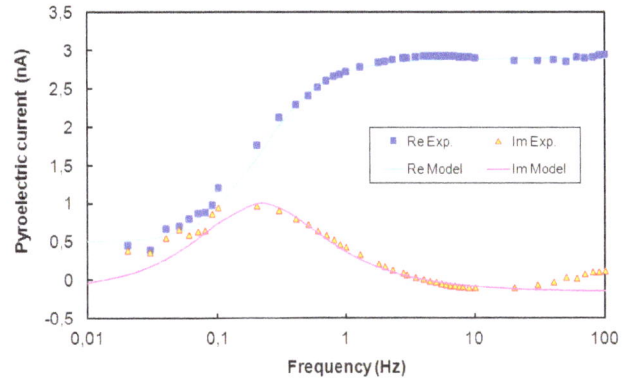

Figure 1. Pyroelectric current spectrum of a PZT plate in comparison to a fit to Eq. (5).

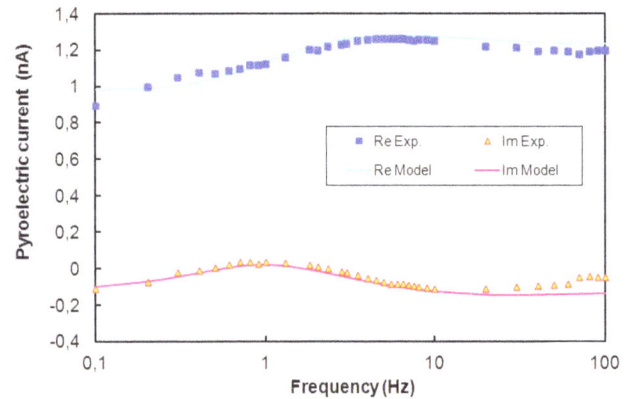

Figure 2. Pyroelectric current spectrum of a LTCC/PZT sensor–actuator in comparison to a fit to Eq. (4) with $p_1 = -p_0/5$.

pyroelectric spectrum is comparable to the results of Bauer and Ploss (1990). Note that the analytical model of Bauer and Ploss (1990) was originally developed for non-embedded samples. The following demonstrates that this model can be successfully applied to embedded piezoelectric transducers.

Figure 2 illustrates the pyroelectric current spectrum of a LTCC/PZT module fitted to Eq. (5). Between 0.1 and 10 Hz the embedded PZT plate loses heat to the LTCC layers with a thermal relaxation time of 0.16 s. A minor decrease of the real part at higher frequencies is attributed to a slightly inhomogeneous polarisation distribution, which is taken into account by $p_1 = -p_0/5$. A detailed analysis was previously reported by Eydam et al. (2014).

The MFC actuator is an example for a broad relaxation time distribution, which is well described by a Cole–Cole function with $\alpha = 0.4$ (Fig. 3). The relaxation time constant with a value of 5.3 s describes the heat loss of the sample surface to the environment with a thermal conductance of about $100 \, \mathrm{W \, m^{-2} \, K^{-1}}$. Here, the pyroelectric current spectrum is satisfactorily modelled by assuming a homogeneous polarisation.

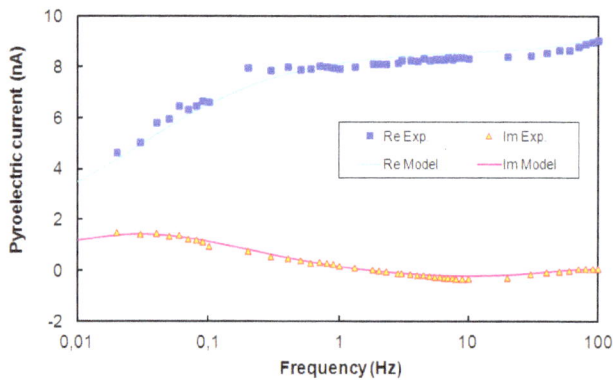

Figure 3. Pyroelectric current spectrum of a MFC actuator in comparison to a fit to Eq. (5).

4.2 Thermal pulse method

4.2.1 Modelling results

The mean temperature of the sample and the resulting pyroelectric current were determined for the PZT plate with the analytical models and the FEM model as described above.

Figure 4 illustrates the mean temperature over a time period of 0.1 s. The analytical models start at the temperature maximum whereas the FEM model shows the increase of the temperature from room temperature to a maximum value since the laser pulse has a given time duration of $0.5\,\mu s$. The model of Bloß et al. (2000) does not reach the maximum immediately but shows a slight increase of the temperature at the beginning due to the thermal mass of the electrode. At longer times, the temperature decreases exponentially to room temperature. The heat diffusion time amounts to 0.01 s for the PZT plate; i.e. it is located shortly after the temperature maximum. At this point, the plate has reached an inner thermal equilibrium. On the other hand, the thermal relaxation time of 0.8 s is the point when the sample reverts to thermal equilibrium with its environment.

The pyroelectric current is proportional to the time derivative of the mean temperature of the sample presented in Fig. 5 (cf. Eq. 2). In the analytical models, the current starts at a very high positive value caused by initial heating and falls down almost immediately to a negative value. On the other hand, the FEM model illustrates additionally the rise of the positive current during the heating period (cf. Fig. 5b). After reaching a minimum the current slowly returns to zero in all models; i.e. the sample returns to steady-state conditions.

4.2.2 Experimental results

Figures 6 and 7 illustrate the measured pyroelectric current of the PZT plate and of the LTCC/PZT module and the MFC actuator. There is a short initial negative peak followed by a large positive signal.

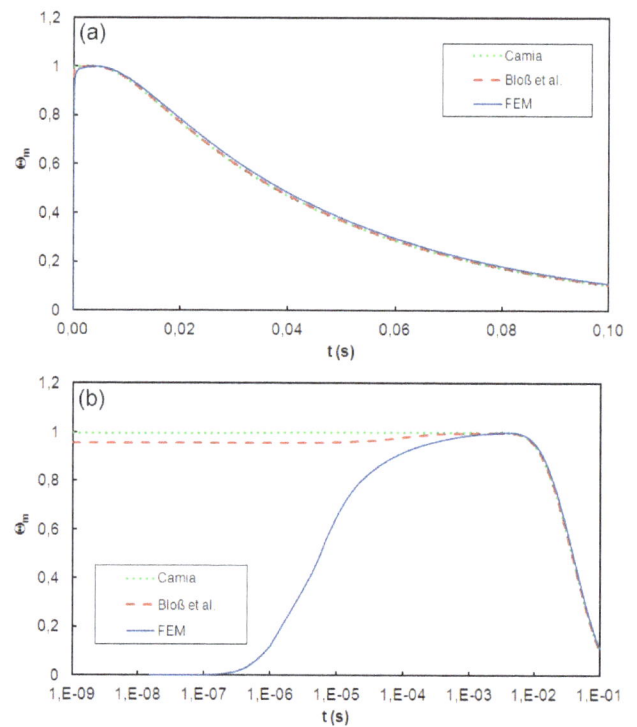

Figure 4. Mean temperature within a PZT plate, determined by three different models (cf. Sect. 2.2) with (**a**) linear and (**b**) logarithmic time axis.

Figure 6 illustrates that a longer pulse width leads to a higher heat input and thus a larger signal. The experimental data in Fig. 6 are in qualitative agreement with the simulation in Fig. 5b. Consequently, the experimental results manifest the initial heating of the piezoelectric plate. However, there is still a time shift between the signal maxima of the model and the experiment. One reason could be the thermal buffering effect of the top electrode (Bloß et al., 2000). It absorbs thermal energy during the short thermal pulse but only slowly transfers it to the piezoelectric since the heat transfer is limited by the piezoelectric's thermal diffusivity. Another reason could be the neglect of the rise and decay times of the laser pulse in the simulation. Both effects are now subject of further research. On the other hand, the signal of the cooling period at longer times is still too noisy for a quantitative analysis. Due to the very high gain of the current amplifier, mainly noise is present in the measured signal for measurement times of more than 0.5 ms. The signal shape is similar to the thermal pulse response of electret polymers obtained in Mellinger et al. (2005), where the high-gain signal decreases to zero for times exceeding 1 ms. For further measurements, a low-pass filter and a 50 Hz notch filter will be used to reduce the noise contribution.

In Fig. 7, the time shift of the maximum signal for the embedded piezoelectric plates is suspected to be caused by the heat transfer time through the embedding top layer. The

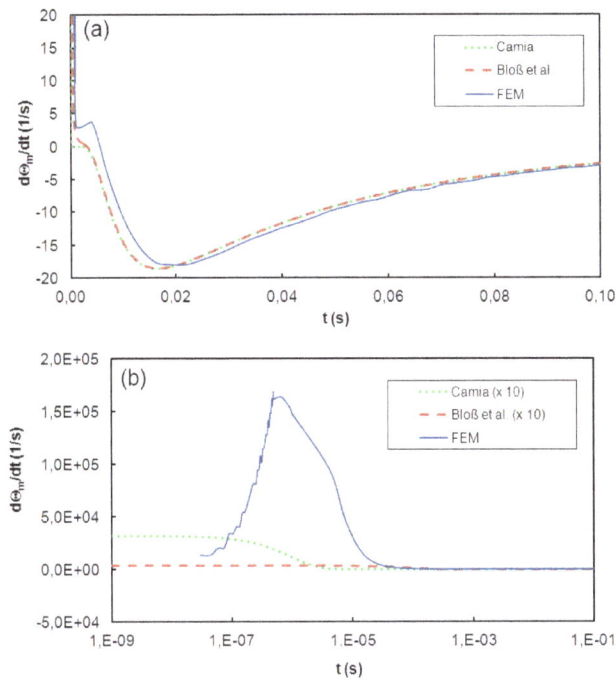

Figure 5. Time derivative of the mean temperature (proportional to the pyroelectric current) within a PZT plate, determined by three different models (cf. Sect. 2.2) with (**a**) linear and (**b**) logarithmic time axis. In (**b**) the values of the models of Camia (1967) and Bloß et al. (2000) were magnified by a factor of 10.

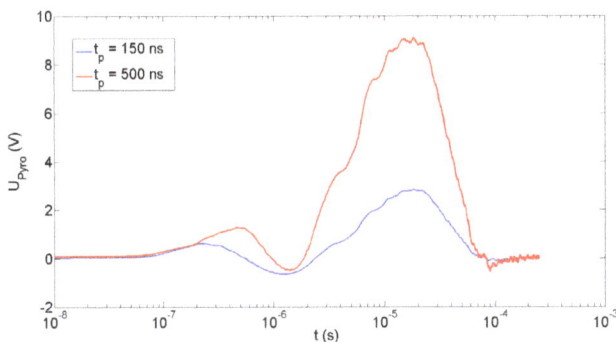

Figure 6. Time dependence of the output voltage of the current amplifier (10^8 V A^{-1}, 7 kHz bandwidth) for the PZT plate for an optical power of 22.5 W and a pulse width of 150 and 500 ns, respectively.

5 Conclusions

The pyroelectric response of an embedded piezoceramic plate for the LIMM has been described by an analytical model, which was successfully applied to integrated sensor–actuator modules. For the thermal pulse method, the tem-

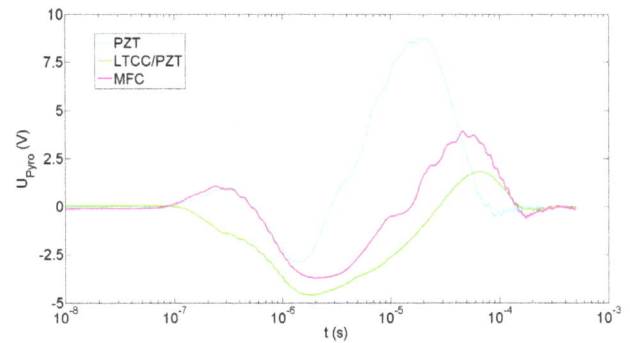

Figure 7. Time dependence of the output voltage of the current amplifier (10^8 V A^{-1}, 7 kHz bandwidth) for the PZT plate, the LTCC/PZT sensor–actuator and the MFC actuator for an optical power of 65 W and a pulse width of 150 ns.

perature distribution and the resulting pyroelectric current have been characterised both by two analytical and one FEM model. Further improvement of the thermal pulse set-up is required to reduce the signal-to-noise ratio in the cooling period. In the next step, a Fourier transform will be performed to analyse the thermal pulse signal in the frequency domain. This enables the application of the LIMM models to a signal that was recorded in a much shorter measuring time.

Acknowledgements. This research is supported by the Deutsche Forschungsgemeinschaft (DFG) in context of the Collaborative Research Centre/Transregio 39 PT-PIESA, subproject C8.

References

Bauer, S. and Ploss, B.: A method for the measurement of the thermal, dielectric, and pyroelectric properties of thin films and their applications for integrated heat sensors, J. Appl. Phys., 68, 6361–6367, 1990.

Bloß, P., DeReggi, A. S., and Schäfer, H.: Electric-field profile and thermal properties in substrate supported dielectric films, Phys. Rev. B, 62, 8517–8530, 2000.

Camia, F. M.: Traité de Thermocinétique impulsionelle, Dunod, Paris, 1967.

Carslaw, H. S. and Jaeger, J. C.: Conduction of Heat in Solids, 2nd edn., Oxford University Press, New York, NY, 1959.

Collins, R. E.: Measurement of charge distribution in electrets, Rev. Sci. Instrum., 48, 83–91, 1977.

Eydam, A., Suchaneck, G., Esslinger, S., Schönecker, A., Neumeister, P., and Gerlach, G.: Polarization characterization of PZT disks and of embedded PZT plates by thermal wave methods, in: AIP Conference Proceedings 1627, Electroceramics XIV, Bucharest, Rumania, 16–20 June 2014, 31–36, 2014.

Flössel, M., Gebhardt, S., Schönecker, A., and Michaelis, A.: Development of a novel sensor-actuator-module with ceramic multilayer technology, Journal of Ceramic Science and Technology, 1, 55–58, 2010.

Lang, S. B. and Das-Gupta, D. K.: Laser-intensity-modulation method, a technique for determination of spatial distributions of polarization and space charge in polymer electrets, J. Appl. Phys., 59, 2151–2160, 1986.

Mellinger, A., Singh, R., and Gerhard-Multhaupt, R.: Fast thermal-pulse measurements of space-charge distributions in electret polymers, Rev. Sci. Instrum., 76, 013903, doi:10.1063/1.1832153, 2005.

Mellinger, A., Singh, R., Wegener, M., Wirges, W., and Gerhard-Multhaupt, R.: Zerstörungsfreie Tomographie von Raumladungs- und Polarisationsverteilungen mittels Wärmepulsen (Non-destructive Space-charge and Polarization Tomography with Thermal Pulses), Tech. Mess., 74, 437–444, 2007.

PI Ceramic GmbH: Werkstoffdaten, Spezifische Parameter der Standardmaterialien, available at: http://www.piceramic.de/download/PI_Ceramic_Werkstoffdaten.pdf, last access: 17 August 2015.

Pham, C.-D., Petre, A., Berquez, L., Flores-Suárez, R., Mellinger, A., Wirges, W., and Gerhard, R.: 3D high-resolution mapping of polarization profiles in thin Poly(vinylidenefluoride-trifluoroethylene) (PVDF-TrFE) films using two thermal techniques, IEEE T. Dielect. El. In., 16, 676–681, 2009.

Stewart, M. and Cain, M.: Use of scanning LIMM (Laser Intensity Modulation Method) to characterise polarization variability in dielectric materials, J. Phys. Conf. Ser., 183, 012001, doi:10.1088/1742-6596/183/1/012001, 2009.

Stewart, M. and Cain, M.: Using High Frequency LIMM to characterize the poling state of piezoelectric ceramic thin films, MRS Proceedings 1805, 2015.

Suchaneck, G., Eydam, A., Hu, W., Krantz, B., Drossel, W.-G., and Gerlach, G.: Evaluation of polarization of embedded piezoelectrics by the thermal wave method, IEEE T. Ultrason. Ferr., 59, 1950–1954, 2012.

Suchaneck, G., Eydam, A., and Gerlach, G.: A laser intensity modulation method for the evaluation of the polarization state of embedded piezoceramics, Ferroelectrics, 453, 127–132, 2013.

19

Signal modeling of an MRI ribbon solenoid coil dedicated to spinal cord injury investigations

Christophe Coillot[1], **Rahima Sidiboulenouar**[1], **Eric Nativel**[2], **Michel Zanca**[1,4], **Eric Alibert**[1],
Maida Cardoso[1], **Guillaume Saintmartin**[1,3], **Harun Noristani**[3], **Nicolas Lonjon**[3,4], **Marine Lecorre**[3,4],
Florence Perrin[3], **and Christophe Goze-Bac**[1]

[1]Laboratoire Charles Coulomb (L2C-UMR5221), BioNanoNMRI group, University of Montpellier,
Place Eugene Bataillon, 34095 Montpellier, France
[2]Institut d'Electronique et des Systèmes (IES-UMR5214), University of Montpellier,
Campus Saint-Priest, 34095 Montpellier, France
[3]Institut des Neurosciences de Montpellier (INSERM U1051), University of Montpellier,
34095 Montpellier, France
[4]Nuclear medicine, CMC Gui de Chauliac, University Hospital Montpellier, 34095 Montpellier, France

Correspondence to: Christophe Coillot (christophe.coillot@univ-montp2.fr)

Abstract. Nuclear magnetic resonance imaging (NMRI) is a powerful tool for biological investigations. Nevertheless, the imaging resolution performance results in the combination of the magnetic field (B_0) and the antenna efficiency. This latter one results in a compromise between the size of the sample, the location of the region of interest and the homogeneity requirement. In the context of spinal cord imaging on mice, a ribbon solenoid coil is used to enhance the efficiency of the MRI experiment. This paper details the calculation of the local magnetization contribution to the induced voltage of MRI coils. The modeling is illustrated on ribbon solenoid antennas used in emitter–receiver mode for the study. The analytical model, which takes into account the emitting mode, the receiving step and the imaging sequence, is compared to the measurement performed on a 9.4 T VARIAN MRI apparatus. The efficiency of the antenna, in terms of signal to noise ratio, is significantly enhanced with respect to a commercial quadrature volumic antenna, given a significant advantage for the study of spinal cord injuries.

1 Context of the study: the spinal cord injuries

Spinal cord injuries (SCIs) are devastating neuropathologies that affect over 2.5 million patients worldwide, yield major handicaps and represent high costs to our society (from about USD 1 million to up than 4 million per patient, National SCI Statistical Care, Sekhon and Fehlings, 2001). Neurological difficulty depends on the spinal level and lesion severity. Unfortunately, there is no effective treatment for any symptoms associated with SCI. MRI is indeed well-established as the most commonly used imaging approach to diagnose and follow-up spinal cord injury patients. In the context of spinal cord injury studies in animals, MRI allows the localization of the region of the lesion and its evolution in order to

understand the fundamental biological mechanisms and the perspective of translation to clinics, to evaluate the effect of therapeutical trials. Even if it is preferentially used for in vivo studies, in vitro imaging of tissue has the advantage of an enhanced resolution because of the acquisition time which is less constrained (Nor et al., 2015).

The aim of the ex vivo MRI study is to deepen the in vivo analysis of altered tissues by means of higher MRI spatial resolution and to evaluate putative correlation with histology. Nevertheless, the imaging resolution performance results from the combination of the magnetic field (B_0), the acquisition time and the antenna efficiency in terms of signal to noise ratio (SNR). By a literal shortcut, MRI experimenters usually define the SNR as the ratio between the

Figure 1. Spinal cord tissue of a CX3CR1 mouse: length is typically about 25–35 mm and diameter is about 2–3 mm.

mean voxel intensity for a given sample (which is related to the magnetization quantity and to the coil sensitivity (in V/T)) divided by the voxel intensity in a region outside the sample (which is closely related to the coil plus preamplifier stochastic noise contribution). In the following, we will consider the SNR of MRI experimenters. In order to enhance the imaging performances a dedicated antenna must be designed. The MRI coil design will result in a compromise between the size of the sample, the location of the region of interest and the homogeneity requirement. The requirement homogeneity on the MRI coil is defined (cf. Mispelter et al., 2006, p. 309) as a $\pm 5\%$ magnetic field variation. Since we will focus on the final induced voltage, this criterion is excluded. Thus, we define the homogeneity zone as the region where the induced voltage variation remains within 10 % of its maximum. The MRI coils can be used either in emitter–receiver mode or solely in one of the two modes. In case of the separation between emitter and receiver mode an active or passive decoupling is mandatory. Next, an impressive variety of MRI coils have been invented and used (Mispelter et al., 2006): solenoid, saddle coil, loop coil, loop gap, scroll coil and bird cage. The configuration of the sample (its size) and the requirements of the experiment (in terms of SNR and homogeneity) could dictate the choice. In the context of spinal cord tissue (as shown in Fig. 1) the choice of the coil is restricted to the solenoid coil, the scroll coil or the loop gap. The solenoid coil appears to be a relevant choice for simplicity of manufacturing and signal to noise ratio efficiency reasons (Hidalgo et al., 2009) even if the scroll coils seems to be competitive (Grant et al., 2010; Mem et al., 2013). The use of ribbon wire instead of round wire is guided by homogeneity considerations over the sample volume.

The homogeneity of the image remains however an important issue for all MRI experiments. MRI coil designers usually anticipate it through a magnetic field intensity mapping (Mispelter et al., 2006; Hidalgo et al., 2009; Mem et al., 2013). Neglecting the MRI pulse sequence dependency a contrario, some authors have deduced the mapping of the radio-frequency coil using sequence dependency (Akoka et al., 1993; Insko and Bolinger, 1993) to correct it a posteriori. The purpose of this work is to derive a simple analytical model of the induced voltage for the solenoid coil used in emitting–receiving mode, which anticipates the effect of the MRI pulse sequence. We believe this analytic model could

offer a useful tool to guide the MRI coil designer by evaluating the signal homogeneity in the longitudinal direction of the solenoid prior to its realization. The method could be applied to other coils and combined with magnetic field numerical simulation, to get it in the whole sample volume.

2 The nuclear magnetic resonance (NMR)-induced voltage: from global to local

When the magnetic field (B_0) is applied to paramagnetic matter a macroscopic nuclear moment (M_0) arises while precessing at the Larmor frequency ω_0 (Bloch, 1946):

$$\omega_0 = -\gamma B_0. \tag{1}$$

The intensity of the net magnetic moment depends on the intensity of the "polarization" magnetic field B_0 (assumed in z direction following Fig. 3). Then, a varying magnetic field at Larmor frequency (B_1) is used to rotate the magnetization transverse to the polarizing magnetic field (cf. Fig. 4). The flip angle of the magnetization (θ) will depend on B_1 magnitude and duration (τ):

$$\theta = -\gamma B_1 \tau. \tag{2}$$

After application of B_1 magnetic field, the spins precess transversally to B_0 and are associated with an electromagnetic field whose magnetic component is classically measured by means of a coil.

The pioneer work on NMR antenna from Hoult and Richards (1976) invokes the Lorentz's reciprocity theorem to give a formulation of the induction law suited to NMR experiments:

$$\xi = -\int_{\text{Sample}} \frac{\partial \left(\frac{B_1}{I} M\right)}{\partial t} \mathrm{d}V_S, \tag{3}$$

where ξ is the electromotive force, B_1 is the varying magnetic field, I is the electrical current, M is the magnetization of the sample and V_S is the sample volume. This formula is a well-known basis for the NMR coil SNR formulation. SNR is one of the most important parameter featuring the antenna efficiency, the other one being the homogeneity of the radio-frequency magnetic field over the sample. However, the equation derived by Hoult and Richard (namely Eq. 3) hides the dependency of the detected signal to the location of the spins while it is the quintessence of NMRI-induced voltage.

For this reason, the formulation of the induced voltage due to local elementary magnetization proposed by Pimmel (1990) in his PhD work (which is unfortunately in French but has been reported in the book of Mispelter et al., 2006) is a well suited approach to describe the NMRI signal dependency on the magnetization location $\mathbf{r} = (x, y, z)$:

$$\delta e(t) = -\frac{\partial}{\partial t}(\delta \mathbf{m}(t) \cdot (\mathbf{B}_1(\mathbf{r})/I)), \tag{4}$$

where $\delta e(t)$ is the contribution to the total electromotive force of the elementary magnetization $\delta m(t)$ when a magnetic field by unit current $B_1(x,y,z)$ is applied also at point (x,y,z). This formulation assumes the elementary magnetization at point (x,y,z) is constant (equal to $M_0\delta V_e$, where δV_e is the volume element), but the effect of the pulse sequence on the local magnetization is not implicit. This 3-D formulation of the local contribution is elegant but confusing since the precessing component of the magnetization only appears in the plane perpendicular to B_0. For this reason, the local contribution to the induced voltage of the NMR signal proposed by Jacquinot and Sakellariou (2011) gives a more precise indication of the problem:

$$\delta e(r) = -\frac{\partial}{\partial t}[\delta m_\perp(r)\cdot(B_{1\perp}(r)/I)], \qquad (5)$$

where $B_{1\perp}(r)$ and $\delta m_\perp(r)$ are the components in the plane perpendicular to B_0.

A generalization of the induced voltage, taking into account the propagative phenomena in the sample, is proposed by Insko et al. (1998). Thus Eq. (3) is generalized to

$$\delta e(r) = -\frac{\partial}{\partial t}[\delta m_\perp(r)\cdot(B'_{1\perp}(r)/I)], \qquad (6)$$

where $B'_{1\perp}(r)$ is the generalized magnetic field retarded potential form:

$$B'_{1\perp}(r) = \frac{\mu_0}{4\pi}\oint e^{ikr}(1-ikr)\frac{dl\times r}{\|r\|^3}, \qquad (7)$$

where $k = \omega\sqrt{\varepsilon_r\varepsilon_0\mu_r\mu_0}$ is the wave number, ε_r is the relative permittivity of the sample and μ_r its relative permeability. As emphasized by Insko et al. (1998), the $e^{ikr}(1-ikr)$ term can be omitted in the near-field approximation (for kr $\ll$ 1 it follows $e^{ikr}(1-ikr)\to 1$).

When we try to feel the nature of the induced voltage by means of Eq. (3), we have to face some inconsistencies. First, at the time where the induced voltage is measured the magnetic field B_1 and the radio-frequency current I are both null, and consequently the term B_1/I is undefined. Second, the scalar product between the magnetization vector and a magnetic field is usually associated with the Zeeman energy, which is confusing. So, even if it is remarkably true from a mathematical point of view, the magnificent intuition of Hoult and Richard, which have gave birth to their famous formula, leads to misunderstanding for beginners. For these reasons, we derive below another way to write the NMRI-induced voltage.

We start from the mathematical form of the vector potential (A) associated with the magnetic dipole moment (δm) corresponding to the magnetization of a small volume $(\delta m = M_0\delta V)$:

$$A = \frac{\mu_0}{4\pi}\frac{\delta m\times r}{\|r\|^3}f(r), \qquad (8)$$

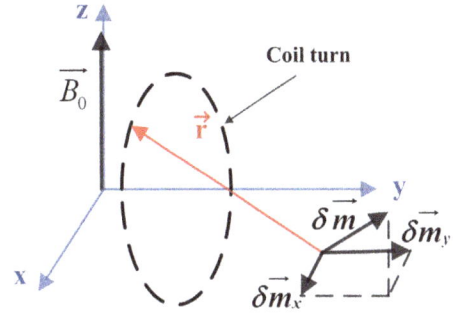

Figure 2. Illustration of the local elementary magnetization (δm) position with respect to the coil turn for the vector potential calculation over the turn.

where r is the distance vector from the elementary magnetization location to the point where the vector potential is computed (cf. Fig. 2) and $f(r) = e^{ikr}(1-ikr)$ summarizes the contribution of the near and far field (Insko et al., 1998).

Next, from the relation between the magnetic field (B) and the vector potential (A),

$$B = \nabla\times A. \qquad (9)$$

From Stokes theorem, for the magnetic flux through the surface, (S)

$$\delta\phi = \iint_{(S)} B dS = \iint_{(S)} \nabla\times A dS = \oint A dl. \qquad (10)$$

By substituting the vector potential by its mathematical equation (as given by Eq. 8),

$$\delta\phi = \oint \frac{\mu_0}{4\pi}\frac{\delta m(r,t)\times r}{\|r\|^3}f(r)dl, \qquad (11)$$

the total flux Φ, which represents the summation over N turns, is

$$\Phi = N\oint \frac{\mu_0}{4\pi}\frac{\delta m(r,t)\times r}{\|r\|^3}f(r)dl, \qquad (12)$$

which allows one to derive the local magnetization contribution to the induced voltage:

$$\delta e(r,t) = -N\frac{\partial}{\partial t}\oint \frac{\mu_0}{4\pi}\frac{\delta m(r,t)\times r}{\|r\|^3}f(r)dl. \qquad (13)$$

By decomposing the components as $\delta m = \delta m_x x + \delta m_y y$, we can write

$$\delta e(r,t) = -\frac{N\mu_0}{4\pi}\oint f(r)\frac{x\times r}{\|r\|^3}dl\frac{\partial\delta m_x(r,t)}{\partial t}$$
$$-\frac{N\mu_0}{4\pi}\oint f(r)\frac{y\times r}{\|r\|^3}dl\frac{\partial\delta m_y(r,t)}{\partial t}. \qquad (14)$$

Finally, we can write the local magnetization contribution to the induced voltage in the standard way of writing a signal coming from a sensor which involves the physical quantity to be measured (here the magnetization) multiplied by the sensor sensitivity (a coefficient or a function S):

$$\delta e(\boldsymbol{r},t) = -\left[S_x(\boldsymbol{r})\frac{\partial \delta m_x(\boldsymbol{r},t)}{\partial t} + S_y(\boldsymbol{r})\frac{\partial \delta m_y(\boldsymbol{r},t)}{\partial t} \right], \quad (15)$$

where $\delta e(\boldsymbol{r},t)$ represents the induced voltage, $\delta m_x(\boldsymbol{r},t)$ and $\delta m_y(\boldsymbol{r},t)$ are respectively the x and y component of the magnetization vector $\delta\boldsymbol{m}$, while $S_x(\boldsymbol{r})$ and $S_y(\boldsymbol{r})$ are the local coil sensitivities in x and y directions related to the coil geometry, defined as

$$S_x(\boldsymbol{r}) = \frac{\mu_0 N}{4\pi}\oint f(r)\frac{\boldsymbol{x}\times\boldsymbol{r}}{\|\boldsymbol{r}\|^3}\mathrm{d}\boldsymbol{l}, \quad (16)$$

$$S_y(\boldsymbol{r}) = \frac{\mu_0 N}{4\pi}\oint f(r)\frac{\boldsymbol{y}\times\boldsymbol{r}}{\|\boldsymbol{r}\|^3}\mathrm{d}\boldsymbol{l}, \quad (17)$$

where $\boldsymbol{x}$ and $\boldsymbol{y}$ are the unit vectors along x and y axis.

By virtue of the scalar triple product, the sensor's sensitivity coefficient can also be expressed:

$$S_x(\boldsymbol{r}) = -\frac{\mu_0 N}{4\pi}\boldsymbol{x}\cdot\oint f(r)\frac{\mathrm{d}\boldsymbol{l}\times\boldsymbol{r}}{\|\boldsymbol{r}\|^3}, \quad (18)$$

$$S_y(\boldsymbol{r}) = -\frac{\mu_0 N}{4\pi}\boldsymbol{y}\cdot\oint f(r)\frac{\mathrm{d}\boldsymbol{l}\times\boldsymbol{r}}{\|\boldsymbol{r}\|^3}, \quad (19)$$

where we definitively recognize the Biot–Savart law at a sign nearby (or the classical $\boldsymbol{B}/I$ term multiplied by the units vector). The sign difference comes from the reverse $\boldsymbol{r}$ direction convention between the usual form of Biot and Savart law with respect to the vector potential writing of Eq. (8) (Insko et al., 1998).

3 1-D NMRI signal modeling of a ribbon solenoid coil

In this section, we will detail the calculation of the induced voltage. We will perform the calculation on N turns of ribbon solenoid of length L and radius R as the one represented in Fig. 3. The helicity of the antenna will be neglected. We assume a perfect homogeneity in the transverse plane (x–z), which is a valid hypothesis for solenoid where the sample is not too close to the coil's wire, as reported in Hidalgo et al. (2009). Next, we assume a homogeneous current distribution flowing through the conductor. Moreover the time propagative phenomenon in the coil can be neglected since we will assume that total wire length will remain much smaller than $\lambda/2$. Then, as discussed in Hoult (2009), we can neglect the far-field contribution (and consequently $f(r)$ is assumed close to 1) even if, according to Insko et al. (1998), it seems to be a rough assumption since kr value at 400 MHz is close to 1. Lastly, the elementary magnetization will be designated as M in the following for simplicity's sake.

The NMRI coil is supposed to be used both in emitter and receiver mode. We will discuss in the following how the magnetization is tilted when the coil is used in emitter mode and

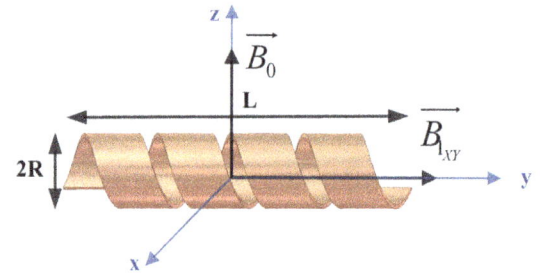

Figure 3. Illustration of the ribbon solenoid coil.

how the signal is detected by the coil when it is used in receiver mode.

3.1 Emitter mode: the magnetization tilt

The magnetic field component generated by a solenoid coil (Fig. 3) on y axis ($B_1(y,t)$) is given by Biot and Savart's law. This one can be formulated using sensitivity equations:

$$B_1(y,t) = -S_y(y)I(t), \quad (20)$$

where $S_y(y)$ is well known for a solenoid while it could be calculated using Eq. (19). It follows that

$$S_y(y) = \quad (21)$$

$$-\frac{\mu_0 N}{2L}\left| \frac{L/2+y}{\sqrt{R^2+(L/2+y)^2}} + \frac{L/2-y}{\sqrt{R^2+(L/2-y)^2}} \right|.$$

In the following, the time dependence of both B_1 and I will be omitted.

Next, by expanding Eq. (2), the distribution of the angle magnetization along the y axis ($\theta(y)$) will be directly related to the magnetic field distribution:

$$\theta(y) = \gamma B_1(y)\tau. \quad (22)$$

Practically, the tilt angle magnetization distribution will be related to the calibration pulse sequence conditions. In this study we assume a calibration pulse performed on a small thickness slice at the center of the antenna (i.e., $y=0$). In case of a different pulse condition (for instance $\pi/2$ pulse obtained over the whole sample volume) the modeling of the magnetization tilt would differ.

Thus, under the hypothesis of a centered pulse calibration, the $\pi/2$ magnetization angle is expressed:

$$\frac{\pi}{2} = \gamma B_1(0)\tau_0, \quad (23)$$

where $B_1(0)$ is the magnetic field at the center of the antenna and τ_0 is the pulse duration.

By combining Eqs. (22) and (23), it appears that tilt angle is proportional to magnetic field ($B_1(y)$) independently to the

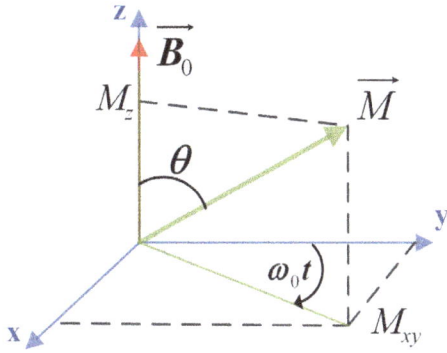

Figure 4. Magnetization vector representation in the static frame.

hardware strategy (adjustment of magnetic field magnitude or pulse duration):

$$\theta(y) = \left(\frac{\pi}{2} \frac{\tau}{B_1(0)\tau_0} \right) B_1(y). \tag{24}$$

Finally, the components of the magnetization vector $M = [M_x; M_y; M_z]$ represented in Fig. 4 are expressed:

$$\begin{pmatrix} M_{xy}(y)\sin(\omega_0 t) \\ M_{xy}(y)\cos(\omega_0 t) \\ M_z(y) \end{pmatrix}, \tag{25}$$

where $M_{xy}(y)$ is the magnetization magnitude in the x–y plane.

3.1.1 Single pulse sequence: the magnetization tilt

In the context of NMR single pulse sequence, the magnetization components will take the following form:

$$\begin{pmatrix} M_0 \sin(\theta(y)) e^{-\frac{T_E}{T_2^*}} (1 - e^{-\frac{T_R}{T_1}}) \sin(\omega_0 t) \\ M_0 \sin(\theta(y)) e^{-\frac{T_E}{T_2^*}} (1 - e^{-\frac{T_R}{T_1}}) \cos(\omega_0 t) \\ M_0 \cos(\theta(y))(1 - e^{-\frac{T_R}{T_1}}) \end{pmatrix}, \tag{26}$$

where (T_1) is the longitudinal relaxation time, (T_R) is the repetition time, (T_2^*) is the transverse relaxation time in a heterogeneous magnetic field and (T_E) is the echo time.

Let us now determine the flip angle distribution for two important MRI pulse sequences, namely gradient echo and spin echo.

3.1.2 Gradient echo sequence: the magnetization tilt

For a gradient echo sequence it is a desirable condition to perform it at the Ernst angle (θ_{ERNST}). Following the same reasoning that has led to Eq. (24), the flip angle distribution ($\theta(y)$) for a gradient echo sequence will follow

$$\theta(y) = \theta_{ERNST} \frac{B_1(y)\tau}{B_1(0)\tau_{ERNST}}, \tag{27}$$

where τ_{ERNST} is the pulse duration required to tilt the magnetization at Ernst angle at the center of the sample.

Thus, the magnetization in x–y plane (component $M_{xy}(y)$, cf. Eq. 25) will be

$$M_{xy}(y) = M_0 \frac{\sin(\theta(y)) \left(1 - e^{-\frac{T_R}{T_1}}\right)}{1 - \cos(\theta(y))e^{-\frac{T_R}{T_1}}} e^{-\frac{T_E}{T_2^*}}, \tag{28}$$

where $\theta(y)$ is the magnetization angle distribution given by Eq. (27).

3.1.3 Spin echo sequence: the magnetization tilt

For the spin echo sequence, the magnetization will be tilted by a $\pi/2$ pulse followed by a π pulse. We use the signal dependency given by Akoka et al. (1993) and Insko and Bolinger (1993) for spin echo sequence: $\sin^3(\theta)$. Thus the magnetization in x–y plane ($M_{xy}(\theta(y))$) will be given by

$$M_{xy}(\theta(y)) = M_0 \sin^3(\theta(y)) \left(1 - e^{-\frac{T_R}{T_1}}\right) e^{-\frac{T_E}{T_2}}, \tag{29}$$

where $\theta(y)$ is the magnetization angle distribution given by Eq. (24) and (T_2) is the true transverse relaxation time.

3.2 Receiver mode: the induced voltage

Once the magnetization flip is determined, we can establish the induced voltage associated with the magnetization precession. The induced voltage created by the elementary magnetization at location y can be simply expressed from Eq. (15) by considering only the sensitivity along y axis:

$$e(t, y) = -S_y(y) \frac{dM_y(y)}{dt}, \tag{30}$$

where $S_y(y)$ is deduced from Eq. (21).

Since $M_y(y) = M_{xy}(y)\cos(\omega_0 t)$, the induced voltage in harmonic regime will be

$$|\underline{e}(y)| = |\omega_0 S_y(y) M_{xy}(y)|, \tag{31}$$

where $M_{xy}(\theta(y))$ is given either by Eqs. (28) or (29) depending of the running pulse sequence.

4 Design of the NMRI ribbon solenoid coil

To design the NMRI ribbon solenoid coil, the first point is to determine the total length of the wire (L_w). In order to neglect the propagative phenomenon into the coil, a length of the solenoid coil about $\approx \lambda/6$ is classically used (Mispelter et al., 2006), while λ is determined by the nuclear frequency of interest into the magnet. In our study, we performed 1 H measurement on a 9.4 T Varian MRI. Thus, the gyromagnetic

Table 1. Solenoid coil design parameter summary.

N	D (mm)	w (mm)	t (mm)
3	8	10	2

frequency (f_0) is about 400 MHz, and it follows that the total length of the solenoid coil should be limited to 12.5 cm:

$$L_{\mathrm{w}} = \frac{\pi N D}{\cos(\psi)} \leq \lambda/6, \qquad (32)$$

where ψ is the pitch angle. The turn number is then deduced from the size of the sample (or its mechanical support). In our study the tissue length is ~ 40 mm while the tube diameter (D) is equal to 8 mm. The copper ribbon has 10 mm width (w) and 50 μm thickness (t_{w}). From Eq. (32) it follows that $N = 3$. Choosing a space between turns (t) equal to 2 mm results in an average length of the solenoid $L \sim 36$ mm (the design parameters are summarized in Table 1).

4.1 NMRI coil electrical model

Basically, the coil can be represented by the electromotive force (given by Eq. 15) in series with an inductance ($L1$) and a resistance which takes into account the occurrence of the skin effect ($R1_{\mathrm{AC}}$). When considering a single ribbon, the current density will tend to flow at the ends of the ribbon: this effect is known as lateral skin effect (Belevitch, 1971). When considering a multiple-turn solenoid (ribbon or round wire), the current density between neighbor conductors will be strengthened especially at the extremities of the coil: this effect is known as the proximity skin effect (Butterworth, 1925). The analytic modeling of these phenomena is beyond the scope of this paper, but they can be efficiently approached by electromagnetic numerical simulations. Finally these effects will dictate the current distribution at high frequencies and thus the homogeneity. In practice, even if it increases the coil's resistance and thus the noise, the use of ribbon wire tends to improve the homogeneity (Grant et al., 2010; Mem et al., 2013). In the case of ribbon solenoid the spacing between turns should be minimized in order to preserve homogeneity on one side but should be sufficient to avoid to strengthen the proximity effect on the other side. The different skins effects are illustrated in Fig. 5.

Lastly, the occurrence of the coil's resonance at a frequency (f_0) where the wavelength (λ) is about twice the wire length (namely $\lambda = c/f_0 \simeq 2L_c$ where c is the vacuum light velocity), will imply $f_0 \simeq c/(2L_c)$) can be interpreted by a capacitance ($C1$) in parallel (cf. Fig. 6) with the previous components (Knight, 2013b).

4.2 Tuning–matching circuit

The intrinsic self-resonance of the coil is much higher than the one to observe; moreover the electrical resonance fre-

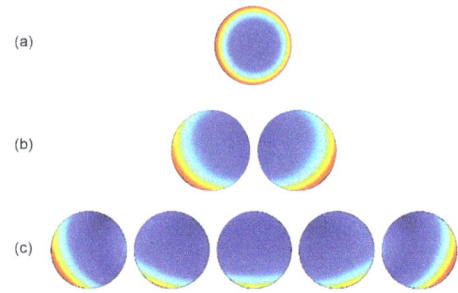

Figure 5. Illustration of the different skin effect regimes in the section of one, two and five turns: (a) usual skin effect, (b) proximity skin effect and (c) lateral skin effect.

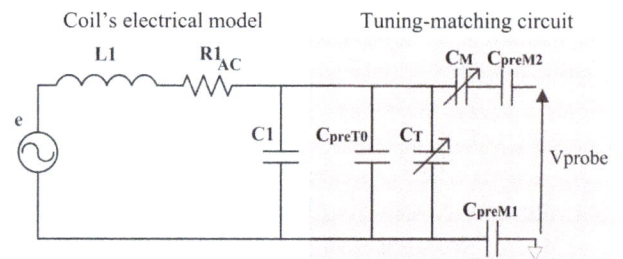

Figure 6. Electrical circuit of the coil plus the tuning–matching circuit.

quency will be affected by the sample dielectric properties. For these reasons, a variable capacitance is usually added in parallel to the coil terminal to adjust the resonance frequency (the tuning circuit is represented by capacitances C_{T} and C_{preT} in Fig. 6). On the other side, the coil must be connected to the radio-frequency power amplifier and impedance of the coil must be matched to the standard 50Ω at the frequency of use. For this purpose the capacitances C_{M}, C_{preM1} and C_{preM2} are used in series (the electrical component values are summarized in Table 2).

5 Experimental results

The experiments have been performed on a 9.4 T magnet from AGILENT. The ribbon solenoid coil has been wound on a glass tube. A small printed circuit board (PCB) is used to realize the tuning–matching circuit. Variable non-magnetic capacitances from VOLTRONICS (Ref. NMKJ10HVE from 0.5 up to 9 pF) are used. Copper foil connected to the ground BNC cable has been used on the reverse face of the mechanical structure to perform an electromagnetic shielding and to prevent the sensibility of the circuit during the manual adjustment. Special care has been given to the connection distance between the ground and the copper foil to prevent occurrence of resonance in the frequency range of interest. A mechanical structure to maintain the glass tube and the PCB, represented in Fig. 7, has been realized using a 3-D printer with polylactic acid material.

Table 2. Electrical parameter summary.

$R1_{AC}$	$L1$	$C1$ (pF)	C_{preT0}	C_{preM1}	C_{preM2}	Q
$\sim 0.3\,(\Omega)$	15 nH	0.55 pF	6.8 pF	0.4 pF	0.4 pF	135

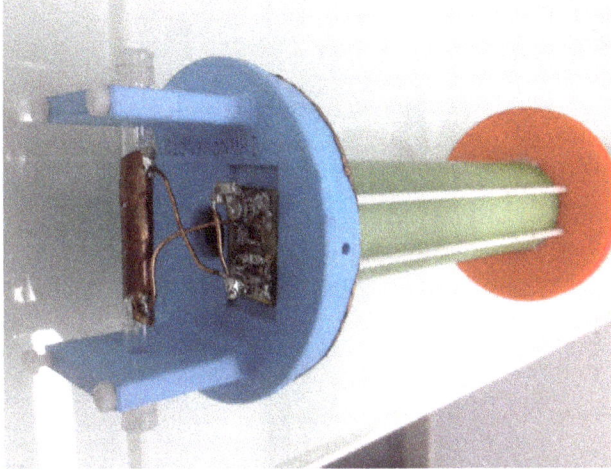

Figure 7. Photograph of the ribbon solenoid coil dedicated to spinal cord injury.

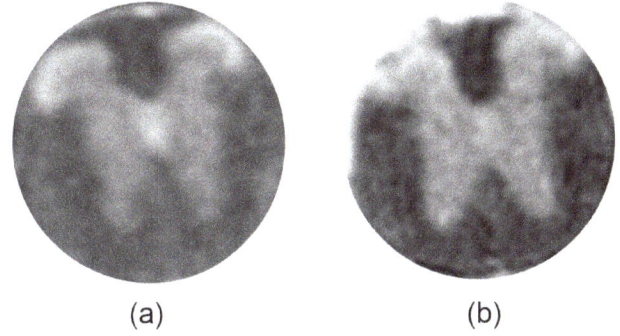

Figure 8. Ex vivo MR images (multi-echo multi-slice sequence: TR = 1155 ms; TE = 14 ms; NE = 1; FOV = 10 mm × 10 mm; 60 slices; thickness = 0.6 mm; resolution = 256 × 256) of spinal cords from adult mice. Hypersignal (butterfly shape) represents the grey matter whereas the surrounding hyposignal corresponds to the white matter: **(a)** image obtained with RAPID Biomedical 43 mm volumic quadrature coil in 14 h 0 min; **(b)** image obtained with the ribbon solenoid coil of Fig. 7 in 1 h 30 min.

The pulse power needed to tilt the magnetization at $\pi/2$ (on a 100 mm length and 4 mm diameter glass tube filled with potable water) in a centered thin slice is about 4.8 dB (the SNR is 1600/45, obtained using a gradient echo sequence with the following parameters: FOV = 10 × 10: TR = 250 ms, TE = 4.32 ms, flip angle = 80°, average = 2, resolution 128 × 128, 20 slices of thickness = 1 mm). For comparison, the pulse power, on the same water sample, needed by a commercial volumetric quadrature antenna with 43 mm inner diameter (from RAPID Biomedical) in the same conditions is about 20 dB (the SNR on the image in the same conditions is 290/45). The enhancement of the SNR between the ribbon solenoid coil and the commercial antenna is ~ 5.5, which is well correlated with the pulse power attenuation. This increase of the SNR allows one either to perform faster acquisition (~ 25 times) for a given resolution or to enhance the resolution for a given acquisition time. In the context of spinal cord injury studies, this improvement in the SNR (as demonstrated on a T2-weighted spin echo sequence in Fig. 8) was crucial. The time acquisition has been divided by 10 while the image quality has been significantly enhanced allowing one to combine high-resolution T2-weighted acquisition and diffusion MRI imaging to investigate accurately the lesion site of the spinal cord.

5.1 Gradient echo sequence: method and experimental results

According to Eq. (27), θ is proportional to $B_1(y)/y$, which is equivalent to the sensitivity term $S_y(y)$, resulting in $\theta(y) \propto S_y(y)$. Then, we normalize the tilt angle distribution to the Ernst angle:

$$\theta(y) = \theta_{ERNST}(y)\frac{S_y(y)}{S_y(0)}, \tag{33}$$

where $S_y(y)$ is given by Eq. (21). Combining Eqs. (28) and (33) into the MRI-induced voltage (Eq. 31) leads to

$$e(y) \propto \theta(y)\frac{\sin(\theta(y))}{(1-\cos(\theta(y))e^{-\frac{T_R}{T_1}})}e^{-\frac{T_E}{T_2^*}}. \tag{34}$$

For a given set of experimental conditions parameters (T_R, T_1, θ_{ERNST}) and given solenoid coil size parameters, the induced voltage $e(y)$ can be plotted. The model (values are normalized) is compared to the normalized experimental data (cf. Fig. 10).

5.2 Spin echo sequence: method and experimental results

For a spin echo sequence, the distribution of the magnetization angle tilt will be

$$\theta(y) = \frac{\pi}{2}\frac{S_y(y)}{S_y(0)}, \tag{35}$$

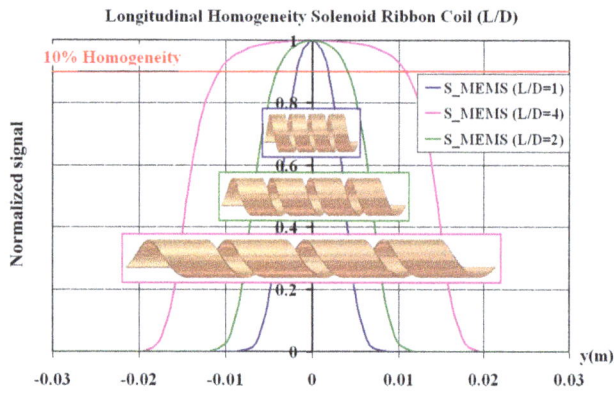

Figure 9. Comparisons of longitudinal homogeneity profiles for different solenoid coil aspect ratios.

where $S_y(y)$ is given by Eq. (21). Combining Eqs. (29) and (35) into the MRI-induced voltage (Eq. 31) leads to

$$e(y) \propto \theta(y)\sin^3(\theta(y))(1 - e^{-\frac{T_R}{T_1}})e^{-\frac{T_E}{T_2}}. \tag{36}$$

The model can be used to anticipate the longitudinal homogeneity range for different aspect ratios (i.e., $L/(2R)$ ratios) of solenoid coils in the context of spin echo sequence. It highlights how to use the model to guide the designer. For instance, it highlights the inefficiency of the short solenoid coil where less than 50 % of the length provides enough signal (considering the homogeneity criterion as 10 % variation of the signal intensity).

Finally, the model so obtained is compared to experimental data in Fig. 10. The measured data have been obtained by inserting a tube of water (100 mm length) into the ribbon solenoid coil. The measurements have been done through the VNMRJ software interface using circular region of interest over the whole diameter.

The comparison between gradient echo and spin echo (Fig. 10a) allows one to illustrate the y axis homogeneity dependency on the pulse sequence. The measurement also shows differences between the two type of sequences, the echo gradient sequence exhibiting a significantly wider homogeneity range. For both sequences we can notice a significant smooth decrease of the signal far from the coil in practice while the decrease is predicted as more abrupt by the model. It is certainly related to the far-field contribution predicted by Insko et al. (1998) conversely to the conclusion given by Hoult (2009).

The real ribbon solenoid coil signal exhibits some magnitude fluctuations which are attributed to the spacing between turns. The occurrence of maxima at the ends of the solenoid coil is attributed to the high-frequency current density distribution discussed above, where the different skin effects tend to distribute the current density at its ends increasing both B_1 and the sensitivity. Finally, the modeling allows the anticipation of the y axis homogeneity tendency at an early stage of the coil design.

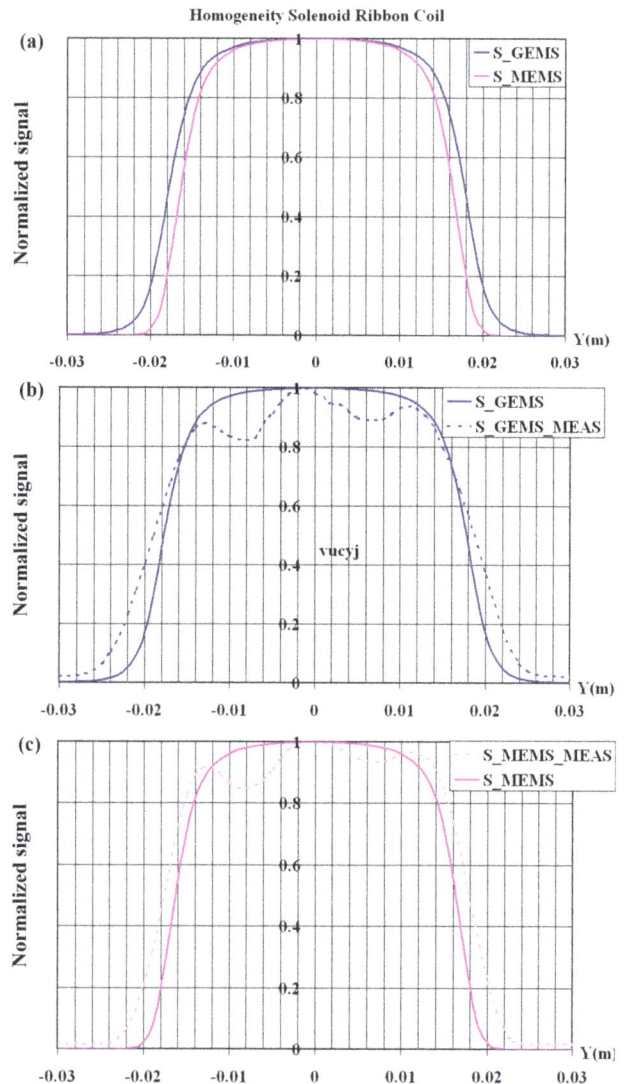

Figure 10. Normalized induced signal comparisons: (**a**) gradient echo sequence (GEMS) versus spin echo sequence (MEMS); (**b**) gradient echo sequence measured (GEMS–MEAS) versus model (GEMS) (TR = 688 ms; TE = 4.5 ms; FOV = 10 mm × 10 mm; 80 slices; thickness = 1 mm; resolution = 128 × 128); (**c**) spin echo sequence measured (MEMS–MEAS) versus model (MEMS) (TR = 10 s; TE = 10 ms; FOV = 10 mm × 10 mm; 80 slices; thickness = 1 mm; resolution = 128 × 128).

6 Conclusions

The step-by-step modeling presented in this paper enables the estimation of the longitudinal signal homogeneity of a solenoid coil depending on the running imaging sequence. The formulation of the local contribution of the elementary magnetization to the induced voltage, using sensitivities coefficients, derived in this paper is well suited for MRI coil designers while it avoids the use of confusing notations. We believe this type of modeling could be applied to other coil shapes in order to guide MRI coil designers in their choice of coils. Two important phenomena reduce the analytic modeling validity, namely, the current density distribution at high frequency and the far-field contribution, which will be more significant for higher frequencies and higher coil and sample sizes. Quantitative modeling of the homogeneity over the volume could be attained by combining classical numerical computations of magnetic field with the MRI pulse sequence conditions.

Finally, the customized antennas are a relevant and cheap way of enhancing the performance of the MRI studies with respect to the commercial antennas.

Acknowledgements. The authors would like to thank Association Verticale, who funded the 9.4 T MRI dedicated to SCI studies, and also thank the Labex NUMEV, who funded the electronic material needed to perform this study.

References

Akoka, S., Franconi, F., Seguin, F., and Le Pape, A.: Radiofrequency map of an NMR coil by imaging, Magnetic Resonance Imaging, 11, 437–441, 1993.

Belevitch, V.: The lateral skin effect in a flat conductor, Philips tech Rev., 32, 221–231, 1971.

Bloch, F.: Nuclear Induction, Physical Review, 70, 460–474, 1946.

Butterworth, S.: On the alternating current resistance of solenoidal coils, Proc. R. Soc. Lon. Ser.-A, 107, 693–715, 1925.

Grant, C. V., Wu, C. H., and Opella, S. J.: Probes for high field solid-state NMR of lossy biological samples, J. Magn. Reson., 204, 180–188, 2010.

Hidalgo, S. S., Jirak, D., Solis, S. E., and Rodriguez, A. O.: Solenoid coil for mouse-model MRI with a clinical 3-Tesla imager: body imaging, Rev. Mex. Fis., 55, 140–144, 2009.

Hoult, D. I.: The Origins and Present Status of the Radio Wave Controversy in NMR, Concept. Magn. Reson. A, 34, 193–216, 2009.

Hoult, D. I. and Richards R. E.: The signal-to-noise ratio of the nuclear magnetic resonance experiment, J. Magn. Reson., 24, 71–85, 1976.

Insko, E. K. and Bolinger, L.: Mapping of the Radiofrequency field, J. Magn. Reson., 103, 82–85, 1993.

Insko, E. K., Elliott, M. A., Schotland, J. C., and Leigh, J. S.: Generalized Reciprocity, J. Magn. Reson., 131, 111–117, 1998.

Jacquinot, J.-F. and Sakellariou, D.: NMR Signal Detection using Inductive Coupling: Applications to Rotating Microcoils, Concept. Magn. Reson. A, 38, 33–51, 2011.

Knight, D. W.: Solenoid Inductance Calculation, version of Sept. 2015, available at: http://www.g3ynh.info/zdocs/magnetics/Solenoids.pdf, 2013a.

Knight, D. W.: The self-resonance and self-capacitance of solenoid coils, version of Sept. 2015, available at: http://www.g3ynh.info/, 2013b.

Meme, S., Joudiou, N., and Szmereta, F.: In vivo magnetic resonance microscopy of Drosophilae at 9.4T, Magn. Reson. Imaging, 31, 109–119, 2013.

Mispelter, J., Lupu, M., and Briguet, A.: NMR probeheads for biophysical and biomedical experiments: theoretical principles and practical guidelines, Imperial College Press, 2006.

Noristani, H. N., Lonjon, N., Cardoso, M., Le Corre, M., Chan-Seng, E., Captier, G., Privat, A., Coillot, C., Goze-Bac, C. and Perrin, F. E.: Correlation of in vivo and ex vivo 1H-MRI with histology in two severities of mouse spinal cord injury, Front. Neuroanat., 9, 24, doi:10.3389/fnana.2015.00024, 2015.

Pimmel, P.: Les Antennes en Resonance Magnétique Nucléaire: Fonctionnement et Réalisation, PhD Thesis, Univ. Claude Bernard, Lyon, 1990.

Sekhon, L. H. and Fehlings, M. G.: Epidemiology, demographics, and pathophysiology of acute spinal cord injury, Spine, 26, S2–12, 2001.

Webb, A. G.: Radiofrequency microcoils for magnetic resonance imaging and spectroscopy, J. Magn. Reson., 31, 55–66, 2013.

Wheeler, H. A.: Simple Inductance Formulas for Radio Coils, P. IRE, 16, 1398–1400, 1928.

20

Selective detection of naphthalene with nanostructured WO$_3$ gas sensors prepared by pulsed laser deposition

Martin Leidinger[1], **Joni Huotari**[2], **Tilman Sauerwald**[1], **Jyrki Lappalainen**[2], **and Andreas Schütze**[1]

[1] Saarland University, Lab for Measurement Technology, Saarbrücken, Germany
[2] University of Oulu, Faculty of Information Technology and Electrical Engineering, Oulu, Finland

Correspondence to: Martin Leidinger (m.leidinger@lmt.uni-saarland.de)

Abstract. Pulsed laser deposition (PLD) at room temperature with a nanosecond laser was used to prepare WO$_3$ layers on both MEMS microheater platforms and Si/SiO$_2$ substrates. Structural characterization showed that the layers are formed of nanoparticles and nanoparticle agglomerates. Two types of layers were prepared, one at an oxygen partial pressure of 0.08 mbar and one at 0.2 mbar. The layer structure and the related gas sensing properties were shown to be highly dependent on this deposition parameter. At an oxygen pressure of 0.2 mbar, formation of ε-phase WO$_3$ was found, which is possibly contributing to the observed increase in sensitivity of the sensor material.

The gas sensing performance of the two sensor layers prepared via PLD was tested for detection of volatile organic compounds (benzene, formaldehyde and naphthalene) at ppb level concentrations, with various ethanol backgrounds (0.5 and 2 ppm) and gas humidities (30, 50 and 70 % RH). The gas sensors were operated in temperature cycled operation. For signal processing, linear discriminant analysis was performed using features extracted from the conductance signals during temperature variations as input data.

Both WO$_3$ sensor layers showed high sensitivity and selectivity to naphthalene compared to the other target gases. Of the two layers, the one prepared at higher oxygen partial pressure showed higher sensitivity and stability resulting in better discrimination of the gases and of different naphthalene concentrations. Naphthalene at concentrations down to 1 ppb could be detected with high reliability, even in an ethanol background of up to 2 ppm. The sensors show only low response to ethanol, which can be compensated reliably during the signal processing. Quantification of ppb level naphthalene concentrations was also possible with a high success rate of more than 99 % as shown by leave-one-out cross validation.

1 Introduction

In order to evaluate and assess indoor air quality (IAQ), different types of gaseous chemical compounds have to be considered. In addition to carbon monoxide (CO), carbon dioxide (CO$_2$) and nitrogen dioxide (NO$_2$), low concentrations of volatile organic compounds (VOCs) play a significant role in deteriorating the quality of breathing air in buildings (World Health Organization, 2010; Bernstein et al., 2008). Exposure to these substances, even at low concentrations, can lead to severe negative effects on human health. For VOCs, health problems mainly include damage to the respiratory system and skin irritations (Jones, 1999). Additionally, some

VOCs are proven to be carcinogenic (e.g., benzene, World Health Organization, 2010) or are suspected to be carcinogenic (e.g., formaldehyde, Guo et al., 2004). Based on toxicity and prevalence, according to the World Health Organization (WHO) and the INDEX project (Koistinen et al., 2008), the highest priority VOCs for IAQ are formaldehyde, benzene and naphthalene. For naphthalene, the WHO guidelines suggest values below 0.01 mg m^{-3} corresponding to 1.9 ppb (World Health Organization, 2010). The main health concerns for naphthalene are lesions in the respiratory tract, including tumors in the upper airways (World Health Organization, 2010).

In order to detect such small concentrations of VOCs without the need for expensive and time-consuming analytical measurements (e.g., GC-MS analysis, Wu et al., 2004), metal oxide semiconductor gas sensors can be applied. Detection of VOCs in the ppb range with such sensors has been successfully demonstrated using temperature cycled operation (TCO) and pattern recognition signal processing with ceramic-based thick film sensors (Leidinger et al., 2014); however, significant cross-sensitivity to ethanol was found mainly for SnO_2-based sensors. In order to reduce this cross-sensitivity, WO_3 layers were investigated. To obtain highly sensitive sensors with small thermal time constants, the MOS thin film layers were produced on microheater substrates by pulsed laser deposition (PLD). We found that these sensors show a high response to naphthalene in the relevant concentration range with high selectivity compared to other gases, especially relevant interferent gases for indoor air quality assessment, e.g., ethanol.

Pulsed laser deposition is a method for depositing a variety of materials ranging from epitaxial thin films (Hussain et al., 2005) to highly porous nanostructured layers (Balandeh et al., 2015). Porous nanostructured layers have been studied especially in the context of gas sensing materials (Caricato et al., 2009; Nam et al., 2006). PLD offers many advantages compared to other deposition methods, for example easily controllable film composition by deposition parameters, and a good repetition of stoichiometry of the target material in the films deposited on the substrate. When using nanosecond laser PLD, as in this study, with a high oxygen partial pressure in the deposition chamber, nanoparticle formation starts during the deposition process leading to a highly porous nanostructured layer (Harilal et al., 2003; Infortuna et al., 2008; Huotari et al., 2015). These types of layers are very suitable for gas sensing purposes because of their high specific surface area.

WO_3 as a material has been widely studied as it offers a large range of possibilities in practical applications, e.g., in gas sensing (Kohl et al., 2000; Wang et al., 2008; Balazsi et al., 2008), and photocatalytic water splitting (Pihosh et al., 2015). There are several methods to produce WO_3 layers ranging from thick film and thin films technologies to chemical methods (Zheng et al., 2011). In this study, PLD was utilized for depositing WO_3 layers on MEMS microhotplates to produce low-cost and high-performance gas sensor devices.

The performance of the PLD sensor layers has been evaluated in test gas measurements. The three high-priority VOCs, benzene, formaldehyde and naphthalene, have been applied in concentrations below, at and above the respective guideline values, and ethanol has been added as an interferent gas in much higher concentrations in order to simulate typical IAQ applications with background gases from, e.g., cleaning agents or alcoholic beverages. The sensors were operated in dynamic operation using temperature cycled operation (TCO), which is a well-known method for increasing

sensitivity and selectivity of gas sensor systems based on MOS sensors (Heilig et al., 1997; Lee and Ready, 1999; Paczkowski et al., 2013; Baur et al., 2014). The resulting signals, after pre-processing and feature extraction, were analyzed by linear discriminant analysis (LDA), a multivariate pattern recognition method which separates different classes of input data while trying to group data sets of the same assigned group (Klecka, 1980; Gutierrez-Osuna, 2002). The combination of TCO and LDA has shown to improve selectivity and sensitivity of gas sensors, both MOS sensors (Gramm et al., 2003; Meier et al., 2007; Reimann and Schütze, 2012; Leidinger et al., 2014) and other types, e.g., GasFET devices (Bur et al., 2012).

2 PLD sensor layer deposition and characterization

2.1 Sensor layer deposition

A XeCl laser with a wavelength of $\lambda = 308$ nm was used to produce WO_3 layers on both Si/SiO_2 substrates and commercial microheater MEMS platforms from a ceramic WO_3 pellet. The laser pulse length was 25 ns and pulse fluence was $I = 1.25$ J cm^{-2}. In all depositions the substrate temperature was kept at room temperature (RT). Two types of samples have been prepared, one at a low O_2 partial pressures of $p(O_2) = 0.08$ mbar, designation PLD0.08O2, and a second type with a higher partial pressure of $p(O_2) = 0.2$ mbar, designation PLD0.2O2. All samples were annealed in a furnace at $400\,^{\circ}$C for 1 h after deposition. The samples deposited on Si/SiO_2 substrates were used as reference samples in structural characterization of the layers, and samples with the MEMS heaters were used in gas sensing measurements.

A Bruker D8 Discover device was used in X-ray diffraction studies, and Raman spectroscopy studies were performed with a HORIBA Jobin Yvon LabRAM HR800 in order to study the crystal structure and symmetry of the layers. The surface morphology and the film composition of the samples were studied with a Veeco Dimension 3100 atomic force microscope (AFM) and with Zeiss Sigma FESEM device.

2.2 Crystal structure characterization of the sensing layers

The grazing incidence diffraction (GID) method of the X-ray diffraction was used to characterize the WO_3 layers annealed at $400\,^{\circ}$C for 1 h. The results are shown in Fig. 1. A clear difference in the crystal structure can be seen. The phase composition of layers deposited at $p(O_2) = 0.08$ mbar is mostly of the monoclinic γ phase of WO_3, but in the samples deposited at $p(O_2) = 0.2$ mbar, also the ferroelectric monoclinic ε phase of WO_3 is present. This is emphasized especially by the (110) and (−112) reflections located at $2\theta \approx 24.0$ and 33.3°, respectively (Johansson, 2012). However, one must remember that the crystal structures of the

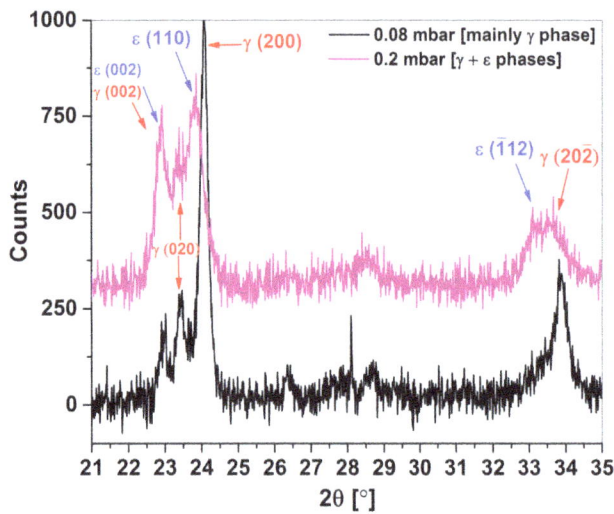

Figure 1. X-ray diffraction spectra of the deposited WO$_3$ layers after annealing in air (400 °C for 1 h).

γ phase and the ε phase are quite similar, and thus both of them have XRD reflections either in the same 2θ angels or very close to each other. The average grain size of both types of samples after annealing was determined to be around 30 nm using the Warren–Averbach method for XRD data (Marinkovic et al., 2001).

In Fig. 2, the Raman spectroscopy studies performed to the WO$_3$ layers are presented. The Raman spectra of the as-deposited samples immediately after deposition without any heat treatment are shown in Fig. 2a, and in Fig. 2b the spectra of the layers after annealing at 400 °C for 1 h are presented. An interesting property of the deposition process can be identified in the non-annealed samples. When the O$_2$ partial pressure is 0.08 mbar, the samples seem to be in an amorphous state after deposition, but when the O$_2$ pressure is 0.2 mbar, some crystallization is already evident during the deposition process at RT, even before any heat treatment to the layers. However, from Fig. 2b, showing the Raman spectra of the samples after the annealing process, it is clearly seen that after heat treatment in a furnace both films have a more crystallized structure. It is also again evident that the layers PLD0.08O2 are composed mostly of γ phase, but the samples PLD0.2O2 have also the ε phase in their crystal structure, verified from the Raman modes at wavenumbers 67, 97, 144, 183, 203, 272, 303, 370, 425, 644, and 680 cm^{-1} (Wang et al., 2008; Johansson et al., 2012; Souza Filho et al., 2000). Similarly as the reflections in XRD measurements, both γ phase and ε phase have Raman modes either at same wavenumbers or very close to each other.

At this point it should be noted that usually the ferroelectric ε phase only exists in temperatures below −40 °C. However, different studies (Wang et al., 2008; Righettoni et al., 2010; Johansson et al., 2012) show that the ε phase can exist in a solid-state form also at temperatures above RT. The

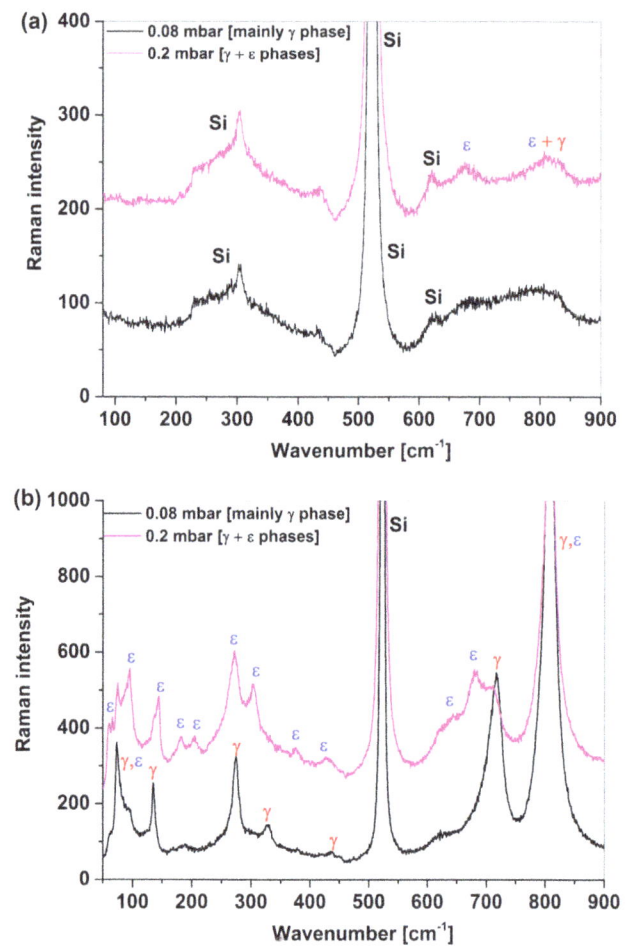

Figure 2. Raman spectra of the deposited WO$_3$ layers **(a)** as-deposited samples, **(b)** after annealing at 400 °C for 1 h.

reason for this is believed to be the small particle size of the samples, similarly as in the samples presented in this study. Also, the existence of the ε phase in the WO$_3$ composition has been proven to enhance WO$_3$ structures sensitivity to acetone (Wang et al., 2008; Righettoni et al., 2010; Sood and Gouma, 2013). The reason was suggested to be the ferroelectricity of the ε phase, namely the spontaneous electric dipole moments it possesses, which are then highly contributing to the chemical reaction between the WO$_3$ surface and the target gas.

2.3 Film composition characterization of the sensing layers

The surface morphology of the as-deposited and annealed WO$_3$ samples was studied by atomic force microscopy, and the results are shown in Fig. 3. The surface micrographs of the as-deposited layer and the annealed layer of sample PLD0.08O2 are shown in Fig. 3a and b, respectively. The sample surfaces consist of small agglomerates of nanopar-

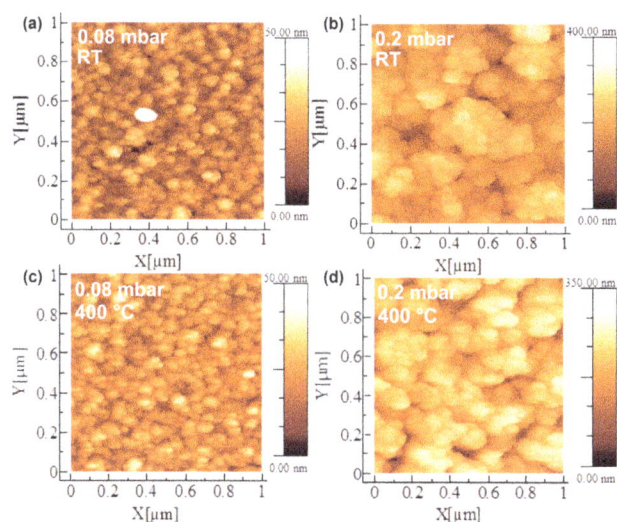

Figure 3. Atomic force microscopy surface micrographs of the deposited WO3 layers showing the influence of oxygen partial pressure during deposition (**a, b**) as-deposited samples, (**c, d**) after annealing at $400\,°C$ for 1 h.

ticles. In Fig. 3c and d, the surface micrographs of the as-deposited layer and the annealed layer of sample PLD0.2O2 are shown, respectively. The layers consist also of small nanoparticles, but agglomerated to bigger clusters. Also, the layer structure is much rougher and more porous than on the samples deposited at 0.08 mbar O_2. It can also be clearly seen that the annealing process at $400\,°C$ for 1 h does not have a great effect on the surfaces of the samples. In both cases, the average surface roughness value R_q was the same before and after the annealing process, being $R_q = 5.5$ nm for sample PLD0.08O2, and $R_q = 42.2$ nm for sample PLD0.2O2. The crystallization, which is observed for the PLD0.08O2 sample seems to be a local process that does not involve larger-scale material transport.

Scanning electron microscopy was used to further study the film composition of the samples. Both surface micrographs and cross-section micrographs were taken from the samples. In Fig. 4a and b the cross-section micrograph and surface micrograph of the as-deposited sample fabricated at $p(O_2) = 0.08$ mbar are shown, respectively. The cross-section graph shows that the film is composed of small nanoparticle agglomerates formed as pillar-like structures, with some porosity in between the columns. The surface graph shows that the film surface is formed of small nanoparticle agglomerates and thus verifies the measurements made with AFM. The cross-section micrograph and surface micrograph of the as-deposited sample fabricated at $p(O_2) = 0.2$ mbar are shown in Fig. 4b and c, respectively. Now the film composition is much more porous and rough compared to the PLD0.08O2 film, and also the nanoparticle agglomerate size is larger. The agglomerates form clearer pillar-like morphology on top of the substrate. The surface of the film is

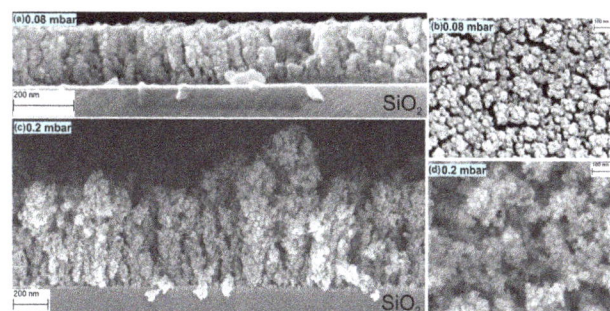

Figure 4. Scanning electron microscopy micrographs (cross sections and top views) of the as-deposited WO3 layers deposited (**a, b**) at $p(O_2) = 0.08$ mbar, and (**c, d**) at $p(O_2) = 0.2$ mbar.

highly porous, concurrent with the AFM measurements performed.

3   Gas sensor performance

3.1   Gas test measurement setup

The gas sensing performance of the two sensor layers was evaluated in an extensive test measurement. The three target VOC gases were applied in three concentrations each, the middle concentrations representing the respective WHO guideline values of 0.1 mg m^{-3} (81 ppb) for formaldehyde and 0.01 mg m^{-3} (1.9 ppb) for naphthalene (World Health Organization, 2010), as well as the European Union guideline value of $5\,\mu$g m^{-3} (1.6 ppb) for benzene (European Parliament, Council of the European Union, 2008). Additionally, ethanol was introduced as a background gas in two concentrations, both much higher than the target gas concentrations. As the third varied parameter, the gas humidity was set in three steps. Table 1 shows all gases and concentrations.

The test gases were generated in a gas mixing system specifically designed for trace gases by Helwig et al. (2014). The gases were mixed into zero air produced by two cascaded zero air generators. Ethanol and formaldehyde test gases were taken from gas cylinders and diluted into the zero air carrier gas stream, either with a one-step dilution (ethanol) or a two-step dilution (formaldehyde). Benzene and naphthalene test gases were generated from permeation tubes in permeation ovens. Each target VOC concentration was set twice during each combination of humidity and ethanol background, first from highest to lowest concentration, then back to the highest concentration. The length of each VOC run was 30 min; between two trace VOC applications the sensors were purged with zero air with the respective ethanol and humidity configuration. In total, 90 different gas mixtures were generated and tested with the sensors; the total length of the measurement was 123 h.

For temperature cycled operation of the sensors, a ramp-up–ramp-down approach was chosen; the temperature of the microheaters was increased from 200 to $400\,°C$ in 20 s and

Table 1. Test gas setup. Each gas concentration was applied at each EtOH background and humidity level for 30 min; between gas exposures sensors were exposed to background for 30 min.

Gas	Concentration (ppb)	EtOH background (ppm)	Humidity (% RH)
Zero air		0; 0.5; 2	30, 50, 70
Formaldehyde	200, 80, 40, 40, 80, 200	0; 0.5; 2	30, 50, 70
Benzene	2.5, 1.5, 0.5, 0.5, 1.5, 2.5	0; 0.5; 2	30, 50, 70
Naphthalene	5, 2, 1, 1, 2, 5	0; 0.5; 2	30, 50, 70

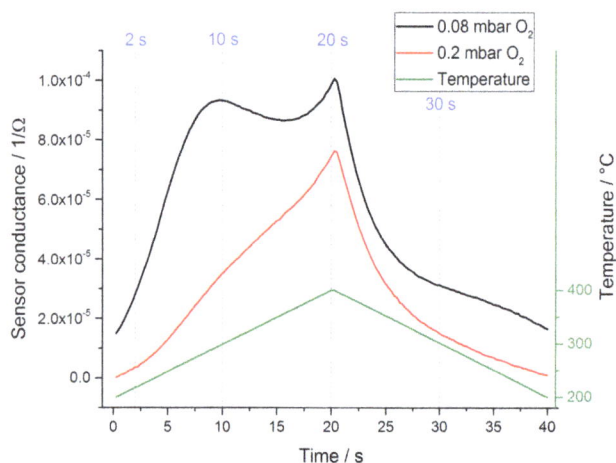

Figure 5. Gas sensor temperature cycle (green) and corresponding sensor signals of the two PLD WO_3 sensors in air at 30 % RH. The dashed lines indicate the selected points for quasi-static sensor signal analysis (cf. Fig. 6).

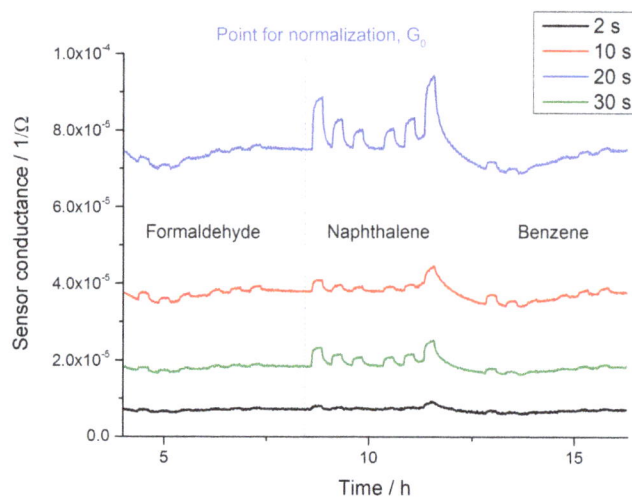

Figure 6. Quasi-static sensor signals for the four points indicated in Fig. 5 during exposure of the PLD0.2O2 sensor to formaldehyde, naphthalene and benzene at relevant ppb levels (Table 1) in air at 30 % RH without EtOH background. For later normalization, the value G_0 was extracted during background before the first exposure to naphthalene as indicated.

then reduced back to 200 °C in the same time, creating a 40 s cycle (see Fig. 5). The two sensor types clearly show differing behavior during the temperature cycle, especially during increase of the heater temperature.

3.2 Gas measurement results

For a first signal evaluation, quasi-static sensor signals were extracted from the raw sensor signal data sets. These signals were generated by plotting the signal value of certain points in the TCO cycle for each cycle, i.e., over the course of the complete measurement. An example of the PLD0.2O2 sensor is given in Fig. 6. Four points of the temperature cycle were selected, indicated in Fig. 5. A section of the measurement was chosen in which all three test VOCs are applied at 30 % RH gas humidity and without ethanol background. The sensor response to all concentrations of naphthalene is clearly visible, as well as the much lower responses to the other VOCs. By normalizing the signals, i.e., calculating the relative change of conductance G/G_0, the sensor response to naphthalene at the different points in the cycle can be determined (cf. Fig. 7). In this plot, it can be seen that the highest sensitivity, of the chosen points in the cycle, is during cooling of the sensor, 30 s after start of the cycle. The

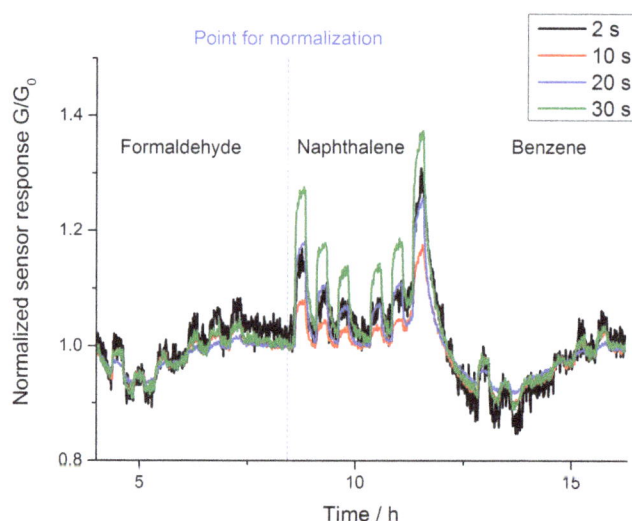

Figure 7. Normalized quasi-static signals for sensor PLD0.2O2 for the four selected points in the temperature cycle (cf. Fig. 5).

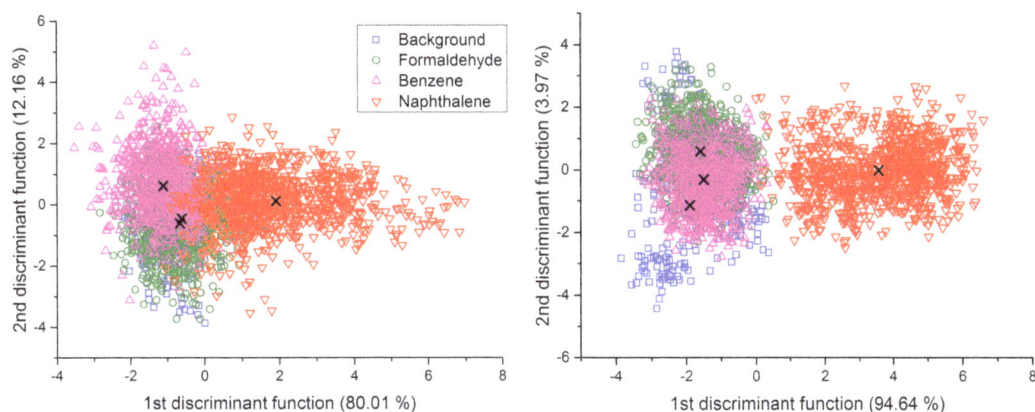

Figure 8. LDA plots for discrimination of different VOCs under varying humidity (30–70 % RH) and changing ethanol background (0–2 ppm), left: PLD0.08O2; right: PLD0.2O2.

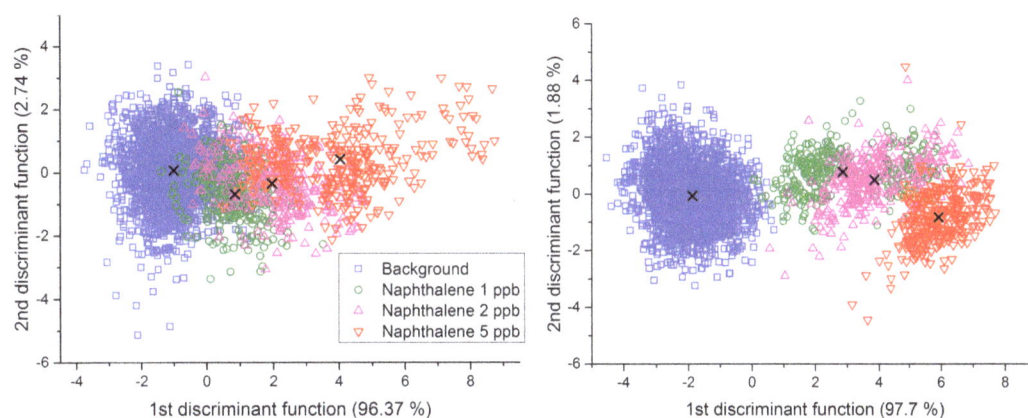

Figure 9. LDA plots for quantification of naphthalene under varying humidity (30–70 % RH) and changing ethanol background (0–2 ppm), left: PLD0.08O2; right: PLD0.2O2.

change of conductance is approx. 15 % for 1 ppb of naphthalene. The reason for the high selectivity to naphthalene is not known in detail and should be studied more closely. However, it was reported earlier that WO$_3$ has a specific response to aromatic compounds (Sauerwald, 2008) and that in general higher molecular weight leads to an increased sensitivity for this material (Sauerwald, 2008; Kohl et al., 2000).

The dynamic sensor signal patterns were analyzed using LDA. The whole procedure of generating data sets for LDA from dynamic gas sensor signals and the options for LDA application were described by Bur et al. (2014). As input data for each sensor, a limited number of features were extracted from the respective sensor signal of each temperature cycle. In this case, the cycle was divided into 20 segments of equal length (2 s). For each segment, the mean value and the slope of the sensor signal were calculated and used as features. This generates a feature vector with 40 values for each temperature cycle. These data sets were grouped into different groups, depending on the desired data analysis.

For the first analysis, the complete data set, containing 2799 feature vectors, was used as input data for the LDA. All cycles which contain a certain VOC were grouped together, regardless of VOC concentration, humidity and ethanol background. This results in one group for each target VOC and a "background" group, which contains the data sets of the TCO cycles which ran when no trace VOC was applied. This analysis checks the performance of the sensors to discriminate the target gases in varying humidity and background conditions. The LDA result plots for the two sensors with the PLD layers are shown in Fig. 8, left for the PLD0.08O2 sensor and right for the PLD0.2O2 type. For both sensors, the background, formaldehyde and benzene groups are overlapping strongly, while the naphthalene group is more separated. Especially for the PLD0.2O2 sensor the naphthalene group is nearly completely split from the other gases.

As a quantitative measure of the discrimination result, leave-one-out cross validation was performed on the data (LOOCV; Gutierrez-Osuna, 2002), with k nearest neighbors (kNN, $k = 5$) as classifier. This method calculates how many

Table 2. Leave-one-out cross-validation (LOOCV) results for all LDA investigations.

LDA no.	Analysis	PLD0.08O2	PLD0.2O2
1a	Gas discrimination (all gases)	66.5 %	71.9 %
1b	Gas discrimination (naphthalene)	86.3 %	99.2 %
2	Naphthalene quantification (full data set)	83.4 %	94.0 %
3	Naphthalene quantification (reduced data set, only 0 ppm EtOH background)	99.3 %	99.7 %
4	Humidity quantification	100 %	99.7 %
5a	Gas discrimination (all gases, only 50 % RH)	85.4 %	86.7 %
5b	Gas discrimination (naphthalene, only 50 % RH)	91.0 %	100 %

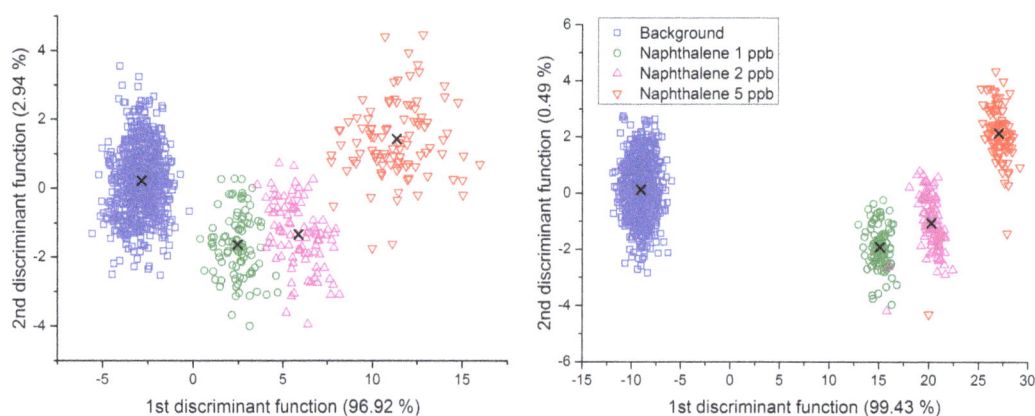

Figure 10. LDA plots for quantification of naphthalene under varying humidity (30–70 % RH) but with 0 ppm ethanol background, left: PLD0.08O2; right: PLD0.2O2.

feature vectors are classified correctly whether the LDA is trained by all other vectors.

Of the two sensors, the one with the PLD0.2O2 layer shows the better discrimination result overall, with 71.9 % correctly classified data points. The PLD0.08O2 achieves 66.5 %. The same order is given for classification of the naphthalene data points. The layer deposited at 0.2 mbar of O_2 shows a much better classification performance (99.2 % correct classifications) compared to the second sensor. The LOOCV results are summarized in Table 2, LDA 1a and 1b.

In the next step, it was checked if a quantification of the naphthalene concentration was possible. The full data set was used again. Each naphthalene concentration was assigned a group (with all humidities and ethanol backgrounds), and the other VOCs were assigned to the background group. The results for both tested sensors are plotted in Fig. 9. In this analysis, the PLD0.2O2 sensor shows the best separation of the groups again. There are some data points from the 1 ppb group located in the 2 ppb group. Otherwise, the three naphthalene concentrations are well lined up along the first discriminant axis. The PLD0.08O2 sensor shows less discrimination of the groups, which is also shown in LOOCV results (see Table 2, LDA 2).

The same analysis was attempted with a reduced data set, which included only the segments of the measurement without the ethanol background (950 sensor cycles in to-

tal). This significantly increases the quality of discrimination and thus naphthalene quantification for the sensors (see Fig. 10). Especially the PLD0.2O2 sensor has excellent separation of the naphthalene concentrations along the first discriminant function; the PLD0.08O2 sensor layer has much wider groups. The LOOCV results, listed in Table 2, LDA 3, show nearly 100 % correct classifications for the better sensor (PLD0.2O2); the second sensor layer also has over 99 % success rate.

Another evaluation of the sensor data was performed regarding quantification of gas humidity. The full data set was split into three groups, for the three gas humidities set during the measurement. This LDA run checks the sensors' cross-sensitivity to humidity, which also shows if the sensor would be able to measure the gas humidity. See Fig. 11 for the LDA result plots. These plots and the corresponding LOOCV results (Table 2, LDA 4) show very good discrimination of the humidities, which means that the sensor layers deposited by PLD have considerable sensitivity to water. However, this also means that the sensors could be used to measure the gas humidity or that the sensor performance could be monitored by comparing the predicted gas humidity with the value of a reference humidity sensor.

If the gas humidity is known, either from the gas sensor itself or from an additional humidity sensor, the quality of the signal processing can be improved by calculating different

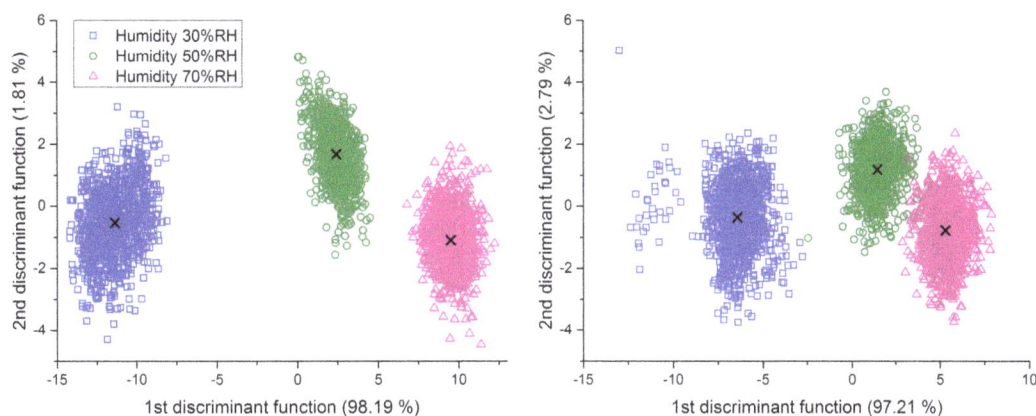

Figure 11. LDA plots for discrimination of ambient humidity levels for all sensor cycles with and without VOCs under varying ethanol background (0–2 ppm), left: PLD0.08O2; right: PLD0.2O2.

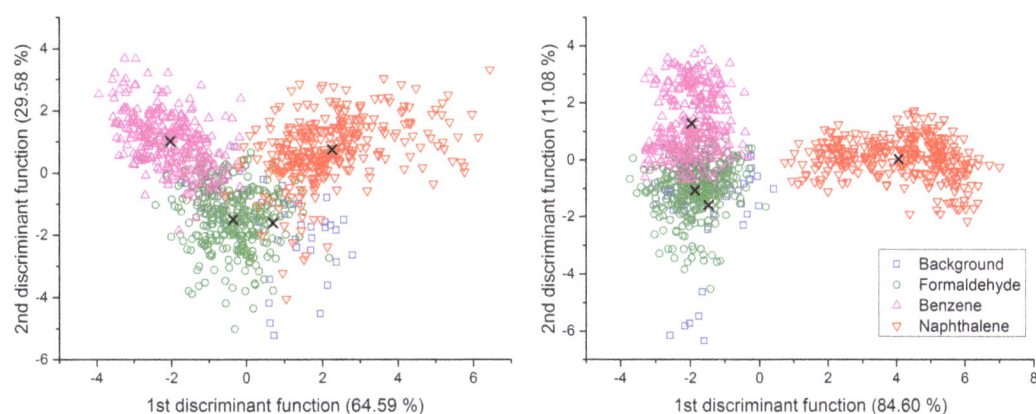

Figure 12. LDA plots for discrimination of different VOCs under constant humidity (50 % RH) and changing ethanol background (0–2 ppm), left: PLD0.08O2; right: PLD0.2O2.

LDA projections for several humidity ranges. A simple version of this approach has been tested by reducing the data set to the sensor signals acquired in one humidity, in this case all cycles measured in 50 % RH. The LDA result for gas identification for this is shown in Fig. 12. Compared to the full data set with three humidity levels (Fig. 8), the separation of the gases is clearly improved, especially for the PLD0.08O2 sensor. LOOCV also shows significant improvement of the classification results (Table 1, LDA 5). For all gases, the ratio of correct classifications was raised from 66.5 to 85.4 % for sensor PLD0.08O2 and from 71.9 to 86.7 % for sensor PLD0.2O2. For the naphthalene group, perfect classification was achieved for the sensor with 0.2 mbar O_2 partial pressure (Table 1, LDA 5b). A hierarchical data processing approach, which in the first step determines the humidity and in the second step classifies the gas, seems promising.

4 Conclusions

Nanoporous WO_3 gas sensing layers have successfully been prepared via nanosecond pulsed laser deposition, characterized, and tested for their gas sensing performance, especially for use in IAQ applications.

The characterization of the PLD sensing layers showed films formed of nanoparticle agglomerates with pillar-like morphology. The films were highly porous in their structure, especially when higher oxygen partial pressure was used during the deposition process. The crystal structure of the films was also dependent on the oxygen partial pressure with higher O_2 pressure resulting in the formation of $WO_3\varepsilon$ phase, which in bulk WO_3 samples is only stable at temperatures below $-40\,°C$. The ε phase in the PLD films was observed to withstand annealing at $400\,°C$ for 1 h probably due to its monocrystalline structure.

The two compared PLD sensor layer samples showed significant differences in their gas sensing performance. The layer deposited at higher oxygen partial pressure displayed improved response and excellent selectivity to naphthalene.

Both discrimination of different target VOCs and quantification of naphthalene were more successful with this sensor.

Naphthalene concentrations down to 1 ppb could be quantified with the sensor layer deposited at 0.2 mbar of O_2 with nearly 100 % success rate as determined by leave-one-out cross validation when no ethanol background was present. Even in varying ethanol background of up to 2 ppm, quantification was still successful for 94 % of all temperature cycles for this sensor. Detection of the presence of naphthalene in concentrations of 1 ppb or more had a success rate of more than 99 %, again determined by LOOCV.

The highly porous structure and possibly the formation of the WO_3 ε phase resulting from the higher oxygen pressure during PLD thus boost the gas sensing performance considerably. With the selected parameters, suitable gas sensing layers for detection and quantification of naphthalene have been obtained. Future investigations will address further improvement of the gas sensitive layers, e.g., by further variations of the deposition parameters as well as introduction of additional nanoparticles for doping and catalytic activation. We are also planning to study PLD deposition based on picosecond laser pulses which allows further optimization of the deposition parameters.

Acknowledgements. This project has received funding from the European Union's Seventh Framework Programme for research, technological development and demonstration under grant agreement no. 604311, Project SENSIndoor.

References

Balandeh, M., Mezzetti, A., Tacca, A., Leonardo, S., Marra, G., Divitini, G., Ducati, C., Medad, L., and Di Fonzo, F.: Quasi-1D hyperbranched WO_3 nanostructures for low-voltage photoelectrochemical water splitting, J. Mater. Chem. A, 3, 6110–6117, doi:10.1039/C4TA06786J, 2015.

Balazsi, C., Wang, L., Zayim, E. O., Szilagy, I. M., Sedlackov, K., Pfeifer, J., Toth, A. L., and Gouma, P.-I.: Nanosize hexagonal tungsten oxide for gas sensing applications, J. Eur. Ceram. Soc., 28, 913–917, doi:10.1016/j.jeurceramsoc.2007.09.001, 2008.

Baur, T., Schütze, A., and Sauerwald, T.: Optimierung des temperaturzyklischen Betriebs von Halbleitergassensoren, tm – Technisches Messen, 82, 187–195, doi:10.1515/teme-2014-0007, 2014.

Bernstein, J. A., Alexis, N., Bacchus, H., Leonard Bernstein, I., Fritz, P., Horner, E., Li, N., Mason, S., Nel, A., Oullette, J., Reijula, K., Reponen, T., Seltzer, J., Smith, A., and Tarlo, S. M.: The health effects of nonindustrial indoor air pollution, J. Allergy Clin. Immun., 121, 585–591, doi:10.1016/j.jaci.2007.10.045, 2008.

Bur, C., Reimann, P., Andersson, M., Schütze, A., and Lloyd Spetz, A.: Increasing the Selectivity of Pt-Gate SiC Field Effect Gas Sensors by DynamicTemperature Modulation, IEEE Sens. J., 12, 1906–1913, doi:10.1109/JSEN.2011.2179645, 2012.

Bur, C., Bastuck, M., Lloyd Spetz, A., Andersson, M., and Schütze, A.: Selectivity enhancement of SiC-FET gas sensors by combining temperature and gate bias cycled operation using multivariate statistics, Sensor. Actuat. B-Chem., 193, 931–940, doi:10.1016/j.snb.2013.12.030, 2014.

Caricato, A. P., Luches, A., and Rella, R.: Nanoparticle thin films for gas sensors prepared by matrix assisted pulsed laser evaporation, Sensors, 9, 2682–2696, doi:10.3390/s90402682, 2009.

European Parliament, Council of the European Union: Directive 2008/50/EC of the European Parliament and of the Council of 21 May 2008 on ambient air quality and cleaner air for Europe, Official Journal of the European Union, 51, 2008.

Gramm, A. and Schütze, A.: High performance solvent vapor identification with a two sensor array using temperature cycling and pattern classification, Sensor. Actuat. B-Chem., 95, 58–65, doi:10.1016/S0925-4005(03)00404-0, 2003.

Guo, H., Lee, S. C., Chan, L. Y., and Li, W. M.: Risk assessment of exposure to volatile organic compounds in different indoor environments, Environ. Res., 94, 57–66, doi:10.1016/S0013-9351(03)00035-5, 2004.

Gutierrez-Osuna, R.: Pattern Analysis for machine olfaction: A review, IEEE Sens. J., 2, 189–202, doi:10.1109/JSEN.2002.800688, 2002.

Harilal, S. S., Bindhu, C. V., Tillack, M. S., Najmabadi, F., and Gaeris, A. C.: Internal structure and expansion dynamics of laser ablation plumes into ambient gases, J. Appl. Phys., 93, 2380–2388, doi:10.1063/1.1544070, 2003.

Heilig, A., Bârsan, N., Weimar, U., Schweizer-Berberich, M., Gardner, J. W., and Göpel, W.: Gas identification by modulating temperatures of SnO_2-based thick film sensors, Sensor. Actuat. B-Chem., 43, 45–51, doi:10.1016/S0925-4005(97)00096-8, 1997.

Helwig, N., Schüler, M., Bur, C., Schütze, A., and Sauerwald, T.: Gas mixing apparatus for automated gas sensor characterization, Meas. Sci. Technol., 25, 055903, doi:10.1088/0957-0233/25/5/055903, 2014.

Huotari, J., Lappalainen, J., Puustinen, J., Baur, T., Alépée, C., Haapalainen, T., Komulainen, S., Pylvänäinen, J., and Lloyd Spetz, A.: Pulsed laser deposition of metal oxide nanoparticles, agglomerates, and nanotrees for chemical sensors, Procedia Eng., 120, 1158–1161, doi:10.1016/j.proeng.2015.08.745, 2015.

Hussain, O. M., Swapnasmitha, A. S., John, J., and Pinto, R.: Structure and morphology of laser-ablated WO_3 thin films, Appl. Phys. A, 81, 1291–1297, 2005.

Infortuna, I., Harvey, A. S., and Gauckler, L. J.: Microstructures of CGO and YSZ thin films by pulsed laser deposition, Adv. Funct. Mater., 18, 127–135, doi:10.1007/s00339-004-3041-z, 2008.

Johansson, M. B., Niklasson, G. A., and Österlund, L.: Structural and optical properties of visible active photocatalytic WO_3 thin films prepared by reactive dc magnetron sputtering, J. Mater. Res., 27, 3130–3140, doi:10.1557/jmr.2012.384, 2012.

Jones, A. P.: Indoor air quality and health, Atmos. Environ., 33, 4535–4564, doi:10.1016/S1352-2310(99)00272-1, 1999.

Klecka, W. R.: Discriminant Analysis, in: Quantitative applications in the social sciences, SAGE University Paper, 72 pp., 1980.

Kohl, D., Heinert, L., Bock, J., Hofmann, Th., and Schieberle, P.: Systematic studies on responses of metal-oxide sensor surfaces to straight chain alkanes, alcohols, aldehydes, ketones, acids and esters using the SOMMSA approach, Sensor. Actuat. B-Chem., 70, 43–50, doi:10.1016/S0925-4005(00)00552-9, 2000.

Koistinen, K., Kotzias, D., Kephalopoulos, S., Schlitt, C., Carrer, P., Jantunen, M., Kirchner, S., McLaughlin, J., Mølhave, L., Fernandes, E. O., and Seifert, B.: The INDEX project: executive summary of a European Union project on indoor air pollutants, Allergy, 63, 810–819, doi:10.1111/j.1398-9995.2008.01740.x, 2008.

Lee, A. P. and Reedy, B. J.: Temperature modulation in semiconductor gas sensing, Sensor. Actuat. B-Chem., 60, 35–42, doi:10.1016/S0925-4005(99)00241-5, 1999.

Leidinger, M., Sauerwald, T., Reimringer, W., Ventura, G., and Schütze, A.: Selective detection of hazardous VOCs for indoor air quality applications using a virtual gas sensor array, J. Sens. Sens. Syst., 3, 253–263, doi:10.5194/jsss-3-253-2014, 2014.

Marinkovic, B., Ribeiro de Avillez, R., Saavedra, A., and Assunção, F. C. R.: A comparison between the Warren-Averbach method and alternate methods for x-ray diffraction microstructure analysis of polycrystalline specimens, Mat. Res., 4, 71–76, doi:10.1590/S1516-14392001000200005, 2001.

Meier, D. C., Evju, J. K., Boger, Z., Raman, B., Benkstein, K. D., Martinez, C. J., Montgomery, C. B., and Semancik, S.: The potential for and challenges of detecting chemical hazards with temperature-programmed microsensors, Sensor. Actuat. B-Chem., 121, 282–294, doi:10.1016/j.snb.2006.09.050, 2007.

Nam, H.-J., Sasaki, T., and Koshizaki, N.: Optical CO gas sensor using a cobalt oxide thin film prepared by pulsed laser deposition under various argon pressures, J. Phys. Chem. B, 110, 23081–23084, doi:10.1021/jp063484f, 2006.

Paczkowski, S., Paczkowska, M., Dippel, S., Schulze, N., Schütz, S., Sauerwald, T., Weiß, A., Bauer, M., Gottschald, J., and Kohl, C.-D.: The olfaction of a fire beetle leads to new concepts for early fire warning systems, Sensor. Actuat. B-Chem., 183, 273–282, doi:10.1016/j.snb.2013.03.123, 2013.

Pihosh, Y., Turkevych, I., Mawatari, K., Uemura, J., Kazoe, Y., Kosar, S., Makita, K., Sugaya, T., Matsui, T., Fujita, D., Tosa, M., Kondo, M., and Kitamori, T.: Photocatalytic generation of hydrogen by core-shell $WO_3/BiVO_4$ nanorods with ultimate water splitting efficiency, Sci. Rep., 5, 11141, doi:10.1038/srep11141, 2015.

Reimann, P. and Schütze, A.: Fire detection in coal mines based on semiconductor gas sensors, Sensor Rev., 32, 47–58, doi:10.1108/02602281211197143, 2012.

Righettoni, M., Tricoli, A., and Pratsinis, S: $Si:WO_3$ Sensors for Highly Selective Detection of Acetone for Easy Diagnosis of Diabetes by Breath Analysis, Anal. Chem., 82, 3581–3587, doi:10.1021/ac902695n, 2010.

Sauerwald, T.: Nachweis von Luftschadstoffen mit Halbleiter-Gassensoren – Möglichkeiten und Einschränkungen, VDI-Berichte 2011, VDI Verlag GmbH, Düsseldorf, 2008.

Sood, S. and Gouma, P.-I.: Polymorphism in nanocrystalline binary metal oxides, Nanomater. Energy, 2, 82–96, 2013.

Souza Filho, A. G., Freire, P. T. C., Pilla, O., Ayala, A. P., Mendes Filho, J., Melo, F. E. A., Freire, V. N., and Lemos, V.: Pressure effects in the Raman spectrum of WO_3 microcrystals, Phys. Rev. B., 62, 3699–3703, doi:10.1103/PhysRevB.62.3699, 2000.

Wang, L., Teleke, A., Pratsinis, S. E., and Gouma, P.-I.: Ferroelectric WO_3 nanoparticles for acetone selective detection, Chem. Mater., 20, 4794–4796, doi:10.1021/cm800761e, 2008.

World Health Organization: WHO Guidelines for Indoor Air Quality: Selected Pollutants, Geneva, 2010.

Wu, C.-H., Feng, C.-T., Lo, Y.-S., Lin, T.-Y., and Lo, J.-G.: Determination of volatile organic compounds in workplace air by multisorbent adsorption/thermal desorption-GC/MS, Chemosphere, 56, 71–80, doi:10.1016/j.chemosphere.2004.02.003, 2004.

Zheng, H., Zhen Ou, J., Strano, M. S., Kaner, R. B., Mitchell, A., and Kalantar-zadeh, K.: Nanostructured tungsten oxide – Properties, synthesis, and applications, Adv. Funct. Mater., 21, 2175–2196, doi:10.1002/adfm.201002477, 2011.

21

A compact readout platform for spectral-optical sensors

Roland Wuchrer[1], Sabrina Amrehn[2], Luhao Liu[1], Thorsten Wagner[2], and Thomas Härtling[1]

[1] Fraunhofer Institute for Ceramic Technologies and Systems IKTS, 01109 Dresden, Germany
[2] Department of Chemistry, University of Paderborn, 33098 Paderborn, Germany

Correspondence to: Roland Wuchrer (roland.wuchrer@ikts.fraunhofer.de)

Abstract. The continuous monitoring of industrial and environmental processes is becoming an increasingly important aspect with both economic and societal impact. So far, spectral-optical sensors with their outstanding properties in terms of sensitivity and reliability have not been considered as a potential solution because of the cost-intensive and bulky readout hardware. Here we present a card-size, inexpensive, and robust readout platform based on a wavelength-sensitive photodiode. In test and characterization experiments we achieved a wavelength shift resolution of better than 0.1 nm and a detection limit of 0.001 AU for ratiometric measurements. We furthermore discuss the capability and current limitations of our readout unit in context with interrogation experiments we performed with a photonic crystal-based fluid sensor. In sum we expect the presented readout platform to foster the exploitation of spectral-optical sensor technology for gas monitoring, chemical analytics, biosensing and many others fields.

1 Introduction

The development of spectral-optical sensors has advanced rapidly over the last decade, and some of these types of sensors are currently finding their way into industrial application. The two most promising working principles are the measurement of wavelength shifts of a single spectral peak and the ratiometric change of two-peak signals. Examples include localized surface plasmon resonance (LSPR) substrates (Steinke et al., 2015; Katzmann et al., 2012) and photonic crystal (PhC) structures (Amrehn et al., 2015) for biological and chemical sensing or gallium arsenide (GaAs) temperature fiber sensors (Willsch et al., 2014) and inorganic phosphor materials for thermometry (Klier and Kumke, 2015). Sensors made of these materials and structures are characterized by high sensitivity, electrical passivity, immunity against intensity fluctuations and applicability under extreme conditions (temperature, humidity, electromagnetic fields, etc.). However, their use is hampered by the available readout hardware.

The hardware providing the necessary spectral resolution (< 0.1 nm) typically lacks applicability for stable, easy-to-use and cost-efficient operation. Conventional interrogators such as grating spectrometers or interferometers pose difficulties in terms of size and weight as they cannot be scaled down without loss of spectral resolution. Furthermore, they need highly demanding measuring conditions like vibration damping, air-conditioning and recalibration. Systems based on a tunable light source are restricted in their measurement range and are susceptible to environmental influences.

Consequently, for exploiting spectral-optical sensors in cost- and energy- efficient sensing systems, robust miniaturized high-resolution interrogation units are needed. One promising approach is photonic integrated circuit technology which exploits planar waveguide structures, e.g., arrayed waveguide gratings, to determine the wavelength information of an optical sensor signal. It is possible to produce small multifunctional chips with picometer resolution in a wafer process (see for example Evenblij and Leijtens, 2014). The major drawback of this technology is the restriction to single-mode waveguide structures. This implies that high-power light sources like superluminescence or laser diodes have to be used for an adequate signal-to-noise ratio, and the compatibility is limited to single-mode fiber sensors like fiber Bragg gratings or ring resonators. Otherwise the intensity loss in the light coupling elements is too high.

In the light of the approaches above, we believe that only with a holistic viewpoint on the entire sensor system can a technically and economically successful sensor application be achieved. In particular, all components from the light source over the actual sensor to the interrogation unit need to be taken into consideration. In the same way, optoelectronic packaging technologies and reliability aspects have to be paid attention to in the design process. Following this credo, we present a compact photocurrent-based interrogation unit based on a wavelength-sensitive photodiode (WSPD) (Braasch et al., 1995; Lu et al., 1996). The key feature of the WSPD is a vertical stack of two photodiodes each having a different spectral sensitivity. This allows the determination of spectral changes in the irradiation light from the ratio of the two generated photocurrents. This approach combines the simplicity of an intensity measurement setup with the robustness of spectral readout and permits application of the interrogation unit for all kinds of spectral-optical sensors.

This article is structured as follows: in the first part we describe the working principle and the readout electronics of the WSPD. We then evaluate the performance of the setup for wavelength shift and ratiometric measurements and highlight advantages and difficulties. Finally, the capability of our readout unit is demonstrated with a photonic crystal-based fluid sensor.

2   Readout platform

The central element of the readout platform is the wavelength-sensitive photodiode WS7.56-TO5i (First Sensor AG). The WSPD is composed of two vertically stacked silicon photodiodes with different spectral response characteristics; see Fig. 1. Through this structure, the central wavelength of monochromatic signals or the centroid of a polychromatic light distribution can be determined by the ratio of the two generated photocurrents I_1 and I_2. Therefore, it is ideally suited for the readout of spectral-optical sensors with their specific spectral changes. For reliable and high-resolution measurements, a low-noise electronics with a flexible amplification was realized. We chose a logarithmic amplifier (LOG112, Texas Instruments), which ensures a highly precise calculation of the current log ratio over a wide current input range of 7.5 decades (100 pA to 3.5 mA). The log ratio voltage V_{LOG} of the two photodiodes is computed by the LOG112 as follows:

$$V_{LOG} = 0.5\,\text{V} \cdot \log \frac{I_1(\lambda, P) + I_{off1}(T)}{I_2(\lambda, P) + I_{off2}(T)}, \qquad (1)$$

where P describes the incident optical power and T the temperature of the photodiodes (formula adapted from the data sheet of the LOG112, Texas Instruments, 2005). The currents I_{off1} and I_{off2} denote offset values which allow considering that the temperature-dependence of the dark current and the noise behavior are different for the two stacked photodiodes. The noise effects of the amplifier electronics are also included. We would like to point out that intensity fluc-

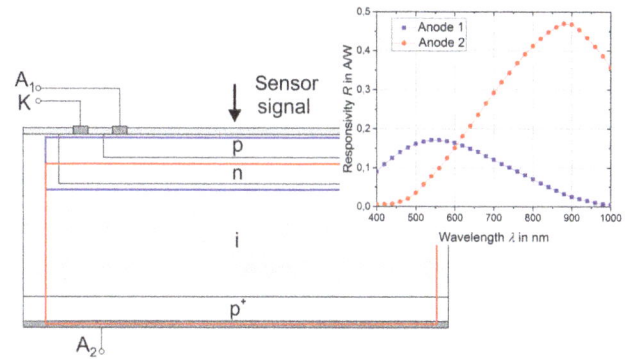

Figure 1. Photodiode structure and spectral responsivity curves of the WSPD (adapted from First Sensor WSPD data sheet, First Sensor AG, 2015).

tuations of the sensor signal have no influence on the readout because of the ratio calculation.

In most cases, a conversion of the calculated log ratio voltage into the centroid wavelength λ_C is not required for the readout of spectral-optical sensors, since the final calibration will be on the measured parameter, e.g., the refractive index. In general, the conversion will be carried out with a scale parameter s, which in the case of λ_C exhibits the unit nm V^{-1}:

$$\lambda_C = s \cdot V_{LOG}. \qquad (2)$$

For many sensor applications, e.g., chemical analytics, a measurement rate in the range of several hertz is sufficient. So, to reduce the noise level of the calculated voltage signal, low-pass filtering with a cutoff frequency of 100 Hz was applied. Moreover, a stable power supply for a minimal noise level is crucial. Several voltage regulators were used to hold the necessary output voltages at a defined value regardless of the changes in the load current or input voltage.

As already mentioned, the dark current behavior of the stacked photodiodes of the WSPD is different. This results in a temperature dependent offset value for the log ratio voltage. For this reason, a temperature sensor (LM45B, Texas Instruments) was placed near the WSPD which enables a recalibration of the log ratio voltage.

In addition, an adjustable LED light source circuit was integrated into the electronics. It reflects the holistic approach we pursue in this work.

A 24 bit analog-to-digital converter, ADS1248 (Texas Instruments), resolves the measured data with high precision. Final processing and controlling of all signals are done with a microcontroller board (Arduino Uno), and the data are displayed with a LabVIEW program. The assembled circuit board is shown in Fig. 2.

A sub-miniature assembly (SMA) fiber adapter was implemented for both the WSPD and the LED, which ensures a stable and easy connection of multimode fibers for sensor irradiation and signal collection. The use of multimode fibers is possible as the WSPD operability is independent of

Figure 2. Photograph of the readout platform.

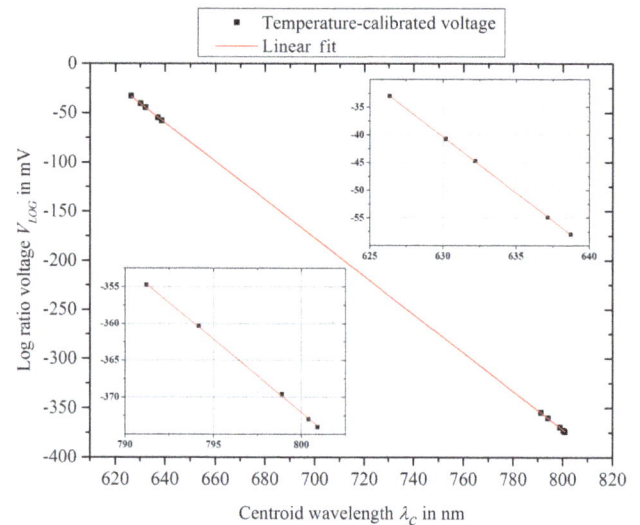

Figure 3. Measured log ratio voltage–centroid wavelength response of the readout platform.

the angle of incidence, polarization, and the spatial intensity distribution of the light.

3   Test measurements

In a first evaluation experiment we tested the performance of the WSPD interrogation electronics for resolving peak wavelength changes and ratiometric response. The experiments were performed under laboratory conditions to guarantee a stable operating temperature for the WSPD. Furthermore, it was ensured that no interfering spectral peaks were present in the test signals and no stray light falsified the measurement. The focus of the test measurements was to characterize the resolution performance of the developed electronics, not the absolute wavelength accuracy.

A free space setup consisting of a white-light source (Xenon lamp) and different bandpass filters was used to determine the wavelength shift resolution. We used two bandpass filters with a central wavelength of 640 and 800 nm, respectively, and a full width at half maximum (FWHM) of 10 nm. By rotating the bandpass filters, their resonance condition was intentionally changed and the transmitted wavelength shifted to shorter wavelengths. The spectra at five rotation angles were analyzed for each filter with a high-resolution grating spectrometer (iHR550, Horiba) so as to obtain a spectral reference. The filtered light was alternately directed to the WSPD and the spectrometer by means of a multimode fiber. The light power at the fiber end was around 20 μW. The acquired data are depicted in Fig. 3.

We observed a highly linear dependence between the centroid wavelength and the voltage ratio of the photodiodes. The measurement uncertainty was assumed as follows: the accuracy of the calibrated spectrometer is 0.07 nm, which represents the data point uncertainty of Fig. 3 in x direction. The temperature dependence of the WSPD output voltage was calculated out using the signal of the LM45B tem-

perature sensor. From the data sheet of the latter we estimated a maximal error of 0.1 K for the temperature monitoring around room temperature (see data sheet of the LM45B, Texas Instruments, 2013). From separate experiments we determined the temperature dependence of the WSPD electronics to be approximately $1\,\text{mV}\,\text{K}^{-1}$, which in sum results in a systematical error of $100\,\mu\text{V}$. To determine the statistical error of the interrogation electronics, the three-fold standard deviation was calculated from 1000 data points per measured wavelength. This represents a measurement time of 20 s. A maximal three-fold standard deviation of $30\,\mu\text{V}$ and a corresponding standard error of $1\,\mu\text{V}$ were obtained. For the linear fit of the 10 measured wavelengths a standard error of $1.2\,\mu\text{V}\,\text{nm}^{-1}$ for the gradient was acquired. Hence, the resulting total error is $102\,\mu\text{V}$ for the y direction of Fig. 3. In the face of these values we refrained from including error bars in the graphs in Fig. 1 since they could not be displayed clearly. In sum, a spectral shift accuracy of around 0.05 nm could be concluded for the WSPD interrogation electronics. The results and the linearly behavior are in agreement with experiments reported elsewhere (Amtor et al., 2011).

In our second experiment, we used two irradiation LEDs simultaneously to carry out a ratiometric measurement. For this, the light of the LEDs was coupled into a Y-type multimode fiber coupler and directed to the WSPD. A 525 nm LED, which was assembled on the circuit board, was varied in its forward current; i.e., its optical power was tuned. At the same time, a 624 nm LED was operated at constant current and worked as a baseline. This ensured that the measured peak ratio was independent of intensity fluctuations. A total light power in the range of 60 to 80 μW was measured at the end of the fiber coupler. As previously, reference spectra were recorded with the spectrometer to corre-

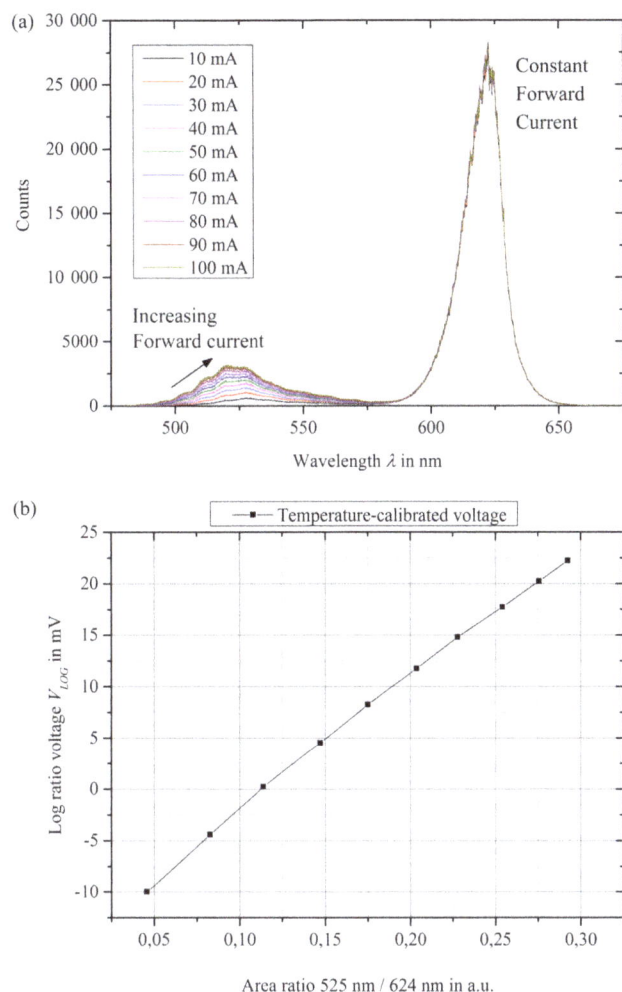

Figure 4. (a) Ratiometric spectra consisting of a baseline at 624 nm and a varying peak intensity at 525 nm. (b) Calculated ratios of the area under the respective emission curves to the measured log ratio voltage of the WSPD readout electronics.

late the LED emission ratio with the measured log ratio voltage. The emission spectra for different currents are depicted in Fig. 4a, while in Fig. 4b we plotted the ratio of the area under the emission curves of the respective LED over the output voltage ratio. In the latter diagram, a slight nonlinearity at small ratios is present, which is due to a nonlinear blueshift of the short-wavelength LED emission upon current tuning. This issue needs to be considered in real sensor applications in which the spectral positions of the peaks need be either constant or well known. As for the determination of the wavelength shift resolution, the temperature uncertainty with 0.1 K was the main error source. In the linear section of the diagram a resolution of LED emission ratio changes of 0.001 AU was achieved.

4 Demonstration of optical sensor interrogation

4.1 Photonic crystal-based fluid sensor

We chose to demonstrate the capability of the readout unit by interrogating a photonic crystal-based fluid sensor with the described electronics. PhCs are periodically ordered transparent nanostructures with a distinct boundary condition for the motion of photons through the crystal structure which result in unique optical material properties. The most well-known example of natural PhCs is probably opals which consist of periodically arranged silica spheres closely packed in 3-D. These gemstones generate iridescent colors (color changes depending on viewing angles). In analogy to the electronic band diagrams, which describe the allowed electronic states in a solid, the optical properties of the PhCs can be described in terms of photonic band structures. In this model the iridescent colors of opals are caused by the formation of photonic stop bands along different crystallographic directions. More details about PhCs can be found in textbooks, e.g., in Joannopoulos et al. (2008).

In the presented work artificially generated inverse opal structures made of tungsten oxide (WO_3) are utilized as sensing layers. They offer similar properties to the opals described above; in particular they exhibit photonic stop bands which can be observed in the reflection spectra (see Fig. 6). Amongst other things the position of the stop bands is very sensitive to the refractive index contrast of the inverse opal material and the material inside the pores, e.g., a fluid (Stein et al., 2008). Therefore a change of the refractive index of a fluid introduced into the pores can be monitored by determining the position of the reflection bands. In a previous work this was demonstrated utilizing a grating-based spectrometer as a readout device (Amrehn et al., 2015), which, however, is quite costly. As will be shown in the following, the diode-based readout unit presented here achieves comparable or even better performance without the need for additional data processing.

4.2 Experimental setup

Tungsten oxide inverse opals films were synthesized by a two-step casting process as follows: firstly, an artificial opal film consisting of poly(methyl methacrylate) (PMMA) spheres was deposited on a glass substrate by drop deposition. In the second step, the pores of the closely packed sphere arrangement were filled with a tungsten salt solution (99.99 % ammonium meta-tungstate hydrate, Sigma–Aldrich). After thermal conversion of the tungsten salt to tungsten oxide and combustion of the PMMA, the resulting tungsten oxide forms the ordered macroporous inverse opal framework. The spherical pores are positioned on the lattice points of the opal spheres. Further details of the synthesis are described elsewhere (Amrehn et al., 2015).

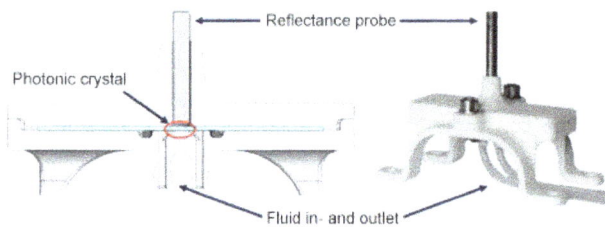

Figure 5. Scheme (left) and photo (right) of the custom-built fluid measurement cell.

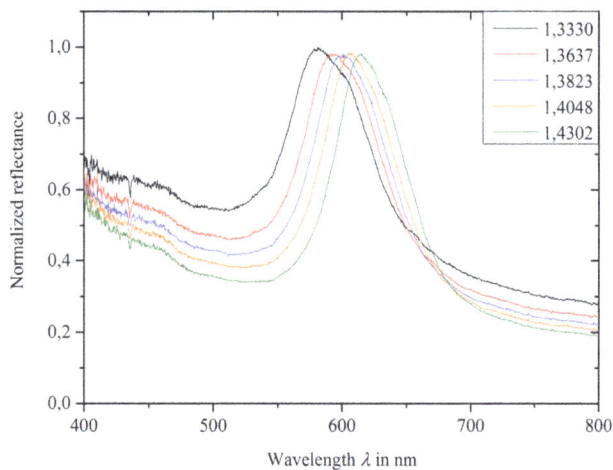

Figure 6. Normalized reflectance spectra of the WO_3 inverse opal structure infiltrated with different water–ethylene-glycol mixtures.

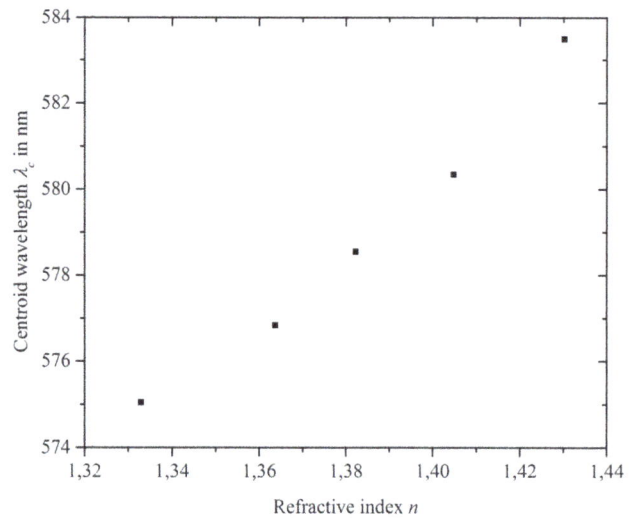

Figure 7. Centroid wavelength of the normalized reflection spectra from Fig. 6.

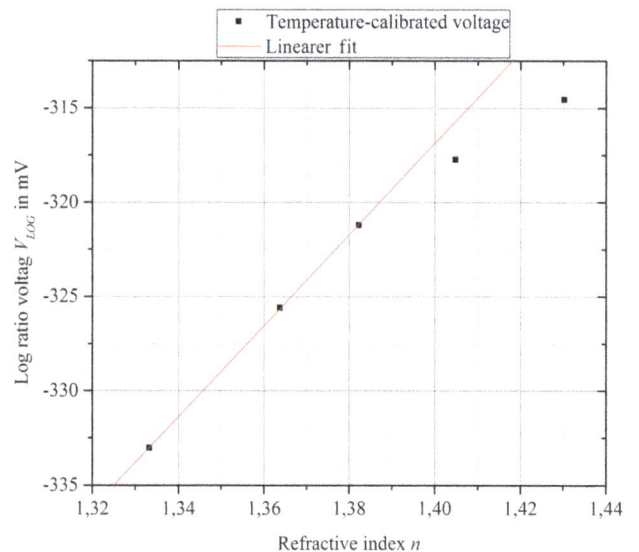

Figure 8. Measured voltages by the WSPD electronics for the PhC reflection spectra at different refractive indexes (Fig. 6).

Utilizing a custom-built measurement cell (Fig. 5) and a reflection/backscattering probe (QR200-7-VIS-BX, Ocean Optics), the optical properties of the inverse opals were characterized. The samples were illuminated by a krypton light source (ecoVis, Ocean Optics) via the probe. Since the reflectance probe is mounted on the backside of the glass slide, it is not in contact with the sample media. The setup allows for continuous fluid measurements. When evaluating the optical properties of the sample, one has to keep in mind that the light has to pass through the glass slide and an additional interface (air/glass). The reflected light is collected by the reflectance probe and analyzed with the readout platform. Different mixtures of ethylene glycol and water were used for testing since this allows the refractive index to be varied over a relatively wide range. The refractive index of each mixture was determined with a refractometer (Kruess DR 201-95).

4.3 Results

Figure 6 shows the normalized reflectance spectra obtained with the described setup. In comparison to the narrowband signal (FWHM = 10 nm) used in the test measurements, the reflection spectra of the PhC structure have a FWHM of around 60 nm. In addition, the signal has an asymmetrical shape and a high baseline. It should be noted that, in con-

trast to the measurements described in Sect. 3, the reflection spectra covered a very broad spectral window as depicted in Fig. 6. Figure 7 shows the centroids of the PhC spectra acquired with the grating spectrometer as reference data. A small nonlinearity of the wavelength shift can be seen. The data obtained with the WSPD readout electronics are illustrated in Fig. 8. From the linear part of the curve (lower refractive indexes) we estimate a limit of detection of 0.001 refractive index units. For higher refractive indices the curve deviates from the behavior of the reference data in Fig. 7. To further evaluate the WSPD data, we simulated the measured data as follows: the normalized reference spectra (Fig. 6) were multiplied with the two WSPD responsiv-

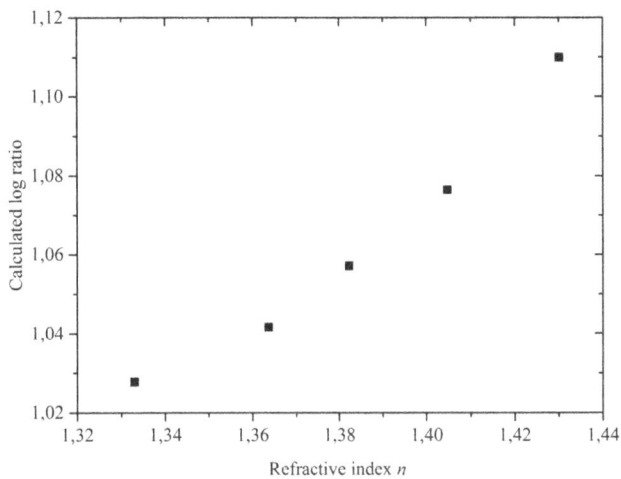

Figure 9. Calculated WSPD-measuring characteristics for the normalized reflection spectra from Fig. 6.

ity curves (Fig. 1). For the two sensitivity-folded spectra, the area under the curves was determined by summarizing the intensity values. Finally, the log ratio of these values was calculated. The result is displayed in Fig. 9. The calculated log ratio values for the WSPD are consistent in their behavior with the centroid wavelengths of the PhC reflection spectra in Fig. 7.

At the moment we cannot provide a satisfying explanation for the observed deviation. Several effects may lead to this behavior, which have to be investigated in depth in future work. These effects include the dependence of the refractive index sensitivity of the PhC on the wavelength, possible deviations of the WSPD element in use from the data sheet (especially at the edges of the broad spectral window), signal drift induced by an electronics artefact, and the higher signal-to-noise ratio induced by the low intensity in the PhC experiments (nW regime) in contrast to the test experiments (μW range). Hence, further work will focus on the redesign of the PhC flow cell as well the WSPD readout electronics to improve the signal evaluation.

5 Conclusion and outlook

In this report we presented a compact electronics platform for reading out spectral-optical sensors and demonstrated its suitability for wavelength shift and ratiometric measurements. The system, based on a wavelength-sensitive photodiode, works independently of intensity fluctuations and permits a simple setup geometry similar to those used for common photodiodes. As the WSPD measures the centroid of the entire spectral distribution of the optical sensor signal, precise knowledge of the shape and spectral behavior of the signal is necessary in order to obtain optimal performance. Furthermore, the operating temperature of the WSPD has to be monitored or stabilized because of the different noise prop-

erties of the stacked photodiodes. In sum it can be concluded that for every sensor application an individual calibration of the WSPD electronic has to be done.

Finally, the applicability of the WSPD setup for the readout of a photonic crystal-based fluid sensor was shown. Although it was possible to unambiguously distinguish different refractive index media, these experiments also produced a slight mismatch between the measured signal and the reference data which is not yet understood. We will focus our future work on a deeper investigation of this observation. Moreover, a redesign of the entire electronics is planned, e.g., the integration of the microcontroller on the circuit board, to realize a compact system and reduce noise effects.

We summarize that, although the WSPD approach still has to overcome some challenges, the application-relevant characteristics like cost efficiency, miniaturization, high spectral resolution, robustness and simple handling are fulfilled with the setup we presented here.

Acknowledgements. The authors thank Hendrik Funke and Peter Blüthgen from Fraunhofer IKTS for remarks and work on the electronics. Also, the authors thank First Sensor AG, Berlin, for providing the responsivity data of the WS7.56. Furthermore, financial support from FhG Internal Programs (grant no. Attract 692271) and the German Federal Ministry of Education and Research (BMBF, grant no. 13N12969) is gratefully acknowledged.

References

Amrehn, S., Wu, X., Schumacher, C., and Wagner, T.: Photonic crystal-based fluid sensors: Toward practical appplication, Phys. Status Solidi A, 212, 1266–1272, doi:10.1002/pssa.201431875, 2015.

Amtor, T., Hofman, C. S., Knorz, J., and Weidemüller, M.: High-precision semiconductor wavelength sensor based on a double-layer photo diode, Rev. Sci. Instrum., 82, 093111, doi:10.1063/1.3640409, 2011.

Braasch, J. C., Holzapfel, W., and Neuschaefer-Rube, S.: Wavelength determination of semiconductor lasers: precise but inexpensive, Opt. Eng., 34, 1417–1420, doi:10.1117/12.201655, 1995.

Evenblij, R. S. and Leijtens, J. A. P.: Space Gator, a giant leap for fibre optic sensing, International Conference on Space Optics, Tenerife, Spain, 7–10 October, 2014.

First Sensor AG: Wavelength-sensitive diode WS7.56-TO5i, available at: http://www.first-sensor.com/en/products/optical-sensors/detectors/wavelength-sensitive-diodes-ws/index.html/ (last access: 19 February 2016), 2015.

Joannopoulos, J. D., Johnson, S. G., Winn, J. N., and Meade, R. D.: Photonic Crystals, in: Molding the Flow of Light, 2nd Edn., edited by: Gnerlich, I., Princeton University Press, Princeton, 2008.

Katzmann, J. and Härtling, T.: Nanorod formation by photochemical metal deposition in nanoporous aluminum oxide templates, J. Phys. Chem. C, 116, 23671–23675, doi:10.1021/jp303896a, 2012.

Klier, D. T. and Kumke, M. U.: Upconversion $NaYF_4$: Yb : Er nanoparticles co-doped with Gd^{3+} and Nd^{3+} for thermometry on the nanoscale, RSC Adv., 5, 67149–67156, doi:10.1039/C5RA11502G, 2015.

Lu, G. N., Chouikha, M. B., Sou, G., and Sedjil, M.: Colour detection using a buried double p-n junction structure implemented in the CMOS process, Electron. Lett., 32, 594–596, doi:10.1049/el:19960337, 1996.

Stein, A., Li, F., and Denny, N. R.: Morphological Control in Colloidal Crystal Templating of Inverse Opals, Hierachical Structures, and Shaped Particles, Chem. Mater., 20, 649–666, doi:10.1021/cm702107n, 2008.

Steinke, N., Wuchrer, R., and Härtling, T.: Aufbau und Biofunktionalisierung einer LSPR-Molekülsensoreinheit, 12th Dresdner Sensor-Symposium, Dresden, Germany, 7–9 December 2015, 111–114, doi:10.5162/12dss2015/P2.2, 2015.

Texas Instruments: Precision Logarithmic and Log Ratio Amplifiers LOG112/LOG2112, available at: http://www.ti.com/general/docs/lit/getliterature.tsp?genericPartNumber=log112&fileType=pdf (last access: 19 February 2016), 2005.

Texas Instruments: LM45 SOT-23 Precision Centigrade Temperature Sensors, available at: http://www.ti.com/general/docs/lit/getliterature.tsp?genericPartNumber=lm45&fileType=pdf (last access: 19 February 2016), 2013.

Willsch, M., Kaiser, J., Bosselmann, T., Wieduwilt, T., and Willsch, R.: Investigation of low-cost two-wavelength interrogation of different fiber optical temperature sensors into electric power facility monitoring systems, 23rd International Conference on Optical Fibre Sensors, Santander, Cantabria, Spain, 2–6 June 2014, 915796, doi:10.1117/12.2059269, 2014.

22

High-speed camera-based measurement system for aeroacoustic investigations

Johannes Gürtler[1], Daniel Haufe[1], Anita Schulz[2], Friedrich Bake[2], Lars Enghardt[2,3], Jürgen Czarske[1], and Andreas Fischer[1]

[1]Chair of Measurement and Sensor System Techniques, Department of Electrical Engineering and Information Technology, TU Dresden, Helmholtzstr. 18, 01069 Dresden, Germany
[2]Institute of Propulsion Technology, German Aerospace Center (DLR), 10623 Berlin, Germany
[3]Institute of Fluid Dynamics and Technical Acoustics, TU Berlin, 10623 Berlin, Germany

Correspondence to: Johannes Gürtler (johannes.guertler@tu-dresden.de)

Abstract. The interaction of sound and flow enables an efficient noise damping. Inevitable for understanding of this aeroacoustic damping phenomenon is the simultaneous measurement of flow and sound fields. Optical sensor systems have the advantage of non-contact measurements. The necessary simultaneous determination of sound levels and flow velocities with high dynamic range has major hurdles. We present an approach based on frequency-modulated Doppler global velocimetry, where a high-speed CMOS camera with data rates over $160\,\mathrm{MSamples\,s^{-1}}$ of velocity samples is employed. Using the proposed system, two-component flow velocity measurements are performed in a three-dimensional region of interest with a spatial resolution of $224\,\mu\mathrm{m}$, based on single-pixel evaluation, and a measurement rate of $10\,\mathrm{kHz}$. The sensor system can simultaneously capture sound and turbulent flow velocity oscillations down to a minimal power density of $40.5\,(\mathrm{mm\,s^{-1}})^2\,\mathrm{Hz}^{-1}$ in a frequency range up to $5\,\mathrm{kHz}$. The presented measurements of the interaction of sound and flow support the hypothesis that the sound energy is transferred into flow energy.

1 Introduction

1.1 Motivation

In order to attenuate the noise of modern jet engines or gas turbines, perforated liners as depicted in Fig. 1 are used (Eldredge and Dowling, 2003). By applying a bias flow through the perforation of such liners it is possible to increase the efficiency and bandwidth of the damping effect (Bechert, 1980). This effect is partially based on the energy transfer from the sound wave into the bias flow as sound-excited oscillations of the flow velocity. In order to be able to design or operate bias flow liners with a high sound damping efficiency, it is desirable to understand the interaction of the sound wave with the flow field. For this reason numerical simulations have been performed (Zhao et al., 2015), but they are often limited because of high calculation efforts, model assumptions or invalid model simplifications and need to be validated. Thus, non-invasive measurement techniques are required that en-

able the simultaneous acquisition of the mean flow velocity and the sound-excited oscillation of the velocity in a three-dimensional (3-D) region of interest above the liner surface. Furthermore, turbulence spectra of the flow velocity need to be analysed in order to quantify the energy transfer from the sound wave to the flow turbulence that contributes to the sound damping. The maximum damping efficiency of the used liner lies at approximately $1\,\mathrm{kHz}$ (Schulz et al., 2015). For this reason, a high measurement rate of $10\,\mathrm{kHz}$ is chosen here in order to sample this maximum as well as flow turbulence up to $5\,\mathrm{kHz}$ properly, according to the Nyquist–Shannon sampling theorem. Additionally, a short experimental time is needed to reduce environmental influences on the experiment, such as ambient temperature and pressure variations. Hence, a planar (2-D) measurement system is needed in order to measure the entire 3-D region of interest for several minutes. The combination of such imaging systems with the number of measurement points N_p and the required mea-

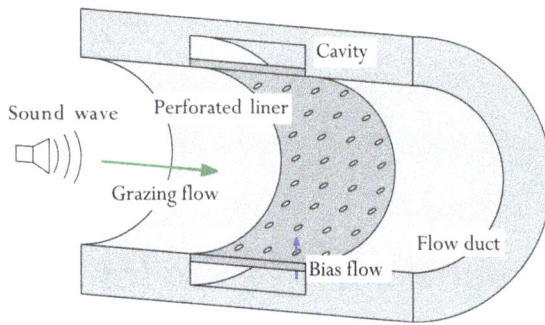

Figure 1. Sketch of a perforated bias flow liner with a cavity behind, installed in a flow duct.

surement rate f_s necessitates a sensor system with high data rates

$$f_d = f_s \cdot N_p \cdot N_c \tag{1}$$

in the range of MSamples s^{-1}, where N_c is the number of simultaneously measured velocity components. These aforementioned requirements are fulfilled by a high-speed camera-based measurement system.

1.2 State of the art

For measurements of small-scale flow velocity fluctuations in the range of several kHz, hot-wire anemometry is a commonly applied technique providing high sensitivity and bandwidth up to 100 kHz (Watmuff, 1995). However, it is a pointwise and invasive technique potentially disturbing the flow and sound field.

Non-invasive, optical measurements of the acoustic particle velocity (APV), i.e. the sound-related particle oscillation velocity (Hann and Greated, 1999; Taylor, 1976) and the flow velocity (Thompson and Atchley, 2005), were realized by using laser Doppler anemometry (LDA). These systems detect the Doppler-shifted frequency of light scattered at single particles moving with the flow with negligible slip and provide data rates in the range of several 10 MSamples s^{-1} (Hann and Greated, 1999). Standard LDA systems yield a point-wise measurement, which objects the requirement of an imaging technique. Enhanced LDA sensor principles with a linear (known as profile sensor) and planar (known as field sensor) measurement volume have been presented (Voigt et al., 2008; Meier and Roesgen, 2012). However, for all LDA principles the measurement rate equals the random particle rate passing the measurement volume observed by one detector element. Hence, a high-speed measurement with a constant measurement and data rate is not possible.

Particle image velocimetry (PIV) offers a simultaneous planar measurement of the flow velocity and the APV (Hann and Greated, 1997; Rupp et al., 2010). It is based on the correlation of interrogation windows, i.e. sections from two subsequently captured images of illuminated particles within a

light sheet. Using PIV systems, measurements of the APV were realized with measurement rates up to 20 kHz (Henning et al., 2013). Furthermore, measurements at a bias flow liner were performed with a PIV system phase-synchronous to the acoustic excitation signal (Schulz et al., 2014). However, standard PIV systems provide only two velocity components to be measured simultaneously, while all three velocity components of the complex flow are required. In addition, the desired measurement system needs a dynamic range of the order of 3×10^3 to measure both typical amplitudes of the APV of the order of 10 mm s^{-1} and superimposed flow velocities up to 30 m s^{-1} (Haufe et al., 2013). Since common PIV systems provide a dynamic range of the order of 10^2 (Adrian, 2005), the application of PIV for aeroacoustic studies requires further attention.

A sufficient dynamic range up to 4×10^3 and high measurement rates up to 50 kHz have been achieved recently using Doppler global velocimetry with sinusoidal frequency modulation (FM-DGV) for aeroacoustic investigations at a bias flow liner (Haufe et al., 2012, 2013). Because of the usage of a linear detector array consisting of eight detectors, an extensive traversing effort is needed to measure a volumetric (3-D) region of interest. Moreover, a low data rate of $8 \cdot 50$ kHz $\cong 0.4$ MSamples s^{-1} is achieved. In order to overcome these drawbacks, the existing FM-DGV system was recently enhanced by implementing a high-speed camera for simultaneous light detection, which allows one velocity component to be measured at a data rate of 800 MSamples s^{-1} (Fischer et al., 2014).

Conventional DGV systems combined with high-speed cameras also provide a high dynamic range over 10^2 as well as high measurement rates in the MHz range (Thurow, 2004). However, neither the camera-based FM-DGV nor the DGV has yet been tested for aeroacoustic studies. It is an open question of whether the sensitivity of the camera sensor is still sufficient to resolve small-scale light intensity fluctuations caused by the sound-excited flow velocity oscillations.

In order to further increase the data rate, the simultaneous measurement of more than one velocity component is desired. Simultaneous planar (2-D) measurements of all three velocity components with a single camera were performed using a hybrid DGV and PIV image processing for the same data set (Willert et al., 2006; Wernet, 2004). Using such systems, measurement rates in the lower Hz range could be achieved. The idea is here to adapt this approach for the high-speed camera-based FM-DGV measurement system, which thus needs to be supplemented by a PIV image processing. It has to be investigated whether such a hybrid FM-DGV–PIV system with a single camera finally allows one to measure all three components of the mean flow field and the sound-excited flow velocity oscillations.

1.3 Aim and structure

A camera-based FM-DGV system using a high-speed camera is demonstrated for volumetric (3-D) field measurements of the flow velocity including the sound-excited oscillations at a bias flow liner. Because of the larger field of view, the higher resolution of the measurement system and the measurement rate of 10 kHz, a data rate of 10 MSamples s^{-1} is achieved according to Eq. (1). Furthermore, the use of the camera allows planar measurements at the bias flow liner, which reduces the traversing effort significantly to one axis. Finally, the high-speed camera offers the perspective of measuring all three velocity components with a single camera by a hybrid FM-DGV–PIV image evaluation.

In Sect. 2 the principle and the setup of the camera-based FM-DGV measurement system are presented. Its validation with respect to aeroacoustic investigations follows in Sect. 3 by performing a measurement of one velocity component at a generic bias flow liner. In addition, the measurement uncertainty is addressed, since a maximum uncertainty of the velocity smaller than 10 mm s^{-1} is needed to resolve the flow velocity oscillations. Using multiple measurements and two different observation directions, the 3-D region of interest is measured for two components of the mean flow velocity and the sound-excited flow oscillations. The measurement results, including the discussion of the acquired turbulence spectra, are presented in Sect. 4. In Sect. 5, the enhancement of the measurement system towards simultaneous three-component measurements with a single camera by using a hybrid FM-DGV–PIV image evaluation is finally discussed as an outlook.

2 Measurement principle

The FM-DGV technique is based on evaluating the velocity-dependent Doppler shift of the frequency of light scattered at moving particles in the flow (Müller et al., 2007; Fischer et al., 2007). In Fig. 2a and b the principle measurement setup is depicted. The scattering particles, which follow the flow velocity v_f with negligible slip, are illuminated from the direction i by a narrow-band laser light with the centre frequency f_c. Light scattered at these particles exhibits a frequency shift by the Doppler frequency

$$f_D = \frac{f_c \|o - i\|}{c} \cdot v, \text{ with } v = \frac{(o - i)}{\|o - i\|} \cdot v_f, \qquad (2)$$

where o is the observation direction, c the speed of light and both vectors i and o are unit vectors. Hence, the Doppler frequency is directly proportional to the velocity component v of v_f along the sensitivity vector $o - i$.

Since (f_c) is 335 THz for the used laser wavelength of 895 nm and f_D is merely in the kHz to MHz range, f_D cannot be measured directly. Therefore, the scattered light is observed through a molecular absorption cell filled with caesium gas. Adequate for the used laser centre frequency

f_c, caesium provides a steep slope in the light-frequency-dependent transmission curve as sketched in Fig. 2c. Thus, the Doppler shift in the frequency of the received light is transformed into a measurable change of the light intensity behind the absorption cell. However, this intensity signal depends not only on the frequency but also on the intensity of the scattered light. In order to eliminate this cross-sensitivity, the laser light frequency is sinusoidally modulated around the absorption minimum of the absorption cell so that

$$f_{Laser}(t) = f_c + f_h \sin(2\pi f_m t), \qquad (3)$$

where f_m is the frequency and f_h the amplitude of the modulation. As can be seen in Fig. 2c, the resulting transmitted intensity signal contains higher-order harmonics of the modulation frequency depending on the shifted laser centre frequency. Finally, the quotient

$$q(f_D) = \frac{A_1(f_D)}{A_2(f_D)} \qquad (4)$$

of the amplitudes A_1 and A_2 of the first- and second-order harmonics is evaluated, which is independent of the mean scattered light intensity and solely depends on the flow velocity component v. After calibrating the system to determine the relationship $v = v(q)$, the desired velocity v can be obtained from the quotient q.

3 Measurement setup and characterization

3.1 Measurement object

Aeroacoustic investigations are realized at a bias flow liner, using the setup shown in Fig. 3. The measurements are performed at the central orifice of $N = 53$ orifices. Each one has the diameter $d = 2.5$ mm and the distance $s = 8.5$ mm to the neighbouring orifices. Above the liner surface, the grazing flow v_g is realized as a suction flow due to a duct fan and is directed in the negative x direction. The mass flow $\dot{m} = 5$ kg h^{-1} is stabilized with a mass flow controller and corresponds to an average bias flow velocity $v_b = \dot{m}/(N\pi(\frac{d}{2})^2\rho\alpha) = 7$ m s^{-1}, with the air density $\rho = 1.2$ kg m^{-3} and the empirical jet contraction factor $\alpha = 0.61$ (Heuwinkel et al., 2010). The jet contraction factor is defined as the ratio of the cross-section area at the smallest jet diameter (vena contracta) behind the orifice and the area of the orifice. Based on previous measurements at the bias flow liner using the same flow parameters, the Kolmogorov length scale of the flow can be estimated to 20 µm (Haufe et al., 2014a). Furthermore the Kolmogorov timescale can be estimated to 24 µs (Pope, 2000). Note that this estimation is based on the assumption of isotropic turbulence and the $k - \epsilon$ turbulence model.

In order to study the damping phenomenon, a sinusoidal acoustic excitation signal at $f_{ac} = 867$ Hz with a sound pressure level of 120 dB is applied using a speaker of the type

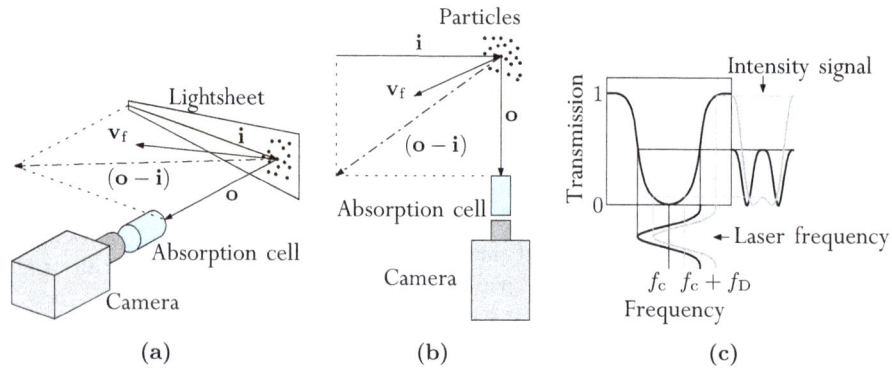

Figure 2. FM-DGV measurement principle: (**a**) and (**b**) the particles are illuminated from the direction i, and the scattered light is detected behind a caesium absorption cell. (**c**) Due to the frequency-dependent transmission curve of the caesium gas, the Doppler frequency shift is transformed into a change of the transmitted light intensity (flow velocity $\boldsymbol{v}_\mathrm{f}$, laser centre frequency f_c, Doppler frequency f_D).

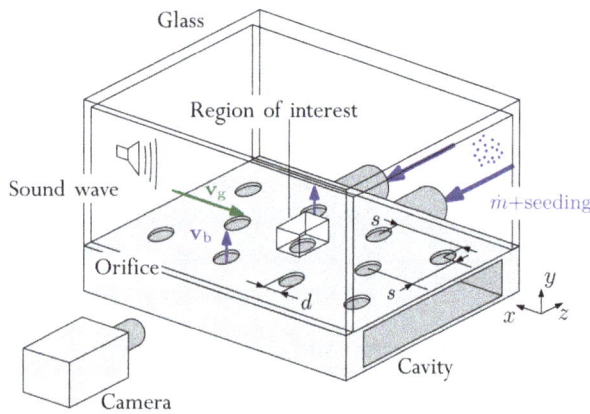

Figure 3. Bias flow liner test rig: a sound wave is excited above the liner facing sheet and is oriented in the direction of the superimposed grazing flow with the velocity $\boldsymbol{v}_\mathrm{g}$. Behind the perforated facing sheet is a cavity, in which an additional mass flow $\dot{m}$ is fed that provides the bias flow with the velocity $\boldsymbol{v}_\mathrm{b}$ through the perforation of the liner. Note that the point of origin of the used coordinate system is located right in the centre of the orifice.

KU-516 from the manufacturer MONACOR. The total harmonic distortion of the speaker is 0.8 % measured with a flush-mounted microphone.

3.2 Illumination and detection setup

The described bias flow liner is expected to exhibit a nonlinear response characteristic, which shall be investigated by analysing the total harmonic distortion of the velocity signal above the liner surface. According to this object, the excitation frequency f_ac and its higher harmonics have to be resolved. Additionally the maximum dissipation of the liner at 1 kHz (Schulz et al., 2015) shall be analysed in the spectrum. Hence, a measurement rate of 10 kHz is necessary in order to properly sample the maximum dissipation range as well as the higher harmonics of the sound-excited oscillations up to

5 kHz, according to the Nyquist–Shannon sampling theorem. For this reason the laser is modulated with $f_\mathrm{m} = 10\,\mathrm{kHz}$, because at least one modulation period is necessary to evaluate the amplitudes A_1 and A_2 according to the FM-DGV measurement principle. As a consequence of the chosen modulation frequency and the Nyquist–Shannon sampling theorem, a high camera frame rate over 40 fps is necessary, in order to resolve the second-order harmonic $2 f_\mathrm{m}$. Here a camera frame rate of 100 fps is used, which yields 10 camera pictures per modulation period. Such a sequence of 10 subsequent raw data pictures showing the modulation of the detected intensity signal is depicted in Fig. 4.

Since light scattering particles are needed, these are added to the bias flow by a particle generator with four Laskin nozzles and seeded into the measurement volume through the orifices of the liner. The liquid seeding particles are made of diethylhexyl sebacate (DEHS) and have a diameter of about 1 μm. They are illuminated in the x–y plane by a laser light sheet, which is produced by a power-amplified diode laser. At a wavelength of 895 nm this laser system provides a maximum of 600 mW output power, which was used for the measurements. By modulating the laser diode current, the laser frequency is modulated around the centre frequency f_c, which is stabilized using a PI controller. Further details of the controlled laser system can be found in Fischer et al. (2013c).

The light scattered at the particles is measured behind the caesium absorption cell with a high-speed camera as shown in Fig. 5a. The cells are temperature stabilized in order to keep the temperature-dependent transmission curve stable. Details of the cell setup and the absorption behaviour of the caesium gas at 895 nm are given by Fischer et al. (2013c). For imaging a high-speed camera of the type Phantom v1610 from the company Vision Research is used. It provides high frame rates up to 1 Mfps at a frame size of 128×16 px, which is reduced to the used frame rate of 100 kfps at a frame size of 128×128 px. Based on the single-pixel resolution of the high-speed camera-based FM-DGV, one velocity value per

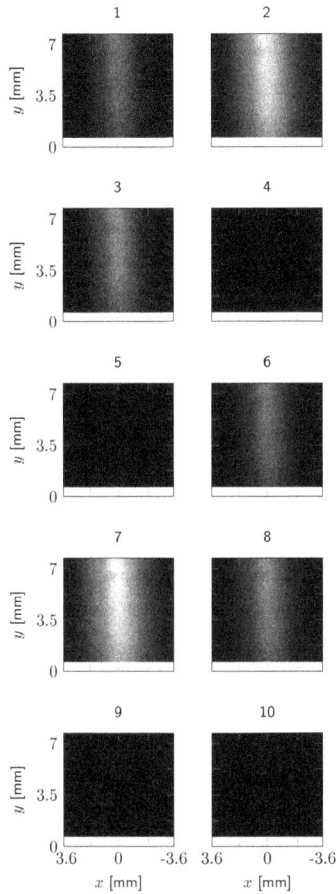

Figure 4. Sequence of 10 subsequent raw pictures, i.e. one modulation period according to $f_\mathrm{m} = 100\,\mathrm{kHz}$. The light intensity is modulated according to the modulation laser frequency (see Fig. 2).

pixel can be measured. This yields a maximum data rate of $128 \cdot 128 \cdot f_\mathrm{m} \approx 160\,\mathrm{MSamples\,s^{-1}}$ for the proposed setup.

The imaging system, which consists of a Keplerian telescope setup, contains the caesium absorption cell between the two lenses and provides a magnification $M = 1:2$. The observed region spans $7.2 \times 7.2\,\mathrm{mm^2}$ in the x–y plane and is located above the central orifice of the liner beginning at $y > 600\,\mu\mathrm{m}$ in order to avoid the detection of reflexes from the liner surface. The volumetric (3-D) velocity field is finally obtained by traversing the light sheet together with the camera along the z direction in eight steps of $450\,\mu\mathrm{m}$, which is depicted in Fig. 5b. For averaging, five repeated acquisitions, each with a duration of $18\,\mathrm{s}$ (i.e. 1.8×10^5 frames), are recorded and evaluated for each z position. Note that the point of origin of the used coordinate system is located right in the centre of the orifice.

3.3 Data evaluation

In principle, the FM-DGV technique combined with a camera allows a high spatial resolution up to the size of a single

Table 1. Specifications of the measurement system including the laser and the high-speed camera.

Max. laser power	$600\,\mathrm{mW}$
Light-sheet thickness	$300\,\mu\mathrm{m}$
Modulation rate f_m	$10\,\mathrm{kHz}$
Pixel size	$28 \times 28\,\mu\mathrm{m^2}$
Camera resolution	$128 \times 128\,\mathrm{px}$
Binned resolution	$34 \times 34\,\mathrm{px}$
Traversing	$8 \times 450\,\mu\mathrm{m}$
Magnification M	$1:2$
Measurement volume	$7.2 \times 7.2 \times 3.6\,\mathrm{mm^3}$
Spatial resolution	$224 \times 224 \times 300\,\mu\mathrm{m^3}$

pixel. However, a binning of 4×4 pixel is applied here to reduce the influence of fluctuations of the scattered light due to spatial averaging (Fischer et al., 2008). The use of this binning reduces the data rate to $10\,\mathrm{MSamples\,s^{-1}}$ and yields a spatial resolution of $224 \times 224 \times 300\,\mu\mathrm{m^3}$. Since its dimension in each direction is about 1 order of magnitude smaller than the orifice diameter, the spatial resolution is sufficient.

With the presented setup it is possible to measure one velocity component. According to Eq. (2), the direction of the measured velocity component v_1 is along the vector $(o_1 - i)$ as sketched in Fig. 5b. In the Sect. 3.4 the results of the one-component measurement setup are used for validation and uncertainty analysis. Since a one-component measurement is not sufficient to understand the flow behaviour at the bias flow liner, a two-component measurement is also performed by adding a second measurement at the same positions but with an anti-parallel observation direction $o_2 = -o_1 = (0, 0, 1)$, which enables the measurement of the velocity component v_2. Through the coordinate transformation

$$\begin{pmatrix} v_y \\ v_z \end{pmatrix} = \begin{pmatrix} \cos\varphi & -\sin\varphi \\ \sin\varphi & \cos\varphi \end{pmatrix} \begin{pmatrix} v_1 \\ v_2 \end{pmatrix} \quad \text{with} \quad \varphi = 45^\circ, \tag{5}$$

the measured velocity components v_1 and v_2 are thereby transformed into the Cartesian components v_y in y direction and v_z in z direction (Fischer et al., 2013b). The results of the two-component measurement are presented in Sect. 4. An overview of the measurement system specifications is given in Table 1.

3.4 Characterization measurement

In order to validate the capability of the measurement system for aeroacoustic investigations at a bias flow liner, the system is examined with respect to two requirements. First, the measurement system has to be able to resolve sound-excited oscillations of the flow velocity. At the same time, it has to allow the detection of flow turbulence down to $10^{-4}\,(\mathrm{m\,s^{-1}})^2\,\mathrm{Hz^{-1}}$ in the case of flow velocities up to $30\,\mathrm{m\,s^{-1}}$ (Haufe et al., 2013). Thus, it also has to provide

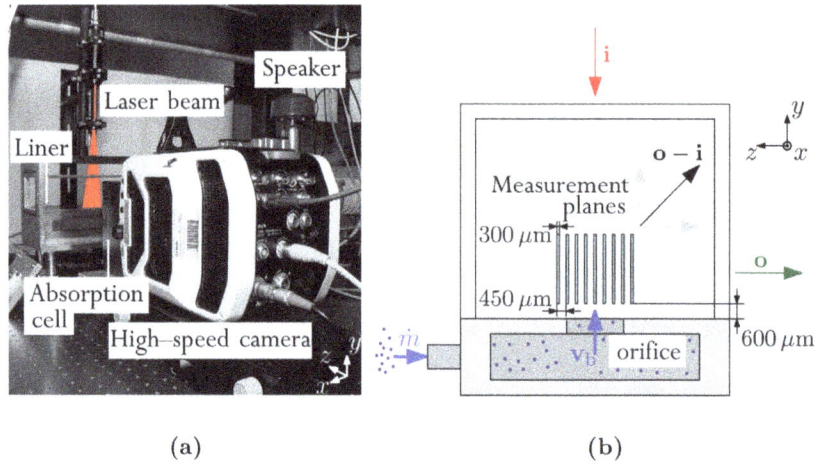

(a)   (b)

Figure 5. Measurement setup: (**a**) the test rig is acoustically excited by a speaker, and the sound–flow interaction at the liner is measured using the FM-DGV system in combination with a high-speed camera. The suction unit for generating the grazing flow v_g is installed behind the speaker. (**b**) The laser light sheet and the imaging system are traversed in z direction in steps of $450\,\mu$m. In the nine resulting measurement planes, the velocity component v in the direction of the vector $(o - i)$ is measured. Note that the sketch is not drawn to scale in order to gain a better clarity.

a small standard deviation of the measured velocity smaller than $10\,\mathrm{mm\,s^{-1}}$, assuming the used measurement time of $5 \times 18\,\mathrm{s}$.

Due to the known sound excitation frequency $f_{ac} = 867\,\mathrm{Hz}$ the fulfilment of the first requirement can be validated by examining the spectrum of the velocity with respect to flow velocity oscillations at f_{ac} and its higher harmonics up to $5 f_{ac}$. Therefore the averaged spectrum at the measurement point at $(x, y, z) = (0, 3.6, 0\,\mathrm{mm})$ is shown in Fig. 6a as an example, which is calculated using a fast Fourier transformation (FFT) of $5 \times 180\,000$ velocity samples corresponding to a measurement time of $90\,\mathrm{s}$. The spectrum shows the expected characteristic peaks at f_{ac} and its higher harmonics up to $5 f_{ac}$, which validates the capability of the measurement system to detect sound-excited oscillations of the flow velocity. Furthermore, the absolute value in the spectrum decreases with increasing frequency. This is typical for flow turbulence indicating the decay of the vortices according to Kolmogorov's theory and is different from the white noise due to the measurement uncertainty (Fischer et al., 2013a). For frequencies over $3\,\mathrm{kHz}$ the spectrum converges towards a constant value $\sqrt{\mathrm{mean}(|\mathrm{FFT}(v)|^2)} = \sigma/\sqrt{900\,000} = 1.5\,\mathrm{mm\,s^{-1}}$, which is a measure of the velocity standard deviation after averaging over $900\,000$ samples, i.e. $90\,\mathrm{s}$. As a result, flow velocity fluctuations down to a power density of $(1.5\,\mathrm{mm\,s^{-1}})^2 \cdot 18\,\mathrm{s} = 40.5\,(\mathrm{mm\,s^{-1}})^2\,\mathrm{Hz^{-1}}$ (based on a window size of $18\,\mathrm{s}$) are successfully resolved in the frequency range up to $5\,\mathrm{kHz}$. Note that σ denotes the standard deviation of all velocity samples in case of no averaging.

The directly calculated standard deviation σ_v of the mean flow velocity from the velocity series is shown in Fig. 6b versus the averaging time N/f_m with N as number of averaged

samples. Further the estimated standard deviation $\sigma/\sqrt{N}$ for the assumption of white noise is shown, too. In comparison σ_v decreases less with increasing averaging time than $\sigma/\sqrt{N}$. This difference is a result of the flow turbulence (see Fig. 6a), which causes non-white noise. However, for $N > 40\,000$ samples σ_v achieves the required uncertainty range $< 10\,\mathrm{mm\,s^{-1}}$. The respective theoretical uncertainty limit for the measurement system is $0.3\,\mathrm{mm\,s^{-1}}$ according to Fischer et al. (2013c). As follows, the measured value of $1.5\,\mathrm{mm\,s^{-1}}$ is 1 order of magnitude larger than the theoretical limit. This is expected, since the measured velocity standard deviation contains contributions from the flow turbulence, which is not a measurement uncertainty. Furthermore the uncertainty limit is based on a FM-DGV system with a higher modulation frequency of $100\,\mathrm{kHz}$ and an avalanche photodiode array as photodetection unit. The uncertainty is increased by fluctuations of the scattered light intensity, i.e. when the scattered light significantly changes during one modulation period of the FM-DGV system. Accordingly, a high modulation frequency (high temporal resolution) and a spatial averaging (low spatial resolution) enable a low uncertainty due to the fluctuations of the scattered light intensity (Fischer et al., 2007). For this reason it is assumed that the higher standard deviation of the camera-based system results from fluctuations of the scattered light, whose influence is increased due the lower modulation frequency of $10\,\mathrm{kHz}$ and the higher spatial resolution. This has to be investigated further in future studies and promises possible system improvements with an optimization of the measurement setup. However, the achieved velocity standard deviation is small enough to fulfil both aforementioned requirements.

Based on the spectrum the total harmonic distortion of the sound-excited periodic flow velocity oscillations is cal-

(a) (b)

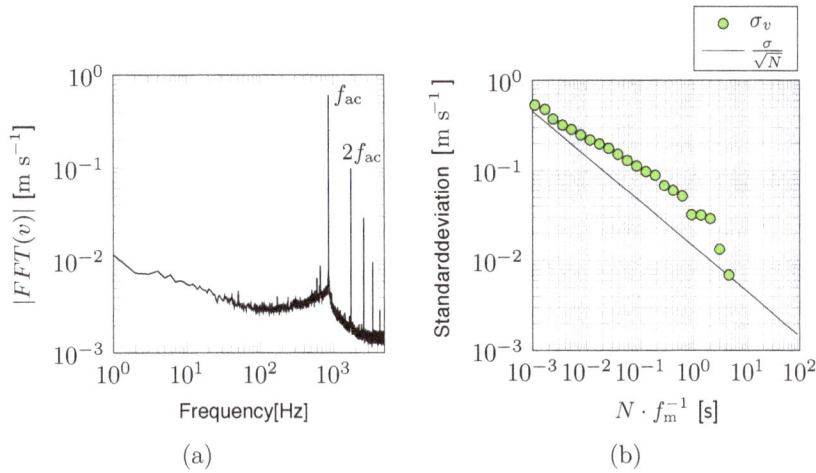

Figure 6. (a) Spectrum of the flow velocity and (b) standard deviation σ_v of the mean flow velocity, calculated directly from the velocity series by averaging over N velocity samples. Estimated standard deviation $\sigma/\sqrt{N}$ for assumption of white noise, calculated using all measured velocity samples without averaging. Both standard deviations depend on the averaging time N/f_m with the measurement rate f_m for the measurement point at $(x, y, z) = (0, 3.6, 0\,\text{mm})$.

culated to 3 %. Hence, the calculated result is larger than the measured total harmonic distortion of 0.8 % of the speaker. Consequently the transfer of the oscillations from the sound wave to the flow velocity has a non-linear response characteristic, as expected from the Navier–Stokes equations being non-linear. Thus, the sound energy is transferred into kinetic energy in the form of flow vortices, which contributes to the damping of the sound like observed by Haufe et al. (2014b) and Schulz et al. (2015). In summary the calculation of the total harmonic distortion also proves the capability of the measurement system to resolve the interaction between sound excitation and flow turbulence. As a result the system is validated to be applicable for aeroacoustic investigations at the bias flow liner.

4 Aeroacoustic measurements

The results of the aeroacoustic investigations are presented by evaluating the data of the two-component measurements. First, in order to investigate the mean flow behaviour, an overview of the mean flow velocities in y and z direction is given in Fig. 7. A more detailed view is given in Figs. 8 and 9. The averaging time is 90 s. Note that in order to improve the visibility of the measurement results the z axis is stretched in Fig. 7. The dissipation rate, i.e. the reduction of the sound wave energy at the used bias flow liner operated with the described parameters, is measured using a microphone and amounts to 0.4 of the insert sound energy.

The maximum flow velocity amounts to $\bar{v}_y = 7.2\,\text{m s}^{-1}$ in y direction and $\bar{v}_z = 3.5\,\text{m s}^{-1}$ in z direction, respectively. The magnitude of the velocity v_y is in good agreement with the estimated value of $v_b = 7\,\text{m s}^{-1}$, calculated from the controlled mass flow $\dot{m}$. Based on the suction flow $\boldsymbol{v}_g$, the mean

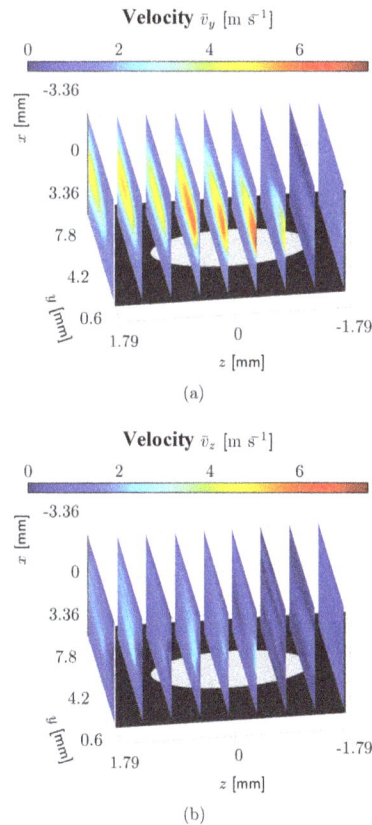

(a)

(b)

Figure 7. Mean flow velocity (a) $\bar{v}_y$ in y direction and (b) $\bar{v}_z$ in z direction.

flow is slightly tilted towards the negative x direction. Furthermore, the flow leaves the orifice slightly oblique in z direction. This is most likely attributed to the asymmetric feed-

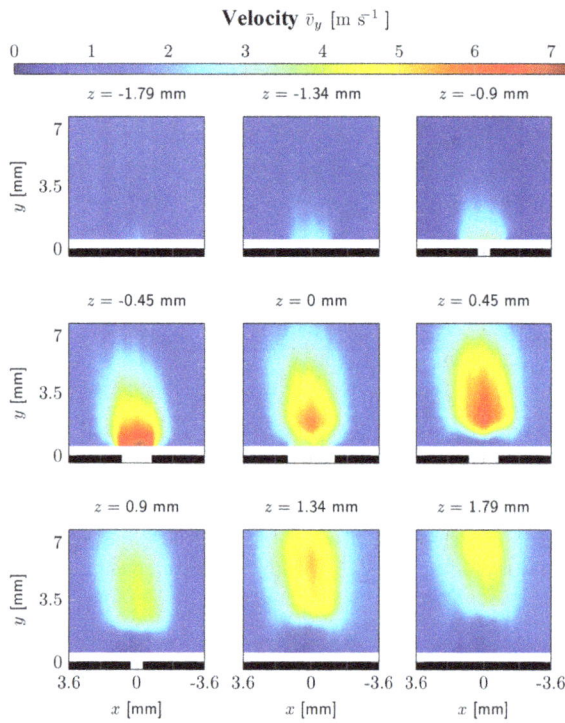

Figure 8. Mean flow velocity $\bar{v}_y$ in y direction.

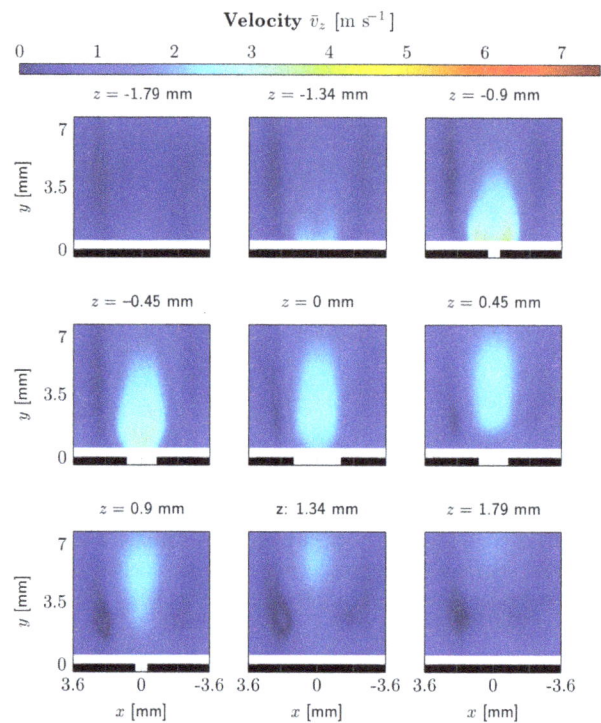

Figure 9. Mean flow velocity $\bar{v}_z$ in z direction.

ing of the bias flow only from the positive z direction and the reflection of the flow at the inner wall of the liner cavity. Because of this, the distance of the velocity maximum to the liner surface increases for each measurement plane with increasing z.

The existence of sound-excited flow oscillations at a single measurement point was already shown in Sect. 3.4 and will be discussed here further with respect to their spatial and temporal behaviour in the 3-D measurement volume. Such investigations of the excited oscillations are possible by using a phase-resolved analysis of the flow oscillation for the known sound signal with the period f_{ac}^{-1}. The phase-averaged velocities in the measurement plane at $z = 0$ mm are shown in Fig. 10 for the phase angle $\varphi_{ac} = 0 \ldots 2\pi$ of the acoustic excitation signal. Therefore the velocity samples are divided into 16 groups of the same acoustic phase angle, according to a phase resolution of $\pi/8$. Then the samples of each group are averaged. As a result, two maxima occur, whose distance of 0.4 cm is 2 orders of magnitude smaller than the acoustic wavelength of 38 cm. Hence, the measured velocity oscillations are dominated by flow oscillations resulting from vortices. The amplitudes of the sound-excited oscillation velocity are presented in Fig. 11 for both velocity components. A more detailed view of both velocities is shown in Figs. 12 and 13. The calculation of the amplitudes is based on the Fourier coefficient of the time series at f_{ac}. The maximum oscillation amplitude in y direction amounts to $3.8\,\mathrm{m\,s^{-1}}$. Accordingly, it is more than 1 order of mag-

nitude larger than the amplitude of the APV of $48\,\mathrm{mm\,s^{-1}}$. Note that the given amplitude of the APV is based on the used sound pressure level of 120 dB, a temperature of 20 °C and the assumption of a plane wave. Consequently, the detected oscillations of the flow velocity are mainly sound-excited and prove the interaction of sound and flow velocity.

Like detected for the mean flow velocity field, regions with a large oscillation amplitude exhibit an increasing distance to the liner surface in positive z direction. Also, sound-excited oscillations are located in the vicinity of the orifice rim, which is a region of high vorticity (Heuwinkel et al., 2010). Thus, an enhancement of the mean flow velocity and the vorticity coincide with an enhancement of the energy transfer from the sound into the flow. Furthermore regions of a high vorticity such as the orifice rim exhibit also a large velocity oscillation amplitude. This coincidence implies the generation of vortices fluctuating with the acoustic excitation frequency. According to the time evolution during one acoustic period in Fig. 10, the vortices detach from the liner surface and decompose continuously with increasing distance to the liner surface. The decomposition into smaller vortices at higher frequencies can also be seen in the velocity spectrum in Fig. 6a due to the decreasing value in the spectrum. This behaviour implies further the proceeding decomposition of the vortices, according to Kolmogorov's theory, and their final dissipation into heat. Hence it is shown that the sound damping performance of the bias flow liner is partially based on these sound-exited flow vortices and their dissipation.

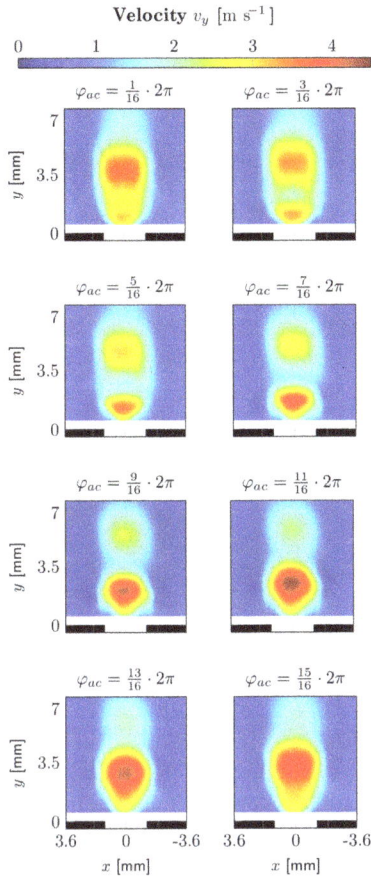

Figure 10. Phase-resolved oscillation of the flow velocity v_y at $z = 0$ mm for different phase angles φ_{ac} of the sound signal.

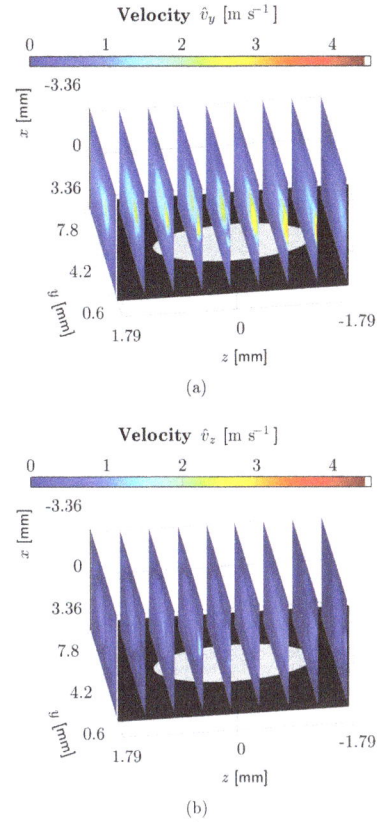

Figure 11. Amplitude of the sound-excited flow velocity oscillations (**a**) in y direction and (**b**) in z direction.

5 Simultaneous three-component measurements with a single camera

In order to understand the complex flow behaviour at the bias flow liner with respect to flow turbulence, an enhancement of the camera-based FM-DGV system is needed, which enables a simultaneous measurement of all three flow velocity components in one measurement plane. Thus, PIV algorithms are additionally applied to determine the in-plane velocity components. This allows a simultaneous three-component measurement of a planar velocity field using a single camera and, thus, increases the data rate when measuring in the 3-D region of interest. Due to this 3-D three-component measurement, the decomposition of the measured superposition of velocity signals resulting from APV and flow velocity oscillations using the Helmholtz–Hodge decomposition could be performed (Petronetto et al., 2010). However, here only a planar measurement at the position $z = 0$ is presented as an example.

The PIV evaluation of the full-resolution images spanning 128×128 px is performed using the open-source software PIVlab. The high seeding concentration, needed for FM-DGV, prohibits a resolution of single particles preferably

used for PIV. However, since the PIV algorithms calculate the cross-correlation function of shifted interrogation windows it is merely needed to resolve moving structures with a sufficient image contrast. The size of the interrogation windows is decreased over three passes in order to increase the precision and the spatial resolution of the PIV results. In the first pass of the evaluation routine the window size is chosen to 35×35 px and then is decreased over two more passes to 15×15 px and 10×10 px. The interrogation window overlap is in all cases adjusted to $50\,\%$, which corresponds to a grid of 5×5 px in the third pass of the evaluation routine. This finally results in a spatial resolution of $280 \times 280 \times 300\,\mu\mathrm{m}^3$ according to the measurement setup in Sect. 3. Due to the symmetric modulation of the laser light frequency nearly around the minimum of the non-linear transmission curve of the caesium, the intensity signal is dominated by the second harmonic at $2 f_\mathrm{m}$ as sketched in Fig. 2c. Based on that, two bright images per modulation period with maximum intensity are present (see Fig. 4). These images can be used for the PIV evaluation, obtaining one measurement result per modulation period. Consequently, the PIV measurement rate equals the measurement rate of the FM-DGV and amounts to $10\,\mathrm{kHz}$.

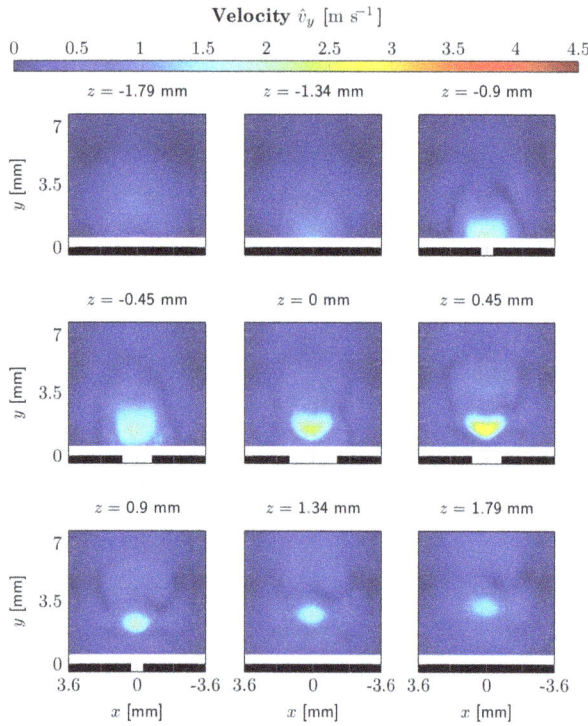

Figure 12. Amplitude of the sound-excited flow velocity oscillations in y direction.

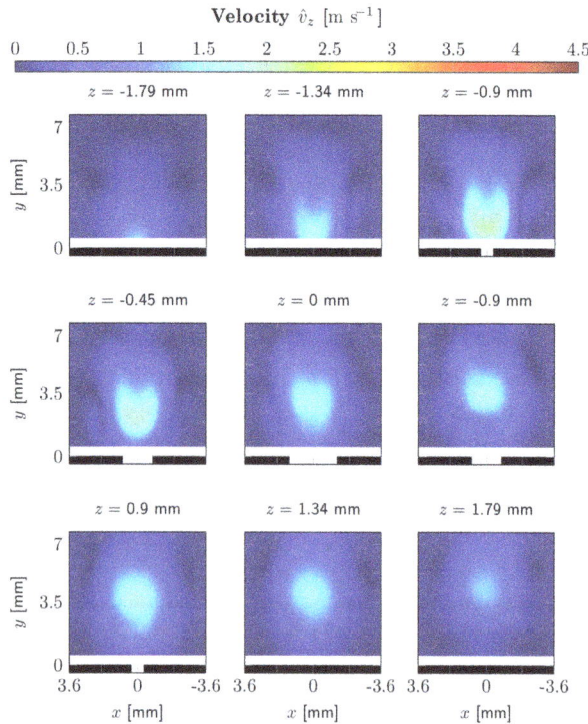

Figure 13. Amplitude of the sound-excited flow velocity oscillations in z direction.

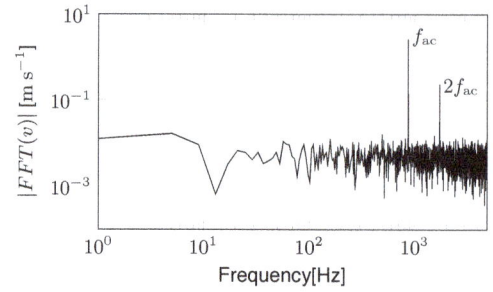

Figure 14. Amplitude spectrum of the flow velocity v_y at $(x, y, z) = (0, 3.6, 0\,\text{mm})$ generated using PIV evaluation.

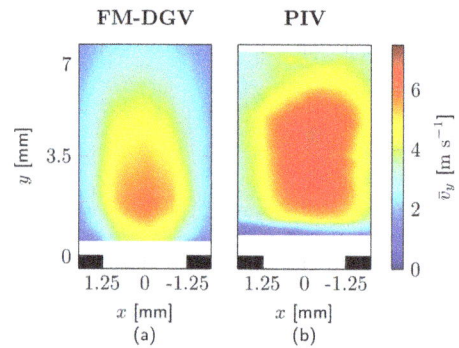

Figure 15. Mean flow velocity $\bar{v}_y$ at $z = 0\,\text{mm}$ measured with (**a**) FM-DGV and (**b**) PIV.

Similar to the validation of the camera-based FM-DGV data in Sect. 3.4, the capability of the PIV evaluation for aeroacoustic investigations needs to be validated. For that reason, the detection of sound-excited flow velocity oscillations is investigated by calculating the averaged spectrum of the velocity $|\text{FFT}(v)|$ at the point $(x, y, z) = (0,3.6,0\,\text{mm})$. This spectrum is shown in Fig. 14. In accordance with the FM-DGV data, the spectrum exhibits the characteristic peaks at $f_{\text{ac}} = 867\,\text{Hz}$ and $2f_{\text{ac}}$. As a result, the capability of the PIV evaluation to resolve sound-excited flow oscillations is proved. However, it is not yet possible to resolve the third or higher harmonics of f_{ac} as well as the flow turbulence. This is due to the velocity standard deviation of $40\,\text{mm s}^{-1}$, which is almost 30 times larger than the standard deviation of the FM-DGV at the investigated measurement point. Note that the uncertainties given here are calculated for two-component velocity data. The larger uncertainty of the PIV system is based on the described interrogation setup, which is optimized to correlate the large particle displacements caused by the mean flow velocity. Hence, the weak velocity gradients of the turbulence cannot be resolved (Keane and Adrian, 1990) properly. In contrast, the mean flow velocity can be resolved, which is depicted in Fig. 15. Here a comparison of the mean flow velocity $\bar{v}_y$ obtained from the FM-DGV measurement (a) and from the PIV evaluation (b) is presented. The results differ in their shape and positioning relative to the orifice.

Only the maximum flow velocities $v_y = 7.1\,\mathrm{m\,s^{-1}}$ due the FM-DGV evaluation and $v_y = 7.2\,\mathrm{m\,s^{-1}}$ due the PIV evaluation are in the same range.

The main reason for the disturbance of the PIV evaluation is considered to be the high seeding particle concentration, which causes a low image contrast and makes correlatable structures barely detectable (see Fig. 4). However, also the out-of-plane movement of particles is problematic, when particles enter or leave the measurement volume during the time of $50\,\mu\mathrm{s}$ between two correlated images (Keane and Adrian, 1990; Nobach and Bodenschatz, 2009). The displacement of particles in z direction is $175\,\mu\mathrm{m}$ due to the velocity $v_z = 3.5\,\mathrm{m\,s^{-1}}$, and thus the displacement is more than half of the light-sheet thickness. Hence, it is most likely that particles occur only in one of the two subsequent images, which leads to a biased estimation of the velocity. Because of the pending issues, the combination of both evaluation techniques yields an increasing uncertainty for all measured velocity components caused by an error propagation. Since two velocity components measured by PIV and one component measured by FM-DGV are combined, the resulting uncertainty calculates to $(\sigma_x, \sigma_y, \sigma_z) = (40, 40, 41)\,\mathrm{mm\,s^{-1}}$ based on a coordinate transformation (Schlüßler et al., 2015).

Consequently, the results of the combined FM-DGV–PIV evaluation are not considered further. The used PIV setup and evaluation parameters need to be optimized in order to perform simultaneous three-component measurements. For instance, the reduction of the seeding particle concentration as well as an enhancement of the light-sheet thickness would improve the quality of the PIV measurements. However, both approaches would yield a higher measurement uncertainty for the FM-DGV data due to light intensity fluctuations and a lower signal-to-noise ratio, respectively. As a third approach the additional use of fluorescent seeding particles is possible. This approach would yield better PIV results without an impact on the FM-DGV results, but it would increase the complexity of the measurement system as well. In summary, the combined FM-DGV–PIV requires further investigation, which is valuable due to the possibility of simultaneous high-speed 2-D three-component measurements using only one camera.

6   Conclusions

The applicability of the high-speed camera-based FM-DGV system for aeroacoustic investigations is successfully proven. The use of a high measurement rate of $10\,\mathrm{kHz}$ and a spatial resolution of $224 \times 224 \times 300\,\mu\mathrm{m}^3$ allows the detection of small-scale velocity oscillations in a frequency range up to $5\,\mathrm{kHz}$ and down to a power density of $40.5\,(\mathrm{mm\,s^{-1}})^2\,\mathrm{Hz^{-1}}$. Furthermore the measurement of two velocity components in a 3-D region of interest ($7.2 \times 7.2 \times 3.6\,\mathrm{mm}^3$) offers investigations of the maximum mean flow velocity of $7\,\mathrm{m\,s^{-1}}$ and the sound–flow interaction. By performing measurements with

this system at a bias flow liner test rig, the detection of sound-excited flow oscillations and of flow turbulence is achieved. Due to this, the periodic (sound-excited flow oscillations) and aperiodic (flow turbulence) flow velocity fluctuations are examined; thus, the energy transfer of the sound into the flow, which contributes to the damping effect of the liner, is shown. Hence, the temporal as well as the spatial behaviour of the flow is investigated.

Additionally an approach for simultaneous three-component measurement is shown using an additional PIV evaluation. However, due to the high seeding particle density the results of the PIV evaluation have high uncertainty, because the setup for the measurement using FM-DGV and PIV in combination needs to be improved.

Acknowledgements. The authors thank the German Research Foundation (DFG) for funding the projects CZ 55/25-3 and EN 797/2-3. Many thanks go to André Döring.

References

Adrian, R. J.: Twenty years of particle image velocimetry, Exp. Fluids, 39, 159–169, 2005.

Bechert, D. W.: Sound absorption caused by vorticity shedding, demonstrated with a jet flow, J. Sound Vib., 70, 389–405, 1980.

Eldredge, J. D. and Dowling, A. P.: The absorption of axial acoustic waves by a perforated liner with bias flow, J. Fluid Mech., 485, 307–335, 2003.

Fischer, A., Büttner, L., Czarske, J., Eggert, M., Grosche, G., and Müller, H.: Investigation of time-resolved single detector Doppler global velocimetry using sinusoidal laser frequency modulation, Meas. Sci. Technol., 18, 2529–2545, 2007.

Fischer, A., König, J., and Czarske, J.: Speckle noise influence on measuring turbulence spectra using time-resolved Doppler global velocimetry with laser frequency modulation, Meas. Sci. Technol., 19, 125402, doi:10.1088/0957-0233/19/12/125402, 2008.

Fischer, A., König, J., Czarske, J., Peterleithner, J., Woisetschläger, J., and Leitgeb, T.: Analysis of flow and density oscillations in a swirl-stabilized flame employing highly resolving optical measurement techniques, Exp. Fluids, 54, 1622, doi:10.1007/s00348-013-1622-3, 2013a.

Fischer, A., König, J., Czarske, J., Rakenius, C., Schmid, G., and Schiffer, H. P.: Investigation of the tip leakage flow at turbine rotor blades with squealer cavity, Exp. Fluids, 54, 1462, doi:10.1007/s00348-013-1462-1, 2013b.

Fischer, A., König, J., Haufe, D., Schlüßler, R., Büttner, L., and Czarske, J.: Optical multi-point measurements of the acoustic particle velocity with frequency modulated Doppler global velocimetry, J. Acoust. Soc. Am., 134, 1102–1111, 2013c.

Fischer, A., Schlüßler, R., Haufe, D., and Czarske, J.: Lock-in spectroscopy employing a high-speed camera and a micro-scanner for volumetric investigations of unsteady flows, Opt. Lett., 39, 5082–5085, 2014.

Hann, D. B. and Greated, C. A.: The measurement of flow velocity and acoustic particle velocity using particle-image velocimetry, Meas. Sci. Technol., 8, 1517–1522, 1997.

Hann, D. B. and Greated, C. A.: The Measurement of Sound Fields Using laser Doppler Anemometry, Acta Acust., 85, 401–411, 1999.

Haufe, D., Schlüßler, R., Fischer, A., Büttner, L., and Czarske, J.: Optical multi-point measurement of the acoustic particle velocity in a superposed flow using a spectroscopic laser technique, Meas. Sci. Technol., 23, 085306, doi:10.1088/0957-0233/23/8/085306, 2012.

Haufe, D., Fischer, A., Czarske, J., Schulz, A., Bake, F., and Enghardt, L.: Multi-scale measurement of acoustic particle velocity and flow velocity for liner investigations, Exp. Fluids, 54, 1569, doi:10.1007/s00348-013-1569-4, 2013.

Haufe, D., Pietzonka, S., Schulz, A., Bake, F., Enghardt, L., Czarske, J., and Fischer, A.: Aeroacoustic near-field measurements with microscale resolution, Meas. Sci. Technol., 25, 105301, doi:10.1088/0957-0233/25/10/105301, 2014a.

Haufe, D., Schulz, A., Bake, F., Enghardt, L., Czarske, J., and Fischer, A.: Spectral analysis of the flow sound interaction at a bias flow liner, Appl. Acoust., 81, 47–49, 2014b.

Henning, A., Kröber, S., and Koop, L.: Acoustic particle velocity measurements: a cross comparison between modern sensor technologies, Meas. Sci. Technol., 24, 085303, doi:10.1088/0957-0233/24/8/085303, 2013.

Heuwinkel, C., Piot, E., Micheli, F., Fischer, A., Enghardt, L., Bake, F., and Röhle, I.: Characterization of a Perforated Liner by Acoustic and Optical Measurements, in: 16th AIAA/CEAS Aeroacoustics Conf., 7 June–9 June 2010, Stockholm, Sweden, p. 3765, 15 pp., 2010.

Keane, R. D. and Adrian, R. J.: Optimization of particle image velocimeters . Part I : Double pulsed systems, Meas. Sci. Technol., 1, 1202–1215, 1990.

Meier, A. H. and Roesgen, T.: Imaging laser Doppler velocimetry, Exp. Fluids, 52, 1017–1026, 2012.

Müller, H., Eggert, M., Czarske, J., Büttner, L., and Fischer, A.: Single-camera Doppler global velocimetry based on frequency modulation techniques, Exp. Fluids, 43, 223–232, 2007.

Nobach, H. and Bodenschatz, E.: Limitations of accuracy in PIV due to individual variations of particle image intensities, Exp. Fluids, 47, 27–38, 2009.

Petronetto, F., Paiva, A., Lage, M., and Lewiner, T.: Meshless Helmholtz-Hodge Decomposition, IEEE Trans. Vis. Comput. Graph., 16, 338–349, 2010.

Pope, S. B.: Turbulent flows, Cambridge University Press, ISBN: 0-521-59125-2, 2000.

Rupp, J., Carrotte, J., and Spencer, A.: Interaction Between the Acoustic Pressure Fluctuations and the Unsteady Flow Field Through Circular Holes, J. Eng. Gas Turbines Power, 132, 061501, doi:10.1115/1.4000114, 2010.

Schlüßler, R., Bermuske, M., Czarske, J., and Fischer, A.: Simultaneous three-component velocity measurements in a swirl-stabilized flame, Exp. Fluids, 56, 183, doi:10.1007/s00348-015-2055-y, 2015.

Schulz, A., Fischer, A., and Bake, F.: Measurement of the acoustic particle velocity under grazing flow using, in: 17th Int. Symp. Appl. Laser Tech. to Fluid Mech., 1–9, Lisbon, Portugal, 2014.

Schulz, A., Haufe, D., Czarske, J., Fischer, A., Bake, F., and Enghardt, L.: Spectral Analysis of Velocity Fluctuations in the Vicinity of a Bias Flow Liner With Respect to the Damping Efficiency, Acta Acust. united with Acust., 101, 24–36, 2015.

Taylor, K. J.: Absolute measurement of acoustic particle velocity, J. Acoust. Soc. Am., 59, 691–694, 1976.

Thompson, M. W. and Atchley, A. A.: Simultaneous measurement of acoustic and streaming velocities in a standing wave using laser Doppler anemometry, J. Acoust. Soc. Am., 117, 1828–1838, 2005.

Thurow: MHz Rate Planar Doppler Velocimetry in Supersonic Jets, in: 42nd AIAA Aerosp. Sci. Meet. Exhib., 17 pp., doi:10.2514/6.2004-23, Reno, 2004.

Voigt, A., Bayer, C., Shirai, K., Büttner, L., and Czarske, J.: Laser Doppler field sensor for high resolution flow velocity imaging without camera., Appl. Opt., 47, 5028–5040, 2008.

Watmuff, J. H.: An investigation of the constant-temperature hot-wire anemometer, Exp. Therm. Fluid Sci., 11, 117–134, 1995.

Wernet, M. P.: Planar particle imaging Doppler velocimetry: a hybrid PIV/DGV technique for three-component velocity measurements, Meas. Sci. Technol., 15, 2011, 18 pp., 2004.

Willert, C. E., Hassa, C., Stockhausen, G., Jarius, M., Voges, M., and Klinner, J.: Combined PIV and DGV applied to a pressurized gas turbine combustion facility, Meas. Sci. Technol., 17, 1670–1679, 2006.

Zhao, D., Ang, L., and Ji, C.: Numerical and experimental investigation of the acoustic damping effect of single-layer perforated liners with joint bias-grazing flow, J. Sound Vib., 342, 152–167, 2015.

23

Enhanced wavelength-selective absorber for thermal detectors based on metamaterials

Astrit Shoshi[1], Thomas Maier[2], and Hubert Brueckl[1]

[1]Center for Integrated Sensor Systems, Danube University Krems, 2700 Wr. Neustadt, Austria
[2]Molecular Diagnostics, Austrian Institute of Technology GmbH, 1220 Vienna, Austria

Correspondence to: Astrit Shoshi (astrit.shoshi@donau-uni.ac.at)

Abstract. The dissipative electromagnetic energy absorption of tailored metamaterials can be exploited to improve the spectral sensitivity and selectivity of thermal detectors. The desired detector characteristics are engineered by tuning the single- or multiband absorption by resonance frequency, magnitude, and spectral bandwidth, strongly depending on the geometrical design of metamaterials. Here, the optical absorption properties of trilayer and multilayer resonant structures are investigated by numerical simulations. We consider isotropic, i.e., polarization-independent, disk-shaped absorber elements consisting of alternating aluminium and aluminium nitride layers of nanometer thicknesses, thus representing low-mass absorbers. Trilayer absorbers show spectral resonances at wavelengths between 2 and 6 μm, reaching near-unity absorption with peak bandwidths ranging from 0.45 to 1.05 μm. The absorption characteristics remain almost unchanged for radiation with an oblique incidence angle up to 40°. Resonant structures of multilayer absorber elements show besides spectral broadening a dual-band perfect absorption, which are suitable for simultaneous multispectral infrared imaging.

1 Introduction

Metamaterials are artificial composite structures which exhibit physical properties different from the intrinsic properties of the individual material components (Veselago, 1968). Unusual optical effects such as the negative refractive index (Shelby et al., 2001) and electromagnetic cloaking (Schurig et al., 2006) have been observed, which are hardly accessible in naturally occurring materials (Pimenov et al., 2007). Such properties are derived from the resonant nature of engineered building units with feature size smaller than the wavelength of interest. In general, periodically arranged metal-dielectric structures with unit cell dimensions in the subwavelength regime are employed to independently tune the electric (Pendry et al., 1996) and magnetic (Pendry et al., 1999) resonances evoked by the incident radiation. The circulating surface currents induced by plasmon resonance in the metallic layer accompanied by the displacement field in the dielectric layer (Dayal and Ramakrishna, 2012) can be manipulated by changing the geometrical and material parameters. As an effective medium, metamaterials are characterized by homogeneous parameters such as the complex

effective electric permittivity $\varepsilon(\omega) = \varepsilon_1 + i\varepsilon_2$ and magnetic permeability $\mu(\omega) = \mu_1 + i\mu_2$ (Smith and Pendry, 2006). In research, much attention has been paid to the real part of ε and μ to create materials with negative refractive index, while at the same time minimizing the undesired losses. Similarly, the imaginary loss terms ε_2 and μ_2 can be engineered to achieve high attenuation and consequently large absorption. By independently manipulating resonances in ε and μ, it is possible to effectively absorb both the incident electric and magnetic field. Moreover, the impedance of the metamaterial, $Z = \sqrt{\varepsilon(\omega)/\mu(\omega)}$, can be matched to free space, giving rise to minimized reflectivity (Landy et al., 2008). The perfect metamaterial absorber (PMA) is defined to have absorption near unity.

The enhanced absorption properties of metamaterials can be exploited to tailor the spectral responsivity and selectivity of thermal sensors (Landy et al., 2008; Maier and Brueckl, 2009, 2010). Thermal sensors such as bolometer, thermopile, and pyroelectric sensors convert temperature changes caused by the absorption of incident radiation into an electrical signal. They usually have a broad spectral response. Exchangeable or fixed optical filter units are used for wavelength selection. In order to achieve a high sensor responsivity, a high absorption is required. Therefore, metamaterials directly integrated on top of thermal sensors are wavelength-selective with an efficient absorption and optimized heat energy transfer (Maier and Brueckl, 2009, 2010). With the integration of low-mass metamaterials, more compact and miniaturized thermal devices can be designed without significantly affecting their response time.

This study focuses on tailoring the optical properties and composition of micron-sized metamaterial structures based on the requirements arising from their integration as wavelength-selective PMAs in thermal sensors.

2   Materials and simulation model

As composite dielectric/metallic materials, highly conductive aluminium (Al) with low ohmic losses and non-dissipative aluminium nitride (AlN) dielectric interlayer were chosen. Al is a good thermal and electrical conductor. AlN has superior thermal and mechanical properties in the infrared. In particular, the high thermal conductivity of AlN is comparable to metals such as Al and is compared to Al oxide (Zhao et al., 2004) about 10 times higher. This is beneficial regarding the heat transfer efficiency to the energy transducer in an integrated sensor. Moreover, these materials are compatible with standard microelectronic production processes.

A simple layout of a near-unity absorber mitigates lithographic demands (Fig. 1). The disc-like top metallic layer (resonator) of an Al–AlN–Al trilayer is located in the center of the square-shaped unit cell consisting of the bottom metal and the dielectric layer of side length p. Due to the circular resonator shape, the excitation of the resonance is expected to be isotropic and, thus, independent of radiation polarization (Dayal and Ramakrishna, 2012). As will be discussed later, the trilayer absorber is additionally extended to a multilayer absorber with alternating dielectric/metallic stacks (Fig. 1). Moreover, a passivation layer, which is usually found in thermal sensors – e.g., SiO_2 in thermopiles, Si_3N_x in bolometers, or Au in pyroelectric detectors – is also considered in the layout (Fig. 1). The disk radius in the unit cell determines the areal density of the metamaterial.

Finite-element simulations based on the commercially available software package COMSOL Multiphysics were performed to analyze the optical properties of a reduced three-dimensional model (Fig. 1). The incident light is an electromagnetic plane wave propagating at normal incidence along the negative z direction. The boundary conditions for the unit cell outer walls perpendicular and parallel to the electric field are a perfect electric and perfect magnetic conductor, respectively. For the top and bottom outer walls scattering boundary conditions with and without an incident wave are used, respectively. The experimental refractive index val-

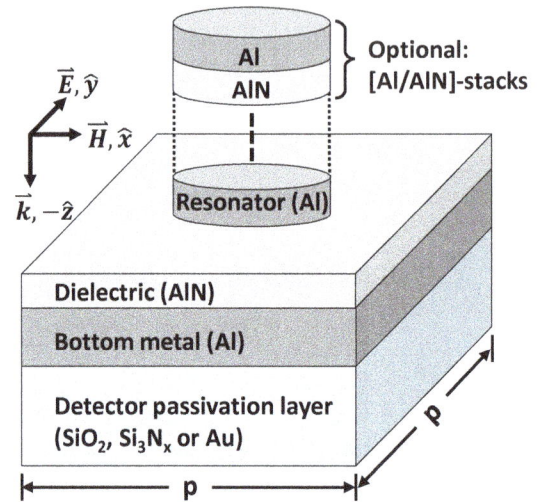

Figure 1. Schematic illustration of the unit cell in the numerical simulations.

ues of the individual materials including the dispersion relation are taken from Kischkat et al. (2012), Ordal et al. (1985), and Palik (1985) and implemented in the model. The field distributions, and the time-averaged power flow in the structure is calculated. Reflection is determined by simulating the same elementary cell geometry with all layer domain properties set to vacuum and normalized to the incident intensity. The difference in the integrated power flow at the individual layer boundaries reveal the absorption values as $A(\omega) = 1 - R(\omega) - T(\omega)$, with the frequency-dependent reflectance $R(\omega)$ and transmittance $T(\omega)$.

3   Simulation results

3.1   Electric and magnetic resonances

The incident electromagnetic plane wave propagating in negative z direction excites plasmon resonances in the top metallic layer, and thus polarization currents depending on the electrical permittivity appear inside the metal. The resonator layer at the top behaves like an electric dipole that serves as a coupler to the electric field of the incident wave. Consequently, the electromagnetic fields are concentrated within sub-wavelength regimes leading to a significant increase of the local field strengths (Fig. 2a). The origins of magnetic resonances are antiparallel currents in the metallic layers, which, together with the displacement field in the dielectric layer, result in circulating currents (Dayal and Ramakrishna, 2012; Tao et al., 2008). The current loop induces a magnetic dipole moment that can resonantly couple to the magnetic field vector of the incident light. For a strong localization of the electromagnetic energy within the metamaterial, both an electric and a magnetic dipole resonant coupling at the same frequency are necessary. This resonant coupling is of destructive nature for the reflected direction, thus eliminating reflection. In a PMA, neither reflections nor transmissions of the incident light at a given wavelength can be observed (Pendry et al., 1999; Zeng et al., 2013).

The simulated electromagnetic field distributions within the absorber element where both resonances appear are exemplarily shown in Fig. 2. Here, typical geometric parameters for resonator radius and thickness, dielectric, bottom metal, and SiO$_2$ substrate thicknesses are 500 and 50, 110, 150, and 250 nm, respectively. The distribution of the electric field in Fig. 2a shows the typical dipole excitation and spatial localization within the dielectric interlayer. The concentration of the magnetic field in the dielectric layer demonstrates the confinement of the magnetic field caused by the oscillating current loop (Fig. 2b). The inhomogeneous distribution of the current density in both metal layers is depicted in Fig. 2c. For smaller wavelengths in close proximity to the resonance, the surface current density displays a rather homogeneous distribution with significantly lower current magnitudes. By approaching the resonance wavelength, the current density experiences a progressive increase at the outer center of the resonator and reaches its maximum value at resonance wavelength. For wavelengths beyond the resonance, the current density steadily decreases. It is noteworthy that the surface current reverses its sign from positive to negative, resulting in a directional change of the circulating current loop. The normalized current density magnitude is slightly asymmetric for wavelengths smaller and larger than the resonance wavelength. The induced image charges in the bottom metal layer behave similar to those observed in the top metal layer. The current sheets in the bottom layer are antiparallel oriented and show slightly lower amplitudes compared to the top layer. In agreement to theoretical predictions, the current sheets in both metallic layers form a circulating current loop, have comparable amplitudes, and are slightly out of phase (Dayal and Ramakrishna, 2012; Zeng et al., 2013). For wavelengths far from the resonance, the whole resonant structure acts like an ordinary material with a current density behavior according to its intrinsic characteristics. Figure 2d displays the corresponding time-averaged energy flow of the electromagnetic wave in the absorber element at resonance. The energy flows from the outer border towards the center of the resonator. The energy flow in minus z direction averaged over the entire unit cell volume demonstrates the wavelength-dependent energy transfer. At resonance wavelength, an enhancement energy transfer by a factor of about 25 is observed (Fig. 2e).

3.2　Trilayer absorbers at normal light incidence

The absorption behavior of circular-shaped trilayer resonators with regard to their geometrical design is discussed. The angle of radiation incidence is 90°, i.e., normal incidence. First, the influence on the optical properties by varying the lateral dimensions as well as the layer thicknesses of the resonator, dielectric, and the bottom metal layer are

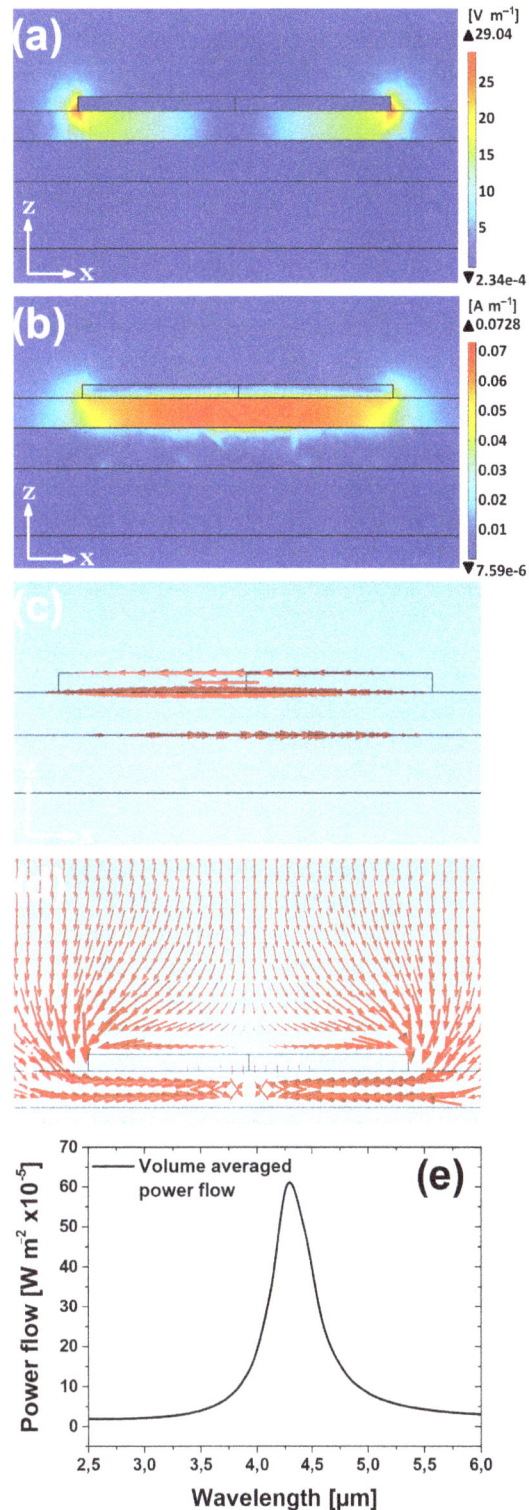

Figure 2. Optical properties of metamaterials for simultaneous electric and magnetic resonances. (**a**) Normalized electric field and (**b**) magnetic field. (**c**) Surface current density. (**d**) Time-averaged energy flow defined by the Poynting vector. (**e**) Volume-averaged energy flow (z component) over the spectral range of interest.

presented. Moreover, the effects arising from changes in the substrate layer thickness and material as well as the resonator density or filling factor are discussed.

Figure 3a shows a selection of absorption spectra for resonator radii varying between 275 and 600 nm, while all other parameters were kept constant. The layer thicknesses of the resonator, dielectric, bottom metal, and SiO$_2$ substrate are 50, 115, 50, and 250 nm, respectively. The unit cell side length is 2.1 μm. A variation of the resonator radius results in a clear shift of the resonance peak, which indicates a change in the effective dielectric permittivity of the metamaterial structure. The resonance wavelength experiences a linear red shift with increasing radius. According to a linear fit, a radius change of 10 nm causes a shift in the resonance wavelength of 80 nm. This is a good reference value in order to estimate peak shifts due to variation in the size of the absorber element, i.e., geometric tolerances caused by fabrication errors. The full width at half maximum (FWHM) of the absorption peak gives information about the coupling strength of excited oscillations in the metamaterial system. There is a linear dependence on the resonator radius ranging from 275 to 600 nm with FWHM values of 100 and 700 nm, respectively. The observed peak broadening is due to a gradual increase of dissipative losses in the metamaterial.

Figure 3b presents the calculated absorption properties for varying dielectric thicknesses between 30 and 210 nm, while all other parameters remain constant: resonator radius and height, bottom metal and substrate thicknesses, and unit cell length are 525 and 50 nm, 50 and 250 nm, and 2.1 μm, respectively. A change in the dielectric thickness influences primarily the inductance or magnetic resonance and, thus, the effective magnetic permeability of the absorber system. A varying dielectric thickness also affects the capacitive coupling to the bottom metal layer, i.e., it changes the effective permittivity, but in a less pronounced manner. Basically, lower layer thicknesses result in higher capacity and lower inductance of the system and vice versa. The resonance frequency remains constant at a resonance wavelength of 4.5 μm in a broad thickness range of 70–210 nm (Fig. 3b). This behavior reflects the relation of the resonance frequency (ω_0) to the capacitance (C) and inductance (L) of the structure, which can be considered as an undamped resonant circuit:

$$\omega_0 = 1\sqrt{LC}. \tag{1}$$

For dielectric thicknesses below 70 nm, there is an imbalance between the inductance and capacitance of the structure, which results in a significant red shift of the resonance along with a reduction in the amplitude. This behavior is characteristic for a damped oscillation. Figuratively speaking, a steady decrease of the dielectric layer thickness means a gradual approach of the antiparallel-oriented magnetic fields in the resonator and bottom metal layer generated by the two corresponding antiparallel-circulating current loops. For large dielectric thicknesses, an antiparallel orientation of the magnetic fields is energetically favored. A steady decrease of

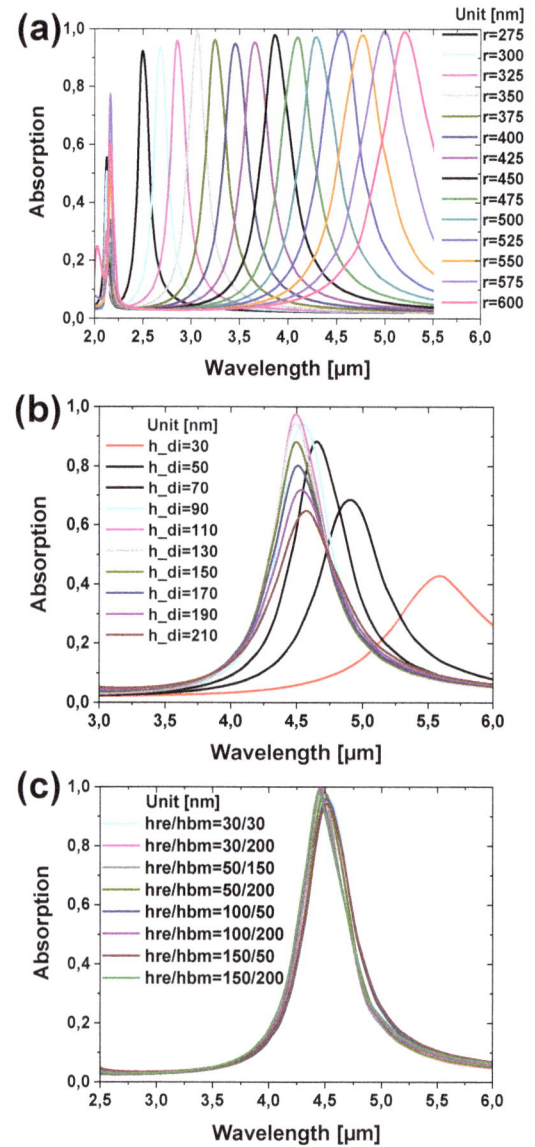

Figure 3. Numerical calculations of absorption spectra for varying **(a)** resonator radius, **(b)** dielectric layer thickness, and **(c)** resonator (hre) and bottom metal (hbm) thickness.

the dielectric layer leads to an increased perturbative interaction. As a consequence, the magnetic fields could perform precession motions when the dielectric thickness falls below a critical value ($\sim$70 nm). A precession means a distortion of flow of the current loops compared to the case for larger dielectric thicknesses. A change in the current flow can be considered as an additional inductance source. The total inductance ($L = L_g + L_p$) of the structure is then the result of the geometrical inductance (L_g) and the inductance induced by current perturbations (L_p). Due to the additional perturbative losses, the magnetic resonance frequency does not solely scale up with the geometrical inductance (L_g). By introducing an effective damping factor to Eq. (1), both a shift of the resonance frequency and a reduction of the amplitude can be explained. However, at very thin dielectric thicknesses, the direction of the magnetic field of the bottom metal layer flips and aligns parallel to the upper resonator magnetic field due to energy minimization reasons. In this configuration no magnetic resonances occur, and the absorption properties are determined by the materials permittivity. The calculations show that the amplitude reaches its maximum value of 0.96 at a dielectric thickness of 110 nm and decreases for larger and smaller layer thickness values similar to a negative parabolic function. The absorption remains within the range 90–130 nm around 0.9. The FWHM value of about 475 nm is lowest at a thickness of 110 nm, and it remains within the range of 90–130 nm almost constant. Smaller or larger layer thicknesses result in a FWHM increase up to 950 nm.

In a PMA, the layer thickness of both the resonator and the bottom metal needs optimization. The penetration depth of an electromagnetic wave into a metallic layer and, thus, the transmission $T(\omega)$ is determined by the wavelength-dependent skin effect. If the thickness of the bottom layer is smaller than the skin depth, image charges are not effectively formed in the bottom metal layer, deteriorating the magnetic resonance. This also implies a dielectric dipole resonance without a proper impedance matching. Therefore, an effective energy transfer is hindered, and unwanted reflections might occur (Dayal and Ramakrishna, 2012). The resonator (bottom layer) thickness range of 30 to 150 nm (30 to 200 nm) was analyzed. The results shown in Fig. 3c are calculated for 525 nm radius, 110 nm dielectric height, 250 nm substrate height, and 2.1 μm unit cell length. For all investigated thickness permutations, the absorption remains always larger than 0.9 and reaches the near-unity absorption value of 0.995 for the pair combination of 50 nm resonator thickness and 150 nm bottom metal thickness. The position of the resonance peak remains constant for all combinations. The FWHM has a slightly oscillatory behavior within the range of 495 ± 25 nm.

Thermal detectors are capped for instance by insulating layers such as SiO_2, Si_3N_x, or conductive Au layers. Their influence on the absorption of integrated metamaterials is also investigated in the simulation model (Fig. 1). In all simulations up to here, a SiO_2 substrate of 250 nm was assumed. SiO_2 thickness variations up to 3 μm show no noticeable changes in the absorption behavior (data not shown). In the case of a gold passivation layer variation (30 to 200 nm), a slightly decreased absorption within the range of 50 to 110 nm from near unity to 0.95 could be observed. Since the transmission $T(\omega)$ in this metamaterial structure is negligible, the marginal absorption deviations can be explained by changes in the image charge formation in the bottom layer and energy flow direction affected by the conductive Au layer.

The resonator density or filling ratio, i.e., resonator base surface area compared to the elementary cell base surface area, affects the absorption. A resonator density of more

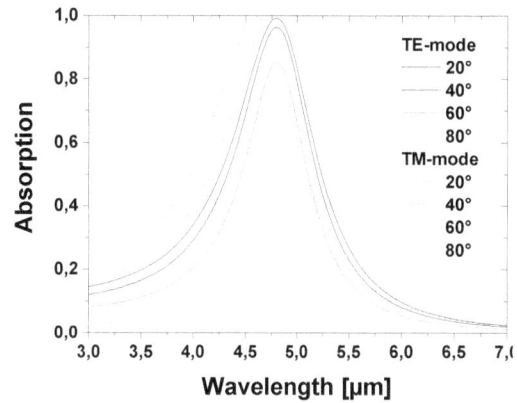

Figure 4. Absorption properties for TE and TM polarization at oblique angle of light incidence.

than 50 % is required for high absorption values, while a decreasing absorption is observed for smaller values (data not shown). Similar to our previous studies (Maier and Brueckl, 2009, 2010), there is no indication of coupling between the absorber elements down to a lateral absorber-to-absorber distance of 600 nm. Thus, for distances larger than 600 nm no influence on the resonance frequency could be observed. Due to the short-range nature of the lateral field distribution at resonant frequency (Fig. 2a and b), coupling phenomena occur typically at distances of about half the resonator diameter.

3.3 Trilayer absorbers at oblique light incidence

Conventional thermal detectors usually respond to a broad spectral range of the incident radiation. A wavelength-selective response is achieved by implementing exchangeable optical filters or microfilters in front of the detector. Such filters possess an angular transmission characteristic and show a limited selective performance above a critical angle of incidence. Here, the angular dependence of the absorption of metamaterial structures are simulated based on the transmission line model. Similar geometric parameters are chosen: the individual thicknesses of the resonator, dielectric, bottom metal are 50, 110, and 150 nm, respectively, with a resonator radius of 500 nm.

For transverse-electric (TE) or transverse-magnetic (TM) polarized light at normal incidence, both the electric and magnetic field are aligned parallel to the resonator plane and are represented by the corresponding in-plane field components. If we consider the case of TE or TM polarization at oblique light incidence, the respective in-plane component of the electric or magnetic field remain unchanged, while the respective in-plane magnetic (H_{ip}) or electric (E_{ip}) field component parallel to the resonator plane decrease with incidence angle α as $H_{ip} = H_0 \cos(\alpha)$ and $E_{ip} = E_0 \cos(\alpha)$, with H_0 and E_0 being the corresponding field magnitudes at normal incidence. In contrast to normal light incidence, we obtain at oblique radiation incidence additional out-of-

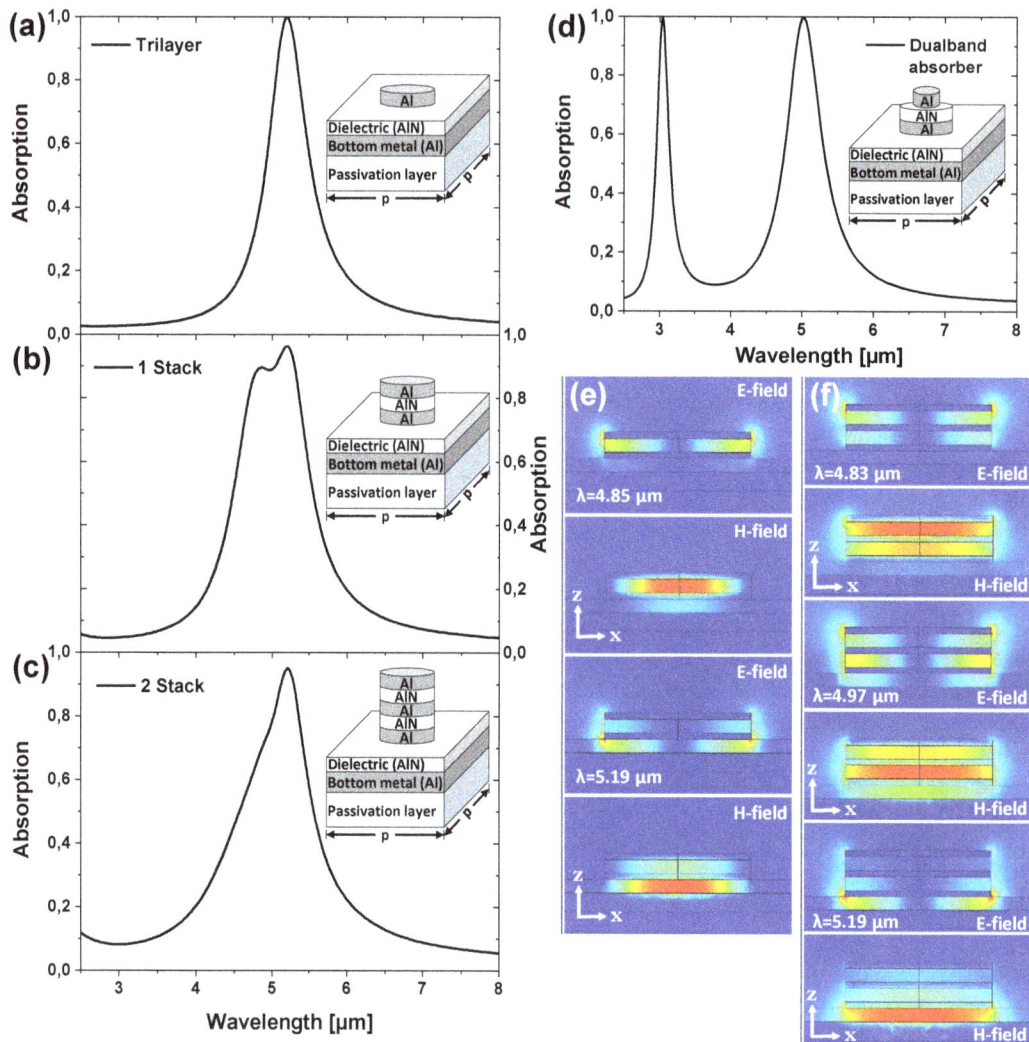

Figure 5. Multilayer absorption spectra for (**a**) trilayer, (**b**) one-stack, and (**c**) two-stack systems. (**d**) Dual-band absorber based on a modified one-stack absorber. All insets show the specific stack configurations. (**e**) and (**f**) represent the electric and magnetic field distribution in the absorber at different resonance wavelengths related to absorption spectra shown in (**b**) and (**c**), respectively.

plane field components (H_{op}, E_{op}) perpendicular to the resonator plane, which increase with incidence angle according to $H_{op}(E_{op}) = H_0(E_0)\sin(\alpha)$. Figure 4 illustrates the angular dependence of the TE and TM mode of the incident radiation with the electric and magnetic field aligned parallel to the long semi-axis, respectively. For both TE and TM polarized light the absorption remains higher than 0.94, and a slight shift in the resonance wavelength can be observed up to an angle of incidence of 40°. For the TE mode even 85 % of the intensity is absorbed at an angle of 60°. The overall nonlinear decrease reaches 45 % absorption at 80°. Remarkable for the TE mode is an unchanged resonance wavelength for all angles with a steady nonlinear decrease in the absorption. Characteristic for the TM mode is the high (> 90 %) absorption up to an angle of 60°. With rising angle, a slight decrease in the intensity of absorption is observed accompa-

nied by a resonance shift of 11 nm/° to smaller wavelengths. Due to Eq. (1), a shift in the resonance wavelength is expected when the inductance and/or capacitance of the systems changes. A gradual angle increase is comparable to a tilt of the resonant structure relative to the incident direction of the electromagnetic wave, and the initial circular shape of the resonator gradually becomes elliptical. This shape transformation from circular to elliptical leads to a decrease of the resonator surface area with increasing angle of incidence. This in turn effectively reduces the capacitance of the resonator which results according to Eq. (1) in a shift of the resonance to smaller wavelengths. However, this implies for TE (TM) polarization a higher impact on the magnetic (electric) resonances, depending on whether the electric or magnetic field is aligned parallel to the resonator plane at all angles. In case of the TE mode, the constant remaining in-plane electric field ensures even at high angles of incidence a continual resonant coupling and, thus, plasmon excitation, while the steadily decreasing in-plane magnetic field leads to a gradual magnetic decoupling. As a consequence, the resonance frequency remains constant, while the amplitude decreases nonlinearly with increasing angle of incidence due to magnetic losses. In contrast to the TM mode, the electric losses caused by a successive decoupling of the in-plane electric field with the angle of incidence result in a nonlinear peak broadening and decrease in absorption (Fig. 4). These findings are also supported by an additional simplified simulation approach considering anisotropic resonators at normal light incidence (data not shown). In this approach, a modulation of the shape from initially circular to elliptical was investigated. The short semi-axis is varied, while the long semi-axis remains constant and corresponds to the initial radius. A linear decrease of the short semi-axis with the electric field oriented parallel to it results in a clear linear resonance shift to smaller wavelengths, while the amplitude remains almost unchanged. For the antiparallel-oriented electric field, a clear amplitude decrease without significant alteration of the resonance frequency is observed. Although the results of the two approaches are not one-to-one transferable, they still share some common basic properties and help to qualitatively interpret results or to recognize tendencies.

3.4   Multilayer absorbers at normal light incidence

A multilayer system with additional dielectric/metallic layers stacked on top of the resonator disc is studied. These disc-shaped, alternating metal and dielectric layers have thicknesses of 50 and 100 nm, respectively. Note that an n-stack system represents the trilayer system with n additional AlN–Al stacks on top of the resonator. The following simulation parameters remain constant during the calculations: resonator radius and height, dielectric, bottom metal, and SiO_2 substrate height are 500 and 50, 115, 150, and 250 nm, respectively. The absorption spectrum of a zero-stack system in Fig. 5a corresponds to the trilayer system at resonance frequency and serves here as a reference. It shows a near-unity absorption at a resonance wavelength of $5.19\,\mu$m with a FWHM value of 650 nm, and the respective maximum electric (magnetic) field value is $29.04\,\mathrm{V\,m^{-1}}$ ($0.0728\,\mathrm{A\,m^{-1}}$). Characteristic for the absorption behavior of a one-stack system is a spectral peak broadening with a FWHM value of 1050 nm, which is due to the excitation of a second resonance at $4.85\,\mu$m and its superposition with the absorption peak of the trilayer at $5.19\,\mu$m (Fig. 5b). The absorption peak at $4.85\,\mu$m results from a resonant coupling of the incident light mainly in the upper stack, which is indicated by the concentration of the electric and magnetic field in the upper circular dielectric layer (Fig. 5e). A steady increase of the wavelength leads to a gradual localization displacement of both fields, resulting in a dominant field concentration in the dielectric of the trilayer at the wavelength of $5.19\,\mu$m.

Compared to the plain trilayer system, both resonances feature a marginally weaker coupling of the electric and magnetic field, which leads to slightly lower absorption peaks. The difference in the excitation frequency arises from lateral geometrical differences. The reduced metallic area in the first stack might result in confined surface currents leading to two different effective permittivities and permeabilities. The calculated resonance peak in a two-stack system basically represents the superposition of three different resonances (Fig. 5c and f). The first, second, and third resonances appear at wavelengths of 4.83, 4.97, and $5.19\,\mu$m, respectively. Similar to the one-stack system, they are due to a main localization of the electric and magnetic field in the top, middle dielectric layer, and the dielectric of the trilayer (Fig. 5f). For comparison, the maximum electric and magnetic field magnitudes of the third resonance are comparable to those calculated in the plain trilayer and one-stack system. In contrast, due to a less efficient electromagnetic coupling the respective maximum electric (magnetic) field values of the first and second resonance are about 38 % (30 %) lower, which lead to an overall nonlinear absorption decrease. Therefore, the peak of the third resonance is well pronounced compared to the first and second resonances, while the spectral peak broadening is still comparable to that observed in a one-stack system. Since the two disc-shaped stacks in the two-stack system hold the same geometry and consist of the same materials, one would expect the first two corresponding resonances to appear at the same wavelength. However, the decisive difference between them lies in their respective boundaries, which are defined by the adjacent layers surrounding the top and the middle stack. The top stack is sandwiched by air and the adjacent layers of the middle stack. Besides the interaction with the top stack, the middle stack is additionally influenced by the underlying trilayer, which finally leads to a slight resonance shift of about $0.14\,\mu$m. In contrast, the resonance wavelengths due to field localizations in the top dielectric layer remain almost constant for the one- and two-stack system (Fig. 5e and f). It should be noted at this point that, so far, only the geometrical parameters of the trilayer system have been optimized towards perfect absorption. Further optimizations by tuning the geometrical parameters of the disc-shaped stacks could minimize unwanted reflections of such multi-stack systems leading to highly efficient electromagnetic coupling at distinct wavelengths. This can result in multiple resonances with perfect absorption as exemplarily demonstrated in the following section for the one-stack system.

A modified design of the shown one-stack system was optimized toward the realization of a wavelength selective "perfect dual-band absorber" (Fig. 5d). The tailored radius of the top disc is 350 nm, while the radius of the underlying dielectric and metallic layer is 575 nm. Except for the radii, all other parameters remain identical. In accordance to Fig. 3a, two absorption peaks appear at a resonance wavelength of 3.0 and $5.0\,\mu$m for absorber radius 350 and 575 nm, respectively, although the disks are stacked in this case. The absorption for both spectral bands reaches values very close to unity (> 99.9 %). The electric and magnetic resonance properties display similar characteristics to the unmodified zero-stack metamaterial structure. As a consequence, a "perfect multiband absorber" can be designed by combining n-stack systems consisting of n resonators of different size (Dayal and Ramakrishna, 2013).

4   Conclusions

Micron-sized composite structures consisting of aluminium and aluminium nitride layer stacks have been investigated as potential wavelength-selective absorbers for thermal detectors by numerical simulations. Electric and magnetic resonances are optimized by tuning the geometrical parameters of the absorber elements, resulting in near-unity absorption (PMA). Over 94 % of the radiation intensity is absorbed at angles of incidence up to 40°. The circularly patterned resonator structures imply polarization-independent absorption. The integration of metamaterial structures in thermal detectors has been addressed by the implementation of typical passivation layers in the simulation model. No significant influences on the absorption properties could be observed. Higher-complexity metamaterials consisting of patterned multilayer resonators show a spectral broadening of the absorption peak due to additional resonances in close proximity to each other. Multilayer structures of different-sized resonators were designed, which can be utilized as multiband PMAs. The spectroscopic characterization of the presented metamaterial structures is part of current investigations.

Acknowledgements. The authors acknowledge funding from BMVIT, BMWFJ, and the Province of Lower Austria, COMET K2 Project, SF-Contract Area 5, project number C150401.

References

Dayal, G. and Ramakrishna, S. A.: Design of highly absorbing metamaterials for Infrared frequencies, Opt. Express, 20, 17503–17508, doi:10.1364/OE.20.017503, 2012.

Dayal, G. and Ramakrishna, S. A.: Design of multi-band metamaterial perfect absorbers with stacked metal–dielectric disks, J. Opt., 15, 055106, doi:10.1088/2040-8978/15/5/055106, 2013.

Kischkat, J., Peters, S., Gruska, B., Semtsiv, M., Chashnikova, M., Klinkmüller, M., Fedosenko, O., Machulik, S., Aleksandrova, A., Monastyrskyi, G., Flores, Y., and Masselink, W. T.: Mid-infrared optical properties of thin films of aluminum oxide, titanium dioxide, silicon dioxide, aluminum nitride, and silicon nitride, Appl. Optics, 51, 6789–6798, doi:10.1364/AO.51.006789, 2012.

Landy, N. I., Sajuyigbe, S., Mock, J. J., Smith, D. R., and Padilla, W. J.: Perfect metamaterial absorber, Phys. Rev. Lett., 100, 207402, doi:10.1103/PhysRevLett.100.207402, 2008.

Maier, T. and Brueckl, H.: Wavelength-tunable microbolometers with metamaterial absorbers, Opt. Lett., 34, 3012–3014, doi:10.1364/OL.34.003012, 2009.

Maier, T. and Brueckl, H.: Multispectral microbolometers for the midinfrared, Opt. Lett., 35, 3766–3768, doi:10.1364/OL.35.003766, 2010.

Ordal, M. A., Bell, R. J., Alexander, R. W., Long Jr., L. L., and Query, M. R.: Optical properties of fourteen metals in the infrared and far infrared, Appl. Optics, 24, 4493–4499, doi:10.1364/AO.24.004493, 1985.

Palik, E. D.: Handbook of Optical Constants of Solids, Academic Press, New York, USA, 1985.

Pendry, J. B., Holden, A. J., Stewart, W. J., and Youngs, I.: Extremely low frequency plasmons in metallic mesostructures, Phys. Rev. Lett., 76, 4773–4776, doi:10.1103/PhysRevLett.76.4773, 1996.

Pendry, J. B., Holden, A. J., Robbins, D. J., and Stewart, W. J.: Magnetism from conductors and enhanced nonlinear phenomena, IEEE T. Microw. Theory, 47, 2075–2084, doi:10.1109/22.798002, 1999.

Pimenov, A., Loidl, A., Gehrke, K., Moshnyaga, V., and Samwer, K.: Negative Refraction Observed in a Metallic Ferromagnet in the Gigahertz Frequency Range, Phys. Rev. Lett., 98, 197401, doi:10.1103/PhysRevLett.98.197401, 2007.

Schurig, D., Justice, J. J., Cummer, S. A., Pendry, J. B., Starr, A. F., and Smith, D. R.: Metamaterial electromagnetic cloak at microwave frequencies, Science, 314, 977–980, doi:10.1126/science.1133628, 2006.

Shelby, R. A., Smith, D., and Schultz, S.: Experimental verification of a negative index of refraction, Science, 292, 77–79, doi:10.1126/science.1058847, 2001.

Smith, D. R. and Pendry, J. B.: Homogenization of metamaterials by field averaging, J. Opt. Soc. Am. B., 23, 391–403, doi:10.1364/JOSAB.23.000391, 2006.

Tao, H., Landy, N. I., Bingham, C. M., Zhang, X., Averitt, R. D., and Padilla, W. J.: A metamaterial absorber for the terahertz regime: Design, fabrication and characterization, Opt. Express, 16, 7181–7188, doi:10.1364/OE.16.007181, 2008.

Veselago, V. G.: The electrodynamics of substances with simultaneously negative values of ϵ and μ, Sov. Phys. Uspekhi, 10, 509–514, doi:10.1070/PU1968v010n04ABEH003699, 1968.

Zeng, Y., Chen, H.-T., and Dalvit, D. A. R.: The role of magnetic dipoles and non-zero-order Bragg waves in metamaterial perfect absorbers, Opt. Express, 21, 3540–3546, doi:10.1364/OE.21.003540, 2013.

Zhao, Y., Zhu, C., Wang, S., Tian, J. Z., Yang, D. J., Chen, C. K., Cheng, H., and Hing, P.: Pulsed photothermal reflectance measurement of the thermal conductivity of sputtered aluminum nitride thin films, J. Appl. Phys., 96, 4563–4568, doi:10.1063/1.1785850, 2004.

Permissions

List of Contributors

A. Dragoneas, L. Hague and M. Grell
Physics and Astronomy, The University of Sheffield, Hicks Building, Hounsfield Road, S3 7RH, Sheffield, UK

S. Fischer
Department of Functional Materials, University of Bayreuth, Bayreuth, Germany
Corporate Technology, Siemens AG, Munich, Germany

D. Schönauer-Kamin and R. Moos
Department of Functional Materials, University of Bayreuth, Bayreuth, Germany

R. Pohle and M. Fleischer
Corporate Technology, Siemens AG, Munich, Germany

C. Coillot
BioNanoNMRI-group, Laboratoire Charles Coulomb (L2C), Universite de Montpellier, Place Eugene Bataillon, 34095 Montpellier, France

M. El Moussalim, A. Rhouni and M. Mansour
Laboratoire de Physique des Plasmas (LPP), Ecole Polytechnique, Route de Saclay, 91128 Palaiseau, France

E. Brun and R. Lebourgeois
3Thales Research and Technology, Palaiseau, France

G. Sou
Laboratoire d'Electronique et d'Electromagnetisme (L2E), Université Pierre et Marie Curie, Paris, France

M. Dietrich, D. Rauch and R. Moos
Bayreuth Engine Research Center (BERC), Zentrum für Energietechnik (ZET), Department of Functional Materials, University of Bayreuth, 95440 Bayreuth, Germany

U. Simon
Institute of Inorganic Chemistry (IAC), RWTH Aachen University, 52074 Aachen, Germany

A. Porch
School of Engineering, Cardiff University, Cardiff CF24 3AA, Wales, UK

M. Liess
RheinMain University of Applied Sciences, Department of Engineering, Am Brückweg 26, 65428 Rüsselsheim, Germany

H. Budzier and G. Gerlach
Technische Universität Dresden, Electrical and Computer Engineering Department, Solid-State Electronics Laboratory, Dresden, Germany

M. Schulz and H. Fritze
Institute of Energy Research and Physical Technologies, Clausthal University of Technology, Goslar, Germany

E. Mayer, I. Shrena, D. Eisele, M. Schmitt and L. M. Reindl
Department of Microsystems Engineering, Albert-Ludwigs-Universität Freiburg, Freiburg, Germany

A. A. Haidry and B. Saruhan
Institute of Materials Research, German Aerospace Center (DLR) Linder Hoehe, 51147 Cologne, Germany

N. Kind
Institute of Materials Research, German Aerospace Center (DLR) Linder Hoehe, 51147 Cologne, Germany
Laboratoire de Tribologie et Dynamique des Systèmes (LTDS) – UMR 5513 Bâtiment D4 Ecole Centrale de Lyon 36 Avenue Guy DE COLLONGUE 69134 Ecull, France

Ali E. Kubba, Ahmed Hasson and Gregory Hall
Fusion Innovations Ltd., Research and Innovation Services, Birmingham Research Park, Vincent Drive, Edgbaston, Birmingham, B15 2SQ, UK

Ammar I. Kubba
School of Engineering, Mechanical Engineering, University of Birmingham, Edgbaston, Birmingham, B15 2TT, UK

B. Schmitt and A. Schütze
Laboratory for Measurement Technology, Saarbrücken, Germany

C. Kiefer
Laboratory for Measurement Technology, Saarbrücken, Germany
Chair of Micromechanics, Microfluidics/Microactuators, Saarbrücken, Germany

C. Weigel and M. Hoffmann
Micromechanical Systems Group, IMN MacroNano®, Technische Universität Ilmenau, Ilmenau, Germany

M. Schneider
Micromechanical Systems Group, IMN MacroNano®, Technische Universität Ilmenau, Ilmenau, Germany
Robert Bosch GmbH, Reutlingen, Germany

J. Schmitt
Micromechanical Systems Group, IMN MacroNano®, Technische Universität Ilmenau, Ilmenau, Germany
Sonceboz Automotive SA, Sonceboz, Switzerland

S. Kahl and R. Jurisch
microsensys GmbH, Erfurt, Germany

Kyle M. Sinding and Bernhard R. Tittmann
Pennsylvania State University, University Park, Pennsylvania, USA

Alison Orr
Misonix Inc. in Farmingdale, New York, USA

Luke Breon
Electric Power Research Institute, Charlotte, North Carolina, USA

I. Motroniuk, R. Stöber and G. Fischerauer
Bayreuth Engine Research Center (BERC), Universität Bayreuth, Bayreuth, Germany

A. Schuler, T. Hausotte and Z. Sun
Institute of Manufacturing Metrology, Friedrich-Alexander-Universität Erlangen-Nürnberg, Naegelsbachstr. 25, 91052 Erlangen, Germany

S. Sachse, U. Enseleit, F. Gerlach, K. Ahlborn and W. Vonau
Kurt-Schwabe-Institut für Mess- und Sensortechnik e.V. Meinsberg, Waldheim, Germany

A. Bockisch, E. Kielhorn, P. Neubauer and S. Junne
Chair of Bioprocess Engineering, Technische Universität Berlin, Berlin, Germany

T. Kuhnke and U. Rother
Rox GmbH, Radeberg, Germany

K. Kulshreshtha and B. Jurgelucks
Institut für Mathematik, Universität Paderborn, Warburger Str. 100, 33098 Paderborn, Germany

F. Bause, J. Rautenberg and C. Unverzagt
Elektrische Messtechnik, Universität Paderborn, Warburger Str. 100, 33098 Paderborn, Germany

Sabrina Amrehn, Xia Wu and Thorsten Wagner
Department of Chemistry, University of Paderborn, Paderborn, Germany

Agnes Eydam, Gunnar Suchaneck and Gerald Gerlach
Solid State Electronics Laboratory, Technische Universität Dresden, Dresden, Germany

Christophe Coillot, Rahima Sidiboulenouar, Eric Alibert, Maida Cardoso and Christophe Goze-Bac
Laboratoire Charles Coulomb (L2C-UMR5221), BioNanoNMRI group, University of Montpellier, Place Eugene Bataillon, 34095 Montpellier, France

Eric Nativel
Institut d'Electronique et des Systèmes (IES-UMR5214), University of Montpellier, Campus Saint-Priest, 34095 Montpellier, France

Michel Zanca
Laboratoire Charles Coulomb (L2C-UMR5221), BioNanoNMRI group, University of Montpellier, Place Eugene Bataillon, 34095 Montpellier, France
Nuclear medicine, CMC Gui de Chauliac, University Hospital Montpellier, 34095 Montpellier, France

Guillaume Saintmartin
Laboratoire Charles Coulomb (L2C-UMR5221), BioNanoNMRI group, University of Montpellier, Place Eugene Bataillon, 34095 Montpellier, France
Institut des Neurosciences de Montpellier (INSERM U1051), University of Montpellier, 34095 Montpellier, France

Harun Noristani and Florence Perrin
Institut des Neurosciences de Montpellier (INSERM U1051), University of Montpellier, 34095 Montpellier, France

Nicolas Lonjon and Marine Lecorre
Institut des Neurosciences de Montpellier (INSERM U1051), University of Montpellier, 34095 Montpellier, France
Nuclear medicine, CMC Gui de Chauliac, University Hospital Montpellier, 34095 Montpellier, France

Martin Leidinger, Tilman Sauerwald and Andreas Schütze
Saarland University, Lab for Measurement Technology, Saarbrücken, Germany

Joni Huotari and Jyrki Lappalainen
University of Oulu, Faculty of Information Technology and Electrical Engineering, Oulu, Finland

RolandWuchrer, Luhao Liu and Thomas Härtling
Fraunhofer Institute for Ceramic Technologies and Systems IKTS, 01109 Dresden, Germany

Sabrina Amrehn and Thorsten Wagner
Department of Chemistry, University of Paderborn, 33098 Paderborn, Germany

Johannes Gürtler, Daniel Haufe, Jürgen Czarske and Andreas Fischer
Chair of Measurement and Sensor System Techniques, Department of Electrical Engineering and Information Technology, TU Dresden, Helmholtzstr. 18, 01069 Dresden, Germany

Anita Schulz and Friedrich Bake
Institute of Propulsion Technology, German Aerospace Center (DLR), 10623 Berlin, Germany

Lars Enghardt
Institute of Propulsion Technology, German Aerospace Center (DLR), 10623 Berlin, Germany
Institute of Fluid Dynamics and Technical Acoustics, TU Berlin, 10623 Berlin, Germany

Astrit Shoshi and Hubert Brueckl
Center for Integrated Sensor Systems, Danube University Krems, 2700 Wr. Neustadt, Austria

Thomas Maier
Molecular Diagnostics, Austrian Institute of Technology GmbH, 1220 Vienna, Austria

Index